# KINETICS OF ENZYME MECHANISMS

# KINETICS OF ENZYME MECHANISMS

J. TZE-FEI WONG

*Department of Biochemistry,*
*University of Toronto,*
*Canada*

1975 ACADEMIC PRESS

London New York San Francisco

*A Subsidiary of Harcourt Brace Jovanovich, Publishers*

ACADEMIC PRESS INC. (LONDON) LTD.
24/28 Oval Road,
London NW1

*United States Edition published by*
ACADEMIC PRESS INC.
111 Fifth Avenue
New York, New York 10003



Library of Congress Catalog Card Number: 74 18511
ISBN: 0 12 762250 0

PRINTED IN GREAT BRITAIN BY
WILLIAM CLOWES & SONS LIMITED
LONDON, COLCHESTER AND BECCLES

To the memory of my mother
who healed the sick and raised her young
in time of war and time of peace

# Preface

Enzyme kinetics can be a difficult subject to learn and to apply. Too simplified a treatment becomes misleading, and is wasteful of good experimental measurements. Too elaborate a treatment becomes incomprehensible, and is wasted on poor experimental measurements. In order to steer between the conflicting demands of simplicity and comprehensiveness, it is essential that we clearly understand the unsimplified principles of kinetic theory without being overwhelmed by mathematical details. The present book examines in the light of this requirement the wide-ranging kinetic concepts and methods which have been developed to interpret the behaviour of different classes of enzyme mechanisms. The aim is that once we become acquainted with the fundamental ideas, further working details will be easy to acquire whenever the need arises. We shall not lose sight of the forest of enzyme kinetics on account of the mathematical trees, and come to appreciate fully the magnificent panorama presented by this forest, across the horizon where the motions of life merge into those of molecules.

Let me record a few words of thanks, although they are not expected to be adequate to my sense of indebtedness. To Professor Charles Hanes, I owe years of sage instruction and delightful collaboration, as well as numerous improvements to this book. Any belief in enzyme kinetics as an appropriate basis for the delineation of biological behaviour this book conveys must be deeply influenced by his steadfast devotion to the subject. There are few others whose contribution to enzyme kinetics would span a period of over four decades, even in the face of such momentous diversions as the discovery of starch-phosphorylase and transpeptidases, and the first enzymic synthesis of a macromolecule (starch).

To Patricia Bronskill, Patricia Gurr and Laszlo Endrenyi I express my appreciation for the enjoyable hours of kinetic endeavour we shared. To Doreen Marks, Theo Hofmann, Gray Scrimgeour and Edward Whitehead, I am thankful for valuable comments. This book was started during a sabbatical leave at Oxford, and it benefited substantially from the gracious hospitality of Professor Joel Mandelstam.

Without the never-failing support and cooperation of my wife and children, the writing of this book would have proceeded at an estimated pace of $3 \times 10^{-7}$ page.sec$^{-1}$.

The literature of enzyme kinetics is rich in quantity as well as quality, and I have only drawn on those most familiar to me amongst the many interesting experimental and theoretical studies. While regretting that space does not allow the inclusion of more studies, I am deeply grateful to those authors and their publishers, upon whose published figures this book has come to rely so much to illustrate the problems and approaches of the subject, for kindly granting me permission to reproduce their figures. Finally, I wish to thank the Medical and National Research Councils of Canada for their generous support over the years of our work on enzymes.

Toronto, Canada
January, 1975

J. Tze-Fei Wong

# Contents

# Symbols

$E$: enzyme
$(E)_0$: total enzyme concentration
$S, A, B, C$: reaction substrates
$P, Q, R$: reaction products
$I, J$: inhibitor, negative effector
$M$: metal activator, positive effector
$C$: metal-substrate complex
$v$: reaction velocity
$v_0$: intial reaction velocity per unit enzyme (in the absence of one or more reaction products)
$v_f$: initial velocity per unit enzyme in the forward direction
$v_r$: initial velocity per unit enzyme in the reverse direction
$V_S$: saturation velocity with respect to substrate $S$
$V_{(S)}$: conditional saturation velocity with respect to substrate $S$ at some fixed concentration of another substrate or modifying ligand
$K_{eq}$: equilibrium constant
$K_m^S$: Michaelis (half-saturation) constant for substrate $S$
$K_{(m)}^S$: conditional Michaelis constant for substrate $S$ at some fixed concentration of another substrate or modifying ligand
$K_S$: dissociation constant for $S$ from the $ES$ complex
$K_S'$: dissociation constant for $S$ from some $ESX$ complex
$\bar{K}_S$: stability constant of the $ES$ complex ($= 1/K_S$)
$k$: elementary rate constant
$\phi$: initial rate parameter (an algebraic combination of rate constants)
$u$: apparent rate constant
$U$: apparent stability constant
$L$: allosteric constant
$T, R$: conformational states of a protein
$H$: slope of Hill plot
$H_m$: maximal slope of Hill plot (Hill coefficient)

# Introduction

The enzymes are protein molecules which catalyse the self-perpetuating systems of chemical reactions which we know as life. Like all proteins, enzymes are subject to long-term modifications through evolution. At the same time their catalytic powers are highly sensitive to ambient conditions, especially as a result of their capacity to form complexes with a wide range of ligands. This ability of the enzymes to respond constantly to their environment has doubtless been basic to the resilience and versatility of the evolving multi-enzyme systems which have so successfully survived the vicissitudes of existence on this planet for perhaps 2,500 million years.

The most momentous of all human enquiries may well be that which seeks to interpret biological phenomena in terms of quantum-mechanical events. Of critical importance to this objective is the understanding of the intimate nature and the mechanism of the enzyme-catalysed reactions. Any biological observation, whether it concerns the colour pattern of a flower, or the growth and aging of man, challenges interpretation first in terms of the regulated participation of individual enzymes, and then in terms of the atomic interactions between each enzyme and its constellation of substrates, products, and other ligands.

Rapid progress is being made in establishing the three-dimensional structures of enzymes, of the sites of interactions with ligands, and even of the complexes formed with various ligands. However, catalysis is fundamentally a rate phenomenon, and the catalytic significance of any enzyme-ligand complex, even if its atomic configuration is elucidated to the last fraction of an angstrom unit, becomes meaningful only on the basis of enzyme kinetic theory. It is the task of kinetic studies to devise and perform observations on the rate behaviour of an enzyme system and to deduce from these observations a quantitative description of the elementary reaction steps which take place during the course of the catalysed reaction. Thus enzyme kinetic theory, which first predicted the existence of the enzyme-substrate complex 70 years ago, long before the chemical nature of the enzymes was suspected, remains a strategic link between the realms of biological behaviour and quantum-mechanical explanations.

# 1 One-Substrate Mechanism

## 1.1. The Enzyme-Substrate Complex

Since both the rate and the specificity of biological processes largely depend on the formation of a specific complex between enzyme and its substrate, the first and foremost contribution of enzyme kinetics has been the discovery of the enzyme-substrate complex.

The kinetic mechanism for a chemical reaction describes the occurrence of the overall reaction through elementary reaction steps. These steps are as a rule monomolecular or bimolecular, since the simultaneous collision of three or more molecules is exceedingly rare. Therefore it is expected that the enzyme mechanism will consist of a combination of monomolecular and bimolecular steps. In 1902 A. J. Brown measured the hydrolysis of sucrose catalysed by the enzyme invertase, and found that the velocity increased with low and moderate substrate concentrations but levelled off at high substrate concentrations. This experiment led Brown to propose that enzyme and substrate must combine to form a complex before catalysis can take place. The essence of this view may be expressed as:

$$E + S \underset{k_{-1}}{\overset{k_1}{\rightleftharpoons}} ES \xrightarrow{k_2} E + P \qquad \text{Mechanism 1.I}$$

This mechanism consists of three elementary reaction steps. Enzyme $E$ and substrate $S$ combine in a bimolecular step with the rate constant $k_1$ to form the enzyme-substrate complex $ES$. Dissociation of the complex may take place before catalysis has occurred, regenerating enzyme and substrate, or after catalysis has occurred, yielding enzyme and product. These latter steps, characterized by the rate constants $k_{-1}$ and $k_2$, are monomolecular in nature.

Before the substrate is added to an enzyme solution, all of the enzyme molecules occur as free $E$. When substrate is added, $E$ and $S$ combine rapidly, and $ES$ accumulates over a *transient phase* to reach a *steady-state* level. The higher the concentration of $S$ added, the higher the initial steady-state concentration of $ES$ will be. However, when most of the enzyme already is bound in $ES$, further increases in the concentration of $S$ will bring about only negligible increases in $ES$ and hence in reaction velocity. Therefore, with increasing concentration of $S$, the velocity will rise toward a well-defined maximum, known as the *saturation velocity*. The existence of this maximum is an inescapable consequence of the enzyme-substrate complex, and provided the original evidence for the formation of the complex, which subsequently in many enzyme systems could be confirmed by direct physical observations on the complex itself, e.g. when Keilin and Mann (1937) discovered that peroxidase changes its hematin absorption spectrum upon combination with hydrogen peroxide.

## 1.2. Steady-State Rate Law

Although the qualitative predictions of the one-substrate mechanism are intuitively evident, analysis of its quantitative predictions requires the derivation of an appropriate rate law that expresses the reaction velocity in terms of the rate constants and the measurable concentrations of total substrate and total enzyme. The concentrations of the molecular species $E$ and $ES$, which are not readily measurable, are usually eliminated from the expression. To begin, we note that reaction product is formed only in the $k_2$-step and the rate of this monomolecular step is

$$v = \frac{d\mathrm{P}}{dt} = k_2(ES). \tag{1.1}$$

To eliminate $(ES)$, we have to examine two other equations in which it also enters. First, the distribution of enzyme between $E$ and $ES$ varies, but the two species always add up to the total enzyme concentration $(E)_0$:

$$(E) + (ES) = (E)_0. \tag{1.2}$$

Secondly, the rate of change of $(ES)$ with time must equal the difference between its formation via the $k_1$ step and its breakdown via the $k_{-1}$ and $k_2$ steps:

$$\begin{aligned}\frac{d(ES)}{dt} &= k_1\mathrm{S}(E) - k_{-1}(ES) - k_2(ES)\\ &= k_1\mathrm{S}[(E)_0 - (ES)] - (k_{-1} + k_2)(ES)\\ &= k_1\mathrm{S}(E)_0 - (k_1\mathrm{S} + k_{-1} + k_2)(ES).\end{aligned} \tag{1.3}$$

$(ES)$ is obtained by rearranging this to

$$(ES) = \frac{k_1\mathrm{S}(E)_0 - \dfrac{d(ES)}{dt}}{k_1\mathrm{S} + k_{-1} + k_2}. \tag{1.4}$$

At the very start of the reaction, substrate and enzyme are suddenly mixed together. The formation of $ES$ begins with the rate of $k_1 \mathrm{S}(E)_0$. Over the transient phase $(ES)$ builds up to a steady-state level appropriate to the concentrates of $E$ and $S$ added originally, and thereafter will fall in response to a fall in $S$. If $\mathrm{S} \gg (E)_0$, this fall in $(ES)$ on account of diminishing substrate will be relatively slow. Therefore during this latter steady-state phase of the reaction $d(ES)/dt$ will be much smaller than $k_1 \mathrm{S}(E)_0$, namely the rapid rate of $ES$ formation at the start of the transient phase. Thus Eqn (1.4) can be simplified to:

$$(ES) = \frac{k_1 \mathrm{S}(E)_0}{k_1 \mathrm{S} + k_{-1} + k_2} . \tag{1.5}$$

Placing this steady-state expression for $(ES)$ back into (1.1) gives us the steady-state rate law first obtained by Briggs and Haldane in 1925:

$$v = \frac{k_2 k_1 \mathrm{S}(E)_0}{k_1 \mathrm{S} + k_{-1} + k_2} = \frac{k_2 \mathrm{S}(E)_0}{\mathrm{S} + \dfrac{k_{-1} + k_2}{k_1}} . \tag{1.6}$$

The rate constants in (1.6) are commonly replaced by two new parameters, the *saturation velocity* $V_S$ and the *Michaelis constant* $K_m^S$:

$$v = \frac{V_S \cdot \mathrm{S}}{\mathrm{S} + K_m^S}$$

$$V_S = k_2 (E)_0, \; K_m^S = \frac{k_{-1} + k_2}{k_1} \tag{1.7}$$

Procedures for determining $V_S$ and $K_m$ are described in Section 1.4.

## 1.3. Quasi-Equilibrium Rate Law

Historically, the steady-state rate law of Briggs and Haldane was not the first usable expression to be obtained. It was foreshadowed by Henri (1901), Michaelis and Menten (1931) and Van Slyke and Cullen (1914) on a more restricted basis. Thus Michaelis and Menten had considered the situation where $k_2$ happens to be much smaller than $k_{-1}$. This would cause the complex formation between enzyme and substrate to approach thermodynamic equilibrium, so that,

$$k_1 \mathrm{S}(E) \simeq k_{-1}(ES),$$

$$(ES) \simeq \frac{k_1 \mathrm{S}(E)_0}{k_1 \mathrm{S} + k_{-1}}$$

and finally,

$$v = k_2(ES) = \frac{k_2 S(E)_0}{S + \frac{k_{-1}}{k_1}}. \tag{1.8}$$

Thus, due to the restrictive assumption that $k_2$ is much smaller than $k_{-1}$, the quasi-equilibrium rate law (1.8) differs from the steady-state rate law (1.7) in that $k_2$ is omitted from its expression for the Michaelis constant:

$$v = \frac{V_S \cdot S}{S + K_m^S}$$

$$V_S = k_2(E)_0 \qquad K_m^S = \frac{k_{-1}}{k_1}. \tag{1.9}$$

The converse assumption of $k_{-1}$ being much smaller than $k_2$ was invoked by Van Slyke and Cullen. Their rate law differs from Briggs and Haldane's only in that $k_{-1}$ is omitted from its expression for the Michaelis constant. The development of these rate laws, like so many other aspects of early enzyme kinetics, has been admirably recalled by Segal (1958). To summarize, the rate law of Briggs and Haldane is general for the simple one-substrate Mechanism 1.I, and the other two rate laws are limiting cases which arise from extreme relative values of the rate constants $k_{-1}$ and $k_2$:

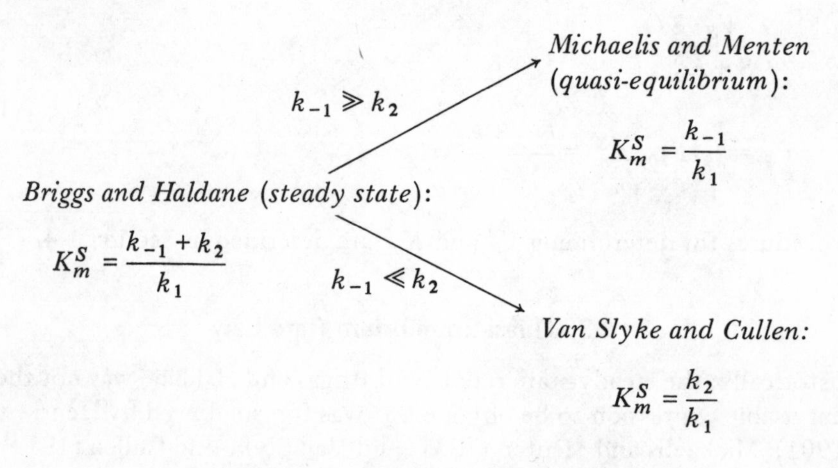


## 1.4. Determination of $V_S$ and $K_m^S$

When enzyme and substrate are mixed together, reaction begins. As reaction progresses, the concentrations of substrate and product will change, and so will the velocity. The typical progress curve, plotting product formation against time, usually starts off with a constant slope which will gradually decrease as substrate and product changes become important. There are two major types of methods for evaluating the relationship between reaction velocity and substrate concentration from progress curves of this shape.

### 1.4.1. Initial rate method

Because concentrations of substrates and other ligands are precisely preset at zero time, analysis of the initial reaction velocity benefits from a minimum of errors along the concentration scale. Estimation of the initial velocity requires an extrapolation of the progress curve back to zero time by visual or computer means (Fig. 1.1).

*(a) Visual estimation*
By sighting along the progress curve toward the origin, the slope of the tangent to the curve at zero time is obtained, and this corresponds to the initial steady-state velocity. A visual aid that offers some assistance in placing the tangent consists of a transparent ruler with a straight line scratched parallel to the edges. By sighting along this line and drawing the estimated slope along an edge, the progress curve itself is left unmarked so that the assessment may be repeated (Hanes *et al.*, 1972).

*(b) Computer estimation*
A series of product measurements are obtained at successive times and the series of points $(\mathbf{P}_1, t_1), (\mathbf{P}_2, t_2), \ldots (\mathbf{P}_i, t_i), (\mathbf{P}_j, t_j) \ldots$ are fitted to an extrapolation formula, e.g. a power series in time as suggested by Booman and Niemann (1956). A simple series is:

$$\mathbf{P} = a_0 + a_1 t + a_2 t^2 + a_3 t^3 + \cdots$$

Fitting the points to the power-series yields the coefficients $a_0, a_1, a_2$, etc. The initial slope is given by the coefficient $a_1$, because

$$\frac{d\mathbf{P}}{dt} = a_1 + 2a_2 t + 3a_3 t^2 + \cdots$$

$$= a_1 \quad \text{when} \quad t = 0$$

For progress curves which are regular in shape and continuously recorded, both types of procedures usually give similar results. In general, computer procedures are less susceptible to systematic visual bias. Visual procedures, on the other hand, involve more direct contact with the raw data, and unusual behaviour is more readily recognized, e.g. Kosow and Rose (1972) observed a biphasic progress curve for hexokinase which could be traced to a slowly developing inhibition of the enzyme by glucose-6-phosphate. In either case, accuracy is improved by a sensitive method of measurement, so that only the earliest portion of the progress curve before the onset of large concentration changes has to be monitored. The use of *differentiators*

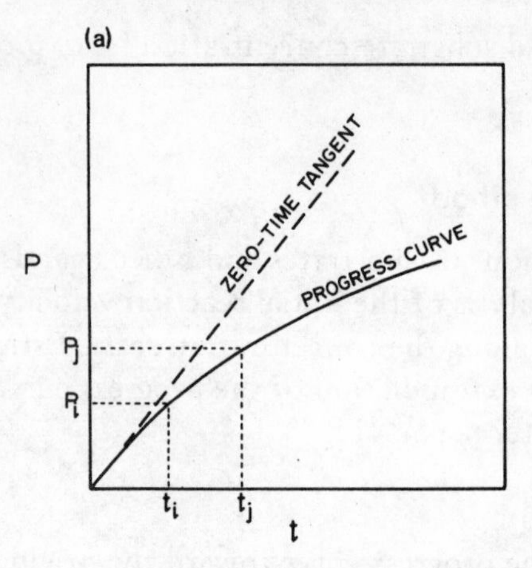


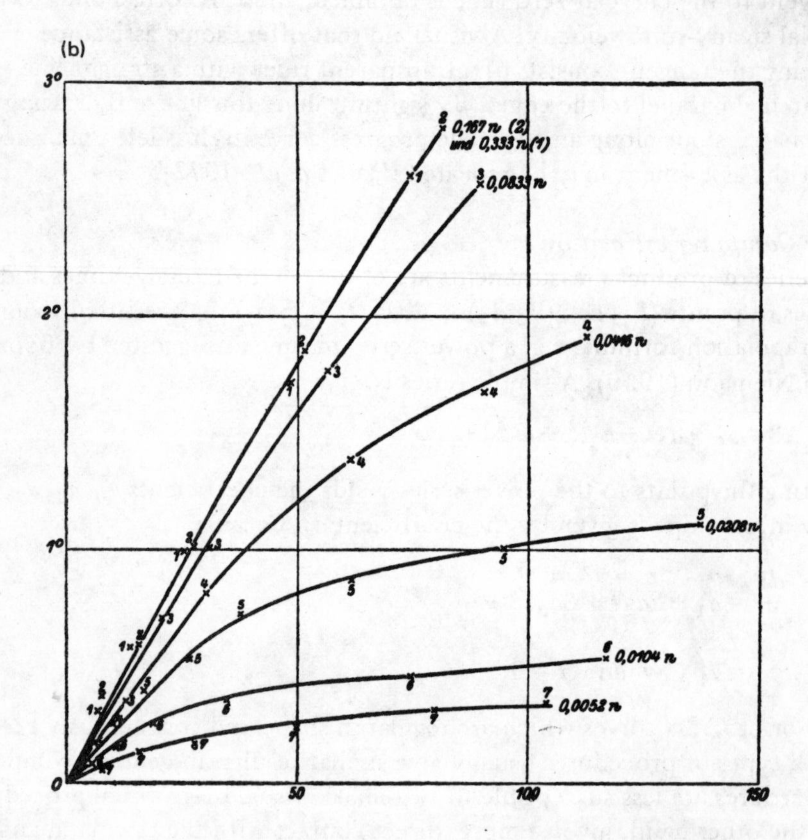


Fig. 1.1. Progress curves. (a) A typical progress curve plotting product concentration versus time. Initial rate can be estimated either by drawing tangent to curve at zero time, or by curve fitting to a series of time points. (b) Invertase progress curves for a series of substrate (sucrose) concentrations ranging from 0.0052 M up to 0.333 M. (From Michaelis and Menten, 1913; ordinate = optical rotation proportional to product concentration, abscissa = time in minutes.)

(Longmuir, 1957; Sargent and Taylor, 1971) and *gradient-flows* (Illingworth and Tipton, 1969; Gurr, Wong and Hanes, 1973) also provides additional procedures for examining the variation of initial velocities with experimental conditions.

After the *initial velocity* $v_0$ has been estimated for a series of reaction mixtures containing different substrate concentrations, a plot of $v_0$ against S can be constructed. The steady-state rate law (1.7) predicts the plot to be a *rectangular hyperbola* specified by the two numerical parameters $V_S$ and $K_m^S$. The form of this hyperbola becomes clear if we consider the velocity predicted for three limiting substrate concentrations:

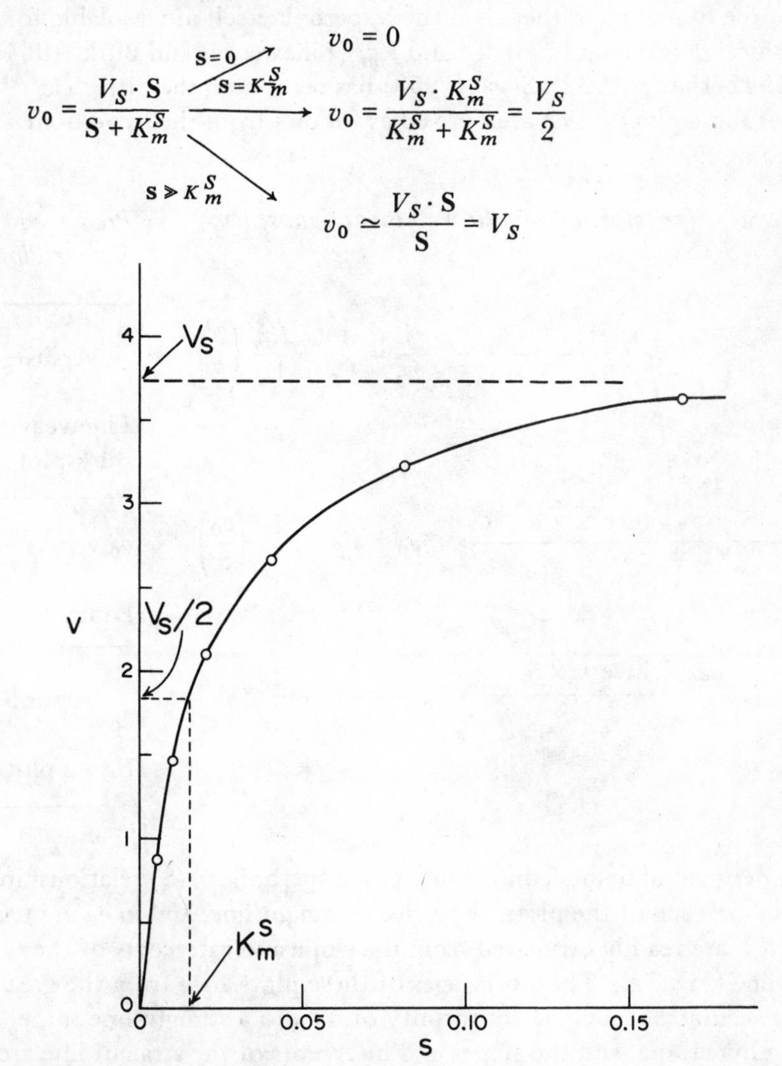


Fig. 1.2. Rectangular hyperbolic variation of velocity with substrate concentration. The initial velocities were estimated by Michaelis and Menten (1913) from the invertase progress curves reproduced in Fig. 1.1(b).

The velocity is zero at zero substrate concentration, and approaches $V_S$ at very high substrate concentrations. It is equal to $V_S/2$ when substrate concentration is equal to the Michaelis constant $K_m^S$. For this reason the Michaelis constant is also referred to as the *half-saturation* constant. Figure 1.2 gives the rates of sucrose hydrolysis by invertase, and illustrates this type of behaviour.

Although velocity approaches $V_S$ at very high substrate concentrations, it is often difficult to decide from an inspection of the hyperbolic curve the exact value of $V_S$. If $V_S$ is imprecisely defined, $V_S/2$ and therefore $K_m^S$ also will be imprecisely defined. To meet this difficulty, three *linear plots* for the hyperbola rather than the hyperbola itself are available for the graphical determination of $V_S$ and $K_m^S$ (Lineweaver and Burk, 1934; Eadie, 1942; Hanes, 1932). Segal (1958) has recounted the historical origins of these plots which are derived as follows from the hyperbolic expression:

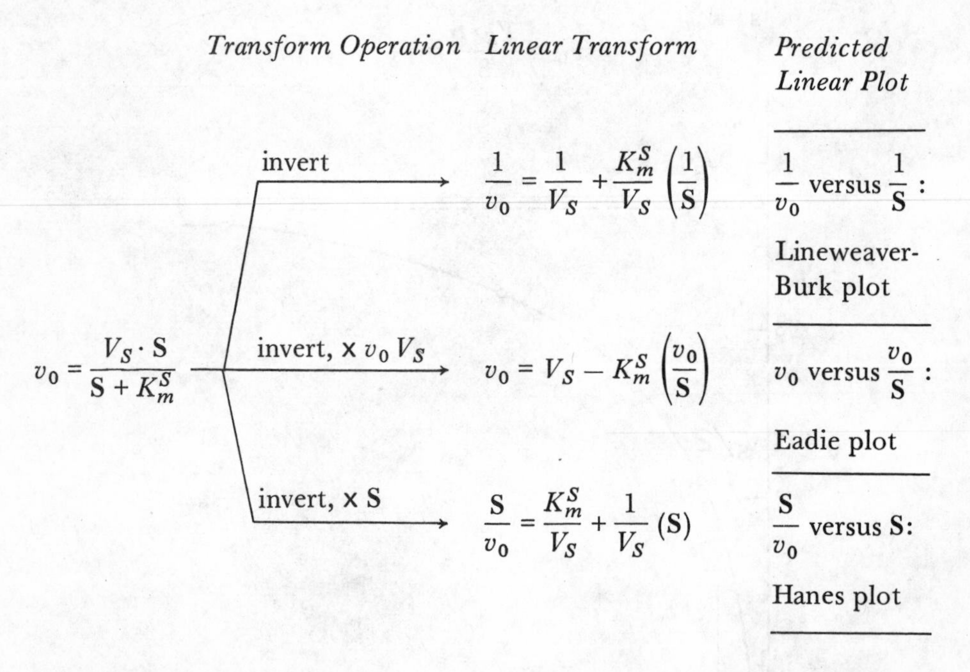


When experimental points conforming to a hyperbolic $v_0(\mathrm{S})$ relationship are placed on each of the plots, they give a straight line, and in each instance $V_S$ and $K_m^S$ are readily estimated from the slope and intercepts of the straight line (Fig. 1.3). The advantages of these plots arise from the greater ease and accuracy of judging conformity of data to a straight line rather than to a hyperbola, and the slopes and intercepts of the straight line are also easy to determine. These advantages become especially important when reliable measurements at high substrate concentrations are prevented by complications such as limited solubility of substrate, physical effects of

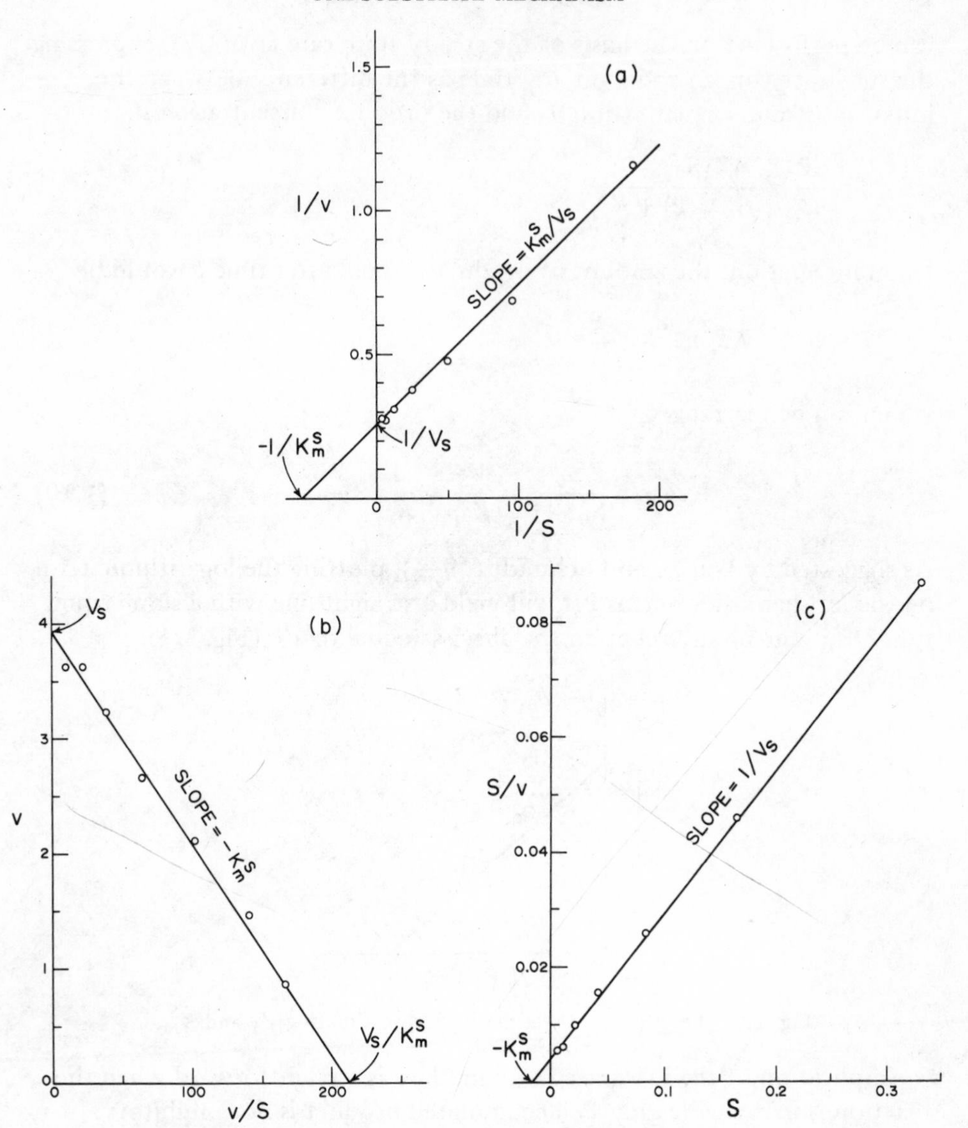


Fig. 1.3. Linear transformations of the rectangular hyperbola. The invertase initial velocities are the same ones used in Fig. 1.2. (a) Lineweaver-Burk plot. (b) Eadie plot. (c) Hanes plot.

concentrated substrate on the enzyme, or opacity of concentrated substrate interfering with optical measurements.

### 1.4.2. Integrated rate method

Since reaction velocity and substrate concentration both vary continuously during the reaction, a single progress curve should supply a sufficient series of velocity-concentration points for the determination of $V_S$ and $K_m^S$. This can be performed on the basis of the steady-state rate law (1.7), expressing the substrate concentration at any time as the difference between the initial substrate concentration $\mathbf{S}_0$ and the product concentration $\mathbf{P}$:

$$v = \frac{d\mathbf{P}}{dt} = \frac{V_S(\mathbf{S}_0 - \mathbf{P})}{(\mathbf{S}_0 - \mathbf{P}) + K_m^S}\,.$$

Upon integration, the amount of product formed after time $t$ would be

$$\mathbf{P} = V_S t + K_m^S \ln \frac{\mathbf{S}_0 - \mathbf{P}}{\mathbf{S}_0}$$

which can be rearranged to

$$\frac{1}{t} \ln \frac{\mathbf{S}_0}{\mathbf{S}_0 - \mathbf{P}} = \frac{V_S}{K_m^S} - \frac{1}{K_m^S} \cdot \frac{\mathbf{P}}{t} \tag{1.10}$$

As suggested by Walker and Schmidt (1944), plotting the logarithmic term on the left-hand side versus $\mathbf{P}/t$ will yield a straight line with a slope equal to $-1/K_m^S$ and an intercept on the abscissa equal to $V_S$ (Fig. 1.4).

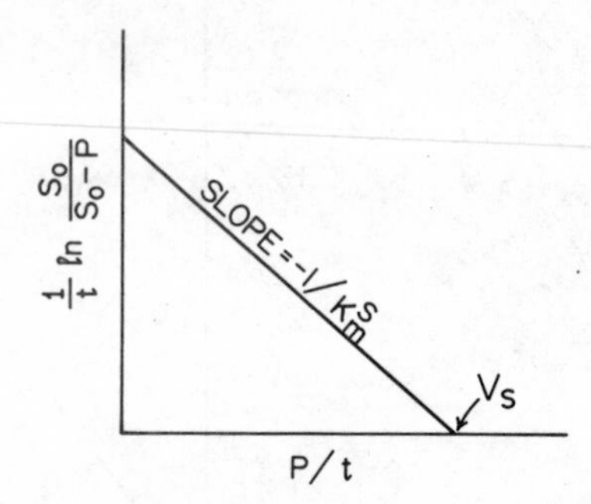


Fig. 1.4. The integrated-rate method for estimating $V_S$ and $K_m^S$.

Application of the integrated rate method is straightforward when the reaction is irreversible and the accumulated product is not inhibitory. Darvey and Williams (1964) and Schwert (1969) succeeded in developing equations for more complex mechanistic situations, but correlations between kinetics and mechanisms are not easily drawn. Schwert suggested that the method may best serve as an auxiliary to the initial rate method.

## 1.5. Validity of Steady State

Since its introduction by Briggs and Haldane, the steady-state method has been a versatile basis for analysing the kinetics of not only the one-substrate mechanism, but also a wide range of other enzyme mechanisms. As Walter and Morales (1964) duly emphasized, the validity of the steady-state method is crucial to the development of enzyme kinetic theory.

### 1.5.1. Nature of the approximation

Substrate concentration decreases during the reaction, and in response the $ES$ complex likewise decreases. The steady-state expression for $d(ES)/dt$ is obtained by differentiation of Eqn (1.5)†:

$$\frac{d(ES)}{dt} = \frac{k_1(k_{-1} + k_2)(E)_0}{(k_1 S + k_{-1} + k_2)^2} \cdot \frac{dS}{dt} . \tag{1.11}$$

The common conception that the steady-state method assumes $d(ES)/dt$ to be zero is clearly unjustified. The approximation of Eqn (1.4) by (1.5) actually does not require $d(ES)/dt$ to be zero, but only requires it to be small relative to $k_1 S(E)_0$. The term $k_1 S(E)_0$ is the rate of change of $(ES)$ at the start of the transient-state immediately upon the mixing of enzyme and substrate; it sets an upper limit to the speed at which enzyme distribution between $(E)$ and $(ES)$ can respond to an abrupt change in substrate concentration. If later in the steady-state phase $d(ES)/dt$ is small relative to this upper limit, enzyme distribution between $(E)$ and $(ES)$ can be regarded as being uniquely dependent upon the substrate concentration, which is in fact what Eqn (1.5) supposes. This supposition, i.e. that the transients of the enzyme distribution process can be neglected within the steady-state phase, is analogous to the one made in using a fast measuring instrument to follow a slow chemical change, i.e. the transients of the instrument can be neglected so that at all times the instrument signal can be regarded as being uniquely dependent upon the chemical concentrations.

### 1.5.2. Two kinds of error

Swoboda (1957) had calculated the time course for $(ES)$ using both the approximate steady-state method and a more precise numerical method (Fig. 1.5a). The steady-state calculation misses entirely the transient-state rise in $(ES)$ immediately following the mixing of enzyme and substrate. During the steady-state fall in $(ES)$, the two calculations approach one another but remain nonidentical. Therefore the discrepancy between the steady-state and numerical calculations is biphasic, larger in the transient-state and smaller in the steady-state, and this reflects the existence of two kinds of inherent errors in the steady-state approximation (Wong, 1965a). The basic differential equation (1.4) has no exact solution. If equations of this type were soluble, the solution would consist of a *complementary function* and a *particular integral* (see Ince, 1944). The steady-state solution in effect neglects the complementary function, thus incurring an error $\delta_c$,

† Briggs and Haldane in fact gave a finite expression for $d(ES)/dt$ in the steady-state. Only an exponent in their expression was incorrect. Later this expression was overlooked, and steady-state theory was misunderstood as requiring that $d(ES)/dt$ be zero.

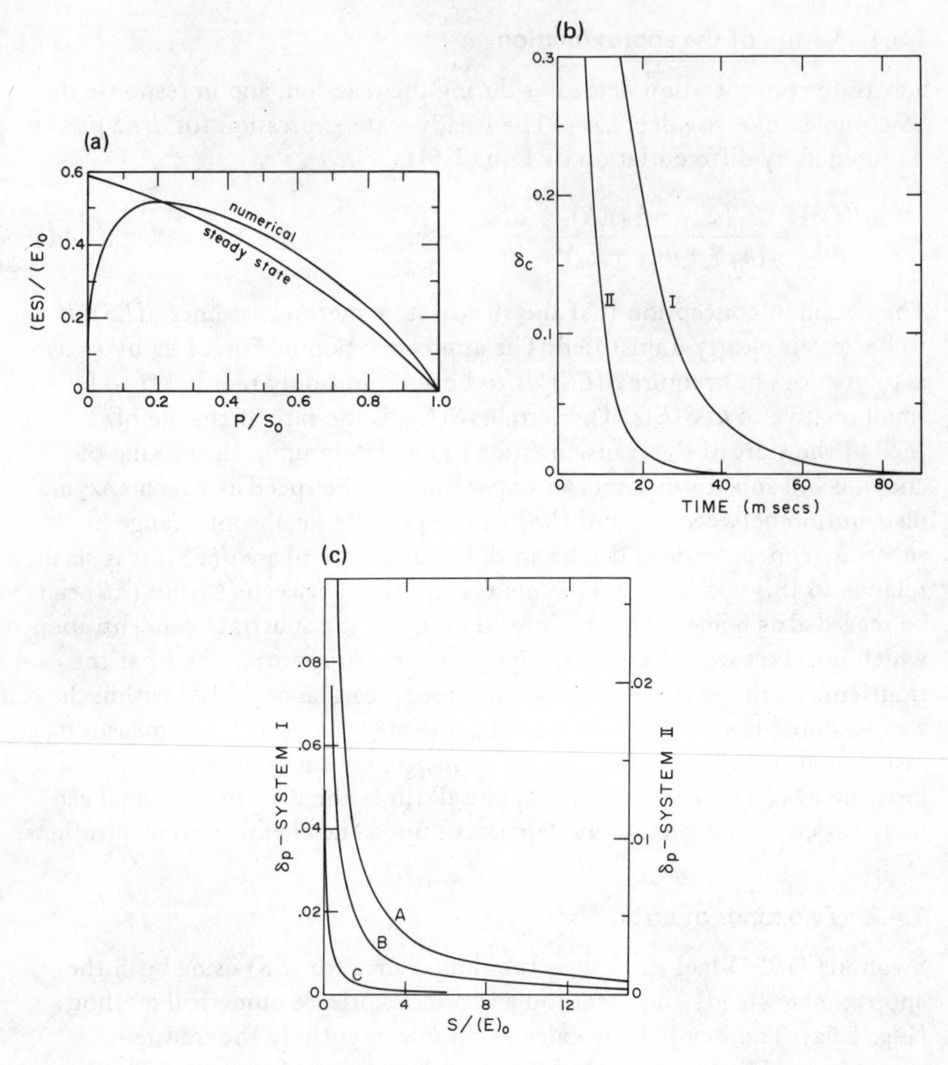


Fig. 1.5. The steady-state approximation. (a) Time course for $(ES)$ calculated by numerical or steady-state method. Substrate/enzyme ratio at the start of the reaction is 2. (After Swoboda, 1957.) (b) Variation of the relative error $\delta_c$ with time; $S_0 = K_m^S$. System I: computed for tryptic hydrolysis of benzoyl-L-arginine ethyl ester. System II: computed for chymotryptic hydrolysis of acetyl-L-phenylalanine ethyl ester. (c) Variation of the relative error $\delta_p$ with substrate/enzyme ratio. Systems I and II are same as in part (b). Substrate concentration equals 0.5 $K_m^S$ for curve A, 2 $K_m^S$ for curve B and 8 $K_m^S$ for curve C. (From Wong, 1965.)

and provides an approximation to the particular integral, thus incurring an error $\delta_p$. The error $\delta_c$ is responsible for the large discrepancy in the transient-state, but vanishes rapidly with time (Fig. 1.5b). The error $\delta_p$ looms large when the experimental substrate/enzyme ratio is low, but vanishes when the ratio is raised (Fig. 1.5c). Of course, a high substrate/enzyme ratio also causes the steady-state to be prolonged relative to the transient-state, so that $\delta_c$ would affect only a negligibly small early portion of the reaction. Therefore the magnitudes of both kinds of errors can be supressed at once by a high substrate/enzyme ratio; and a practical criterion can be formulated to test the validity of the steady-state treatment:

*Steady-state theory predicts proportionality between reaction velocity and enzyme concentration. Since the relative magnitudes of the errors $\delta_c$ and $\delta_p$ vary with enzyme concentration, establishment of this predicted proportionality would indicate that the errors are unimportant under the conditions of the experiment.*

Although this predicted proportionality is very often fulfilled, efforts should not be spared to confirm it at both high and low substrate concentrations. In the striking example of lactate dehydrogenase studied by Wurster and Hess (1970), it proved valid over more than a $10^4$-fold range of enzyme concentrations (Fig. 1.6).

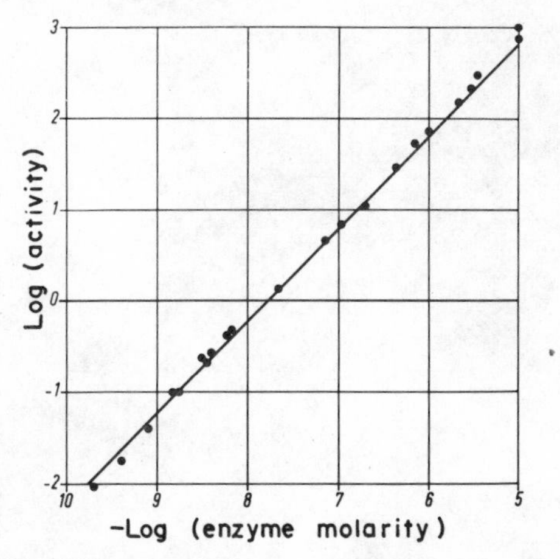


Fig. 1.6. Proportionality between reaction velocity and concentration of lactate dehydrogenase. (From Wurster and Hess, 1970.)

In summary, the steady-state method of Briggs and Haldane gives only an approximate treatment of the steady-state phase of the reaction, and is subjected to two kinds of analysable errors, but its combination of simplicity and sufficient accuracy is not easily replaced. Conveniently, both kinds of errors can be suppressed merely by raising the substrate/enzyme ratio.

## Problem 1.1

A hydrolase hydrolyses an ester with a $K_m$ of 100 mM. It also hydrolyses an amide with a $K_m$ of 0.01 mM. In what way would the progress curve obtained with 1 mM of the ester differ in shape from that obtained with 1 mM of the amide, assuming there is no significant inhibition by reaction products in either case?

# 2 Steady-State Rate Law

The steady-state method, simple as it is, rapidly becomes unwieldy when the rate predictions of complex enzyme mechanisms are to be analysed. However, the mathematical complexities are easily eliminated by a topological approach to analysis that applies to all classes of steady-state enzyme mechanisms. The fundamentals of such an approach may be illustrated by considering the reversible one-substrate mechanism.

## 2.1. Reversible One-Substrate Mechanism

When a one-substrate reaction is thermodynamically reversible, the enzyme catalyst must promote the reaction rate equally in both directions. It becomes necessary to include explicitly in the mechanism both enzyme-substrate and enzyme-product complexes. The mechanism can be represented diagrammatically either in an open form, or a folded form that shows more clearly the topological arrangement of the elementary reaction steps:

$$E \underset{k_{-1}}{\overset{k_1\mathrm{S}}{\rightleftharpoons}} ES \underset{k_{-2}}{\overset{k_2}{\rightleftharpoons}} EP \underset{k_{-3}\mathrm{P}}{\overset{k_3}{\rightleftharpoons}} E$$

(open form)

or

$$\begin{array}{ccc} & E & \\ {\scriptstyle k_{-1}} \nearrow\!\!\swarrow {\scriptstyle k_1\mathrm{S}} & & {\scriptstyle k_3} \nwarrow\!\!\searrow {\scriptstyle k_{-3}\mathrm{P}} \\ ES & \underset{k_{-2}}{\overset{k_2}{\rightleftharpoons}} & EP \end{array}$$

(folded form)

Mechanism 2.I

There are in this mechanism six reaction steps, each represented by an arrow, connecting the three enzyme-containing molecular species $E$, $ES$, and $EP$. The probability of a *monomolecular arrow* is given by its rate constant, and that of a *bimolecular arrow* is given by its rate constant times the concentration of the non-enzymic ligand participating in the step, e.g.,

probability of conversion of $ES$ to $E = k_{-1}$

probability of conversion of $E$ to $ES = k_1\mathrm{S}$

The rate of change of each enzyme species equals the difference between elementary steps leading to its formation, and elementary steps leading to its breakdown. The steady-state method postulates that this rate of change is negligible relative to the rates of the elementary steps themselves. Consequently,

$$\frac{d(E)}{dt} = -(k_1\mathbf{S} + k_{-3}\mathbf{P})(E) + k_{-1}(ES) + k_3(EP) \simeq 0$$

$$\frac{d(ES)}{dt} = k_1\mathbf{S}(E) - (k_{-1} + k_2)(ES) + k_{-2}(EP) \simeq 0 \qquad (2.1)$$

$$\frac{d(EP)}{dt} = k_{-3}\mathbf{P}(E) + k_2(ES) - (k_{-2} + k_3)(EP) \simeq 0$$

The sum of the three enzyme species is equal to the total enzyme:

$$(E) + (ES) + (EP) = (E)_0 \qquad (2.2)$$

The simultaneous algebraic Eqns (2.1) and (2.2) are soluble by the determinant method. To use this method, a determinant, or $\Delta$, is constructed for each enzyme species in matrix form. Expansion of the matrix in accordance with determinant rules yields a series of positive and negative terms. Its final expression is obtained after cancellation of identical positive and negative terms. Application of these operations to our three enzyme species yield:

$$\begin{aligned}
\Delta_E &= k_{-1}k_{-2} + k_2k_{-2} + k_{-1}k_3 + k_2k_3 - k_2k_{-2} \\
&= k_{-1}k_{-2} + k_{-1}k_3 + k_2k_3 \\
\Delta_{ES} &= k_1\mathbf{S}k_{-2} + k_1\mathbf{S}k_3 + k_{-2}k_{-3}\mathbf{P} + k_3k_{-3}\mathbf{P} - k_3k_{-3}\mathbf{P} \\
&= k_1\mathbf{S}k_{-2} + k_1\mathbf{S}k_3 + k_{-2}k_{-3}\mathbf{P} \qquad (2.3) \\
\Delta_{EP} &= k_1\mathbf{S}k_{-1} + k_1\mathbf{S}k_2 + k_{-1}k_{-3}\mathbf{P} + k_2k_{-3}\mathbf{P} - k_1\mathbf{S}k_{-1} \\
&= k_1\mathbf{S}k_2 + k_{-1}k_{-3}\mathbf{P} + k_2k_{-3}\mathbf{P}
\end{aligned}$$

The relative concentrations of the various enzyme species are proportional to their determinants:

$$(E) = \frac{\Delta_E}{\Delta_E + \Delta_{ES} + \Delta_{EP}} \cdot (E)_0$$

$$(ES) = \frac{\Delta_{ES}}{\Delta_E + \Delta_{ES} + \Delta_{EP}} \cdot (E)_0 \qquad (2.4)$$

$$(EP) = \frac{\Delta_{EP}}{\Delta_E + \Delta_{ES} + \Delta_{EP}} \cdot (E)_0$$

Equations (2.4) are called *distribution equations* because they express the distribution of total enzyme amongst the various enzyme species. The rate law is derived readily from these distribution equations:

$$v = \frac{d\mathrm{P}}{dt} = (\text{rate of } k_3 \text{ step}) - (\text{rate of } k_{-3} \text{ step})$$

$$= k_3(EP) - k_{-3}\mathrm{P}(E)$$

$$= \frac{k_3 \Delta_{EP} - k_{-3}\mathrm{P}\Delta_E}{\Delta_E + \Delta_{ES} + \Delta_{EP}} \cdot (E)_0 \tag{2.5}$$

## 2.2. Schematic Rule for Enzyme Distributions

The expansion of a determinant into algebraic terms is performed in accordance with determinant rules. For complex enzyme mechanisms the method is wasteful, because the expansion generates a large number of identical positive and negative terms which disappear from the final expression through cancellations. This wastage of the determinant method is not unique to enzymic problems. About a century ago, Kirchhoff and Maxwell encountered a comparable difficulty with the simultaneous equations describing flow of electrical currents through networks. By systematically characterizing the terms which would disappear through cancellations from those which would not, these distinguished workers succeeded in replacing the determinant rules by more selective rules. The new rules are topological in nature: the *cancellable* and *uncancellable* terms were identified by virtue of the topological network patterns they represent.

A corresponding topological approach was introduced into enzyme kinetics when King and Altman in 1956 devised a *schematic rule* for obtaining directly the uncancellable determinant terms for any enzyme species in a mechanism. Its one requirement that the enzyme species must react only with nonenzymic ligands and not with one another is satisfied by most enzymes (except for those that undergo reversible association of subunits; see Section 8.1.4).

*Schematic rule:*

For an enzyme mechanism with $n$ enzyme species, the determinant expression $\Delta_i$ for any enzyme species $E_i$ will be a series of positive terms. Every term is the product of $n - 1$ arrow-probabilities† from the mechanism. Each of the $n - 1$ arrows must leave from a different enzyme

† The arrow-probabilities are also called *kappas*. The arrow-probability of a monomolecular reaction step is simply the monomolecular rate constant. That of a bimolecular reaction step is given by the bimolecular rate constant times the concentration of the nonenzymic ligand involved in the step.

species other than $E_i$ itself. Altogether, these $n-1$ arrows form a noncyclic and confluent pattern converging at $E_i$. All arrow patterns which are acceptable to this requirement, and none other, will be included in $\Delta_i$. All the $n$ different determinants, $\Delta_1, \Delta_2, \ldots, \Delta_n$, can be similarly obtained, and the fractional concentration of $E_i$ is given by:

$$\frac{(E_i)}{(E)_0} = \frac{\Delta_i}{\Delta_1 + \Delta_2 + \cdots \Delta_i + \cdots \Delta_n}$$

In the reversible one-substrate Mechanism 2.I, there are three enzyme species: $E$, $ES$, and $EP$. In the determinant $\Delta_E$, every term will contain $3 - 1 = 2$ arrows. One leaving from $ES$ and one from $EP$, these arrows will have to converge singly or in sequence at $E$. The only topological patterns acceptable to this requirement are†:

E, ES, EP with arrows $k_{-1}$ (ES → E), $k_{-2}$ (EP → ES): $k_{-1} \cdot k_{-2}$

E, ES, EP with arrows $k_{-1}$ (ES → E), $k_3$ (EP → E): $k_{-1}k_3$

E, ES, EP with arrows $k_2$ (ES → EP), $k_3$ (EP → E): $k_2 \cdot k_3$

For the determinant $\Delta_{ES}$, the two arrows will have to leave from $E$ and $EP$ and converge at $ES$. The acceptable patterns are:

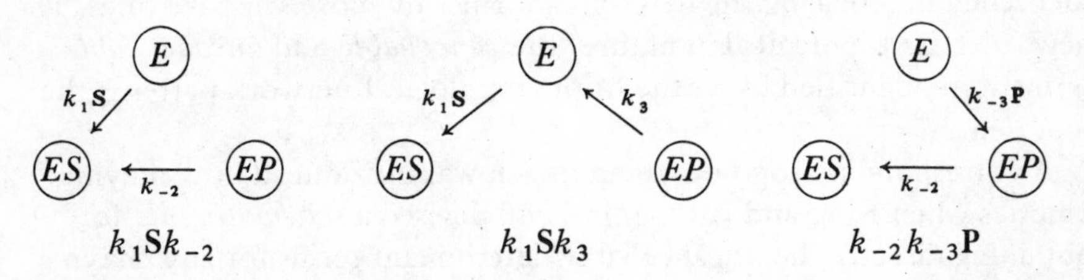


Finally, for the determinant $\Delta_{EP}$, the two arrows will have to leave from $E$ and $ES$ and converge at $EP$. The acceptable patterns are:

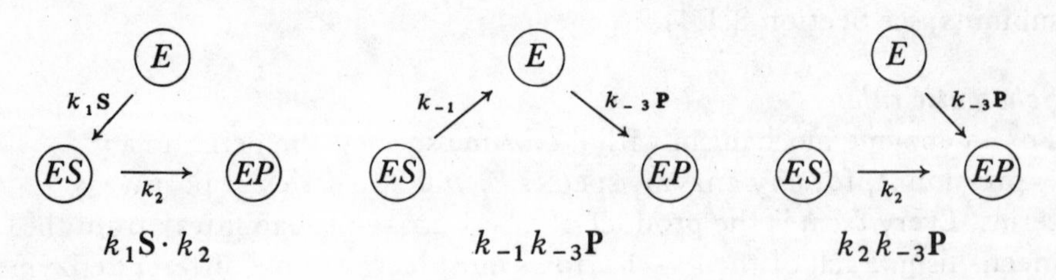


† Another two-arrow combination with one arrow leaving from $ES$ and one from $EP$ is $k_2 k_{-2}$. It is unacceptable because the $k_2$ and $k_{-2}$ arrows form a cycle; it also does not converge to $E$.

In this way, the determinant expansions given in Eqn (2.3) may be written out by inspecting the topology of reaction steps in the mechanism. The entire formal machinery for expanding determinants and cancelling positive and negative terms is no longer necessary for deriving enzyme distribution equations.

## 2.3. Structural Rule for Rate Laws

The schematic rule of King and Altman permits the easy recognition of denominator terms of the rate laws by eliminating the need to cancel terms in expanding the determinant matrices. However, the schematic rule could not eliminate the need to cancel numerator terms during the derivation of the rate laws. For instance, in deriving the numerator for rate law (2.5), four of the six terms generated are cancelled out:

$$\begin{aligned} k_3 \Delta_{EP} - k_{-3}\mathbf{P}\Delta_E &= k_3(k_1\mathbf{S}k_2 + k_{-1}k_{-3}\mathbf{P} + k_2k_{-3}\mathbf{P}) \\ &\quad - k_{-3}\mathbf{P}(k_{-1}k_{-2} + k_{-1}k_3 + k_2k_3) \\ &= k_1\mathbf{S}\cdot k_2\cdot k_3 - k_{-1}\cdot k_{-2}\cdot k_{-3}\mathbf{P} \end{aligned} \tag{2.6}$$

and the rate law for Mechanism 2.I is:

$$v = \frac{(k_1\mathbf{S}k_2k_3 - k_{-1}k_{-2}k_{-3}\mathbf{P})(E)_0}{\begin{aligned} &k_{-1}k_{-2} + k_{-1}k_3 + k_2k_3 + k_1\mathbf{S}k_{-2} + k_1\mathbf{S}k_3 + k_{-2}k_{-3}\mathbf{P} \\ &+ k_1\mathbf{S}k_2 + k_{-1}k_{-3}\mathbf{P} + k_2k_{-3}\mathbf{P} \end{aligned}} \tag{2.7}$$

The importance of eliminating all cancellations in deriving rate laws goes far beyond economy of mathematical procedure. As long as the uncancellable terms could not be recognized in some way from amongst the larger number of terms generated from the determinant matrices, there could be no direct insight into the nature of the rate law. On the other hand, if the uncancellable terms could be directly identified, the nature of the rate law and hence the predicted rate behaviour would be deducible from the topology of an enzyme mechanism without reliance on the apparently fortuitous outcome of cancellations. Toward this end, Wong and Hanes (1962) formulated a *structural rule* for writing steady-state rate laws, which defines the topological properties of all the uncancellable terms in both the numerator and the denominator:

*Structural rule*

(a) The rate expression consists of three groups of terms:

$$\frac{v}{(E)_0} = \frac{(\textit{forward numerator terms}) - (\textit{reverse numerator terms})}{(\textit{denominator terms})}$$

(b) For an enzyme mechanism with $n$ enzyme species, the forward numerator terms all have $n$ arrows, one leaving from each enzyme species. Taken together, these $n$ arrows contain one, and only one, cyclic reaction pathway capable of accomplishing the catalysed reaction in the forward direction. If this pathway does not include all the $n$ arrows, the extra arrows must flow confluently into this pathway.

(c) The reverse numerator terms have the same properties as the forward ones, except that the cyclic reaction pathway they contain accomplishes the catalysed reaction in the reverse instead of the forward direction.

(d) The denominator terms all have $n - 1$ arrows. Being the sum of all the determinants already specified by the schematic rule, they represent all the noncyclic patterns constructible for the mechanism. Every pattern converges at the enzyme species that does not contribute an arrow to the pattern.

Mechanisms with branching reaction sequences give more than one acceptable forward or reverse numerator terms. Mechanisms consisting of an unbranched, linear arrangement of enzyme species can give only one acceptable forward numerator term, containing all the forward arrows in the linear sequence, and only one acceptable reverse-numerator term, containing all the reverse arrows in the linear sequence. Accordingly the rate-law numerator for Mechanism 2.I is:

forward numerator term reverse numerator term

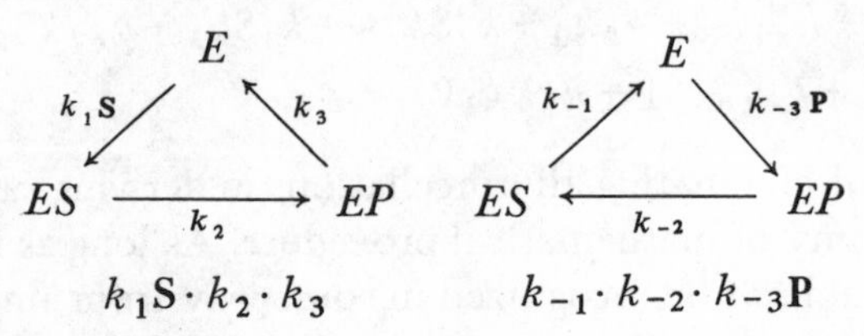


The denominator for the rate law is the same as the denominator for the distribution equations, namely $(\Delta_E + \Delta_{ES} + \Delta_{EP})$. Therefore the entire rate-law (2.7) can be written simply by inspecting the mechanism and applying the structural rule.

For complex mechanisms the problem is to ensure the complete recognition of all the acceptable arrow patterns. Since the acceptable patterns for $\Delta_i$ contain $n - 1$ arrows, one from each of the other $n - 1$ enzyme species besides $E_i$, a simple algorithm used by Wong and Hanes (1962) to generate the patterns is to multiply together the sums of arrows leaving from the other $n - 1$ enzyme species, and select only those arrow combinations which do not include a cycle. (In order not to include a cycle, they must also converge at $E_i$):

$$\Delta_i = \not{c}\ (\Sigma \text{ arrows from } E_1)(\Sigma \text{ arrows from } E_2) \ldots$$
$$\ldots (\Sigma \text{ arrows from } E_{i-1})(\Sigma \text{ arrows from } E_{i+1}) \ldots$$
$$(\Sigma \text{ arrows from } E_n)$$

The symbol $\not{c}$ stands for "noncyclic combinations". For instance, all the determinant patterns for Mechanism 2.I can be generated in this way:

$$\Delta_E = \not{c}(k_{-1} + k_2)(k_{-2} + k_3) = k_{-1}k_{-2} + k_{-1}k_3 + k_2k_3$$

$$\Delta_{ES} = \not{c}(k_1\mathbf{S} + k_{-3}\mathbf{P})(k_{-2} + k_3) = k_1\mathbf{S}k_{-2} + k_1\mathbf{S}k_3 + k_{-2}k_{-3}\mathbf{P}$$

$$\Delta_{EP} = \not{c}(k_1\mathbf{S} + k_{-3}\mathbf{P})(k_{-1} + k_2) = k_1\mathbf{S}k_2 + k_{-1}k_{-3}\mathbf{P} + k_2k_{-3}\mathbf{P}$$

More involved and powerful algorithms for generating such patterns have been developed by Volkenstein and Goldstein (1966) and Seshagiri (1972), and computer programs for the task by Hurst (1969) and Fisher and Schulz (1969).

## 2.4. Topological Reasoning

Mathematically speaking, the steady-state rate law is plainly not a haphazard entity. As defined by the structural rule, it has the form of a ratio between two sums of arrow patterns, and every pattern conforms to a rigid topological description. For any given mechanism, the structural rule easily identifies what categories of arrow patterns will or will not appear in the rate law. Conversely, from the experimentally-determined form of the rate law, we know immediately what kinds of model mechanisms will or will not prove to be compatible with the data. Thus features of rate behaviour and features of reaction mechanism may be correlated through a process of deduction, or *topological reasoning*, based on our knowledge of the structural rule. To begin with, it becomes a simple matter to deduce by this means the useful restrictions imposed on the rate law by the initial rate condition, by the number of enzyme species reacting with each ligand in a reaction mechanism, and by the essential or non-essential nature of the variable ligand.

### 2.4.1. Initial one-way condition

Negative terms in the numerator of the rate law require a complete cyclic pathway for the reverse reaction, and therefore the presence of all the reaction products. For example, in a catalysed reaction of the type $A + B + C \rightleftharpoons P + Q + R$, all three products $P$, $Q$ and $R$ must be present at finite concentrations before a complete reverse pathway can be drawn. Accordingly, if one or more of the reaction products is absent from the reaction mixture at the start of the reaction, the initial-velocity expression will be free of negative numerator terms, regardless of the mechanism. This condition under which there can be a flow of reaction only in the forward but not the reverse direction, may be referred to as the *initial one-way condition.* Easy to satisfy experimentally, it ensures mathematical simplicity through the banishment of negative terms. Analysis of mechanisms to be presented in the ensuing chapters will be based mostly on initial velocities, and the symbols employed will be

$v_0$: *initial reaction velocity per unit enzyme*

$v_f$ and $v_r$: when it becomes necessary to distinguish between the two directions of reaction, $v_f$ will represent the *initial forward velocity per unit enzyme*, and $v_r$ the *initial reverse velocity per unit enzyme.*

### 2.4.2. Concept of "degree"

Whenever a ligand $X$ (substrate, product, inhibitor, modifier, effector, etc.) reacts with an enzyme species, its concentration enters into the arrow for the reaction step, and eventually appears in the arrow-patterns of the rate law. If there are $z$ different enzyme species in the mechanism which react with $X$, no arrow-pattern can be raised to higher than the $z$th power in X, because all arrow-patterns can include only one arrow from any enzyme species. Consequently, when reaction velocity is studied as a function of X, the rate law would be expressible as a ratio between two polynomials in X at most of the $z$th degree:

$$v_0 = \frac{m_z \mathrm{X}^z + m_{z-1}\mathrm{X}^{z-1} + \cdots + m_0}{d_z \mathrm{X}^z + d_{z-1}\mathrm{X}^{z-1} + \cdots + d_0} \,. \tag{2.8}$$

The value of $z$ defines the *degree* of the mechanism with respect to $X$. The $m$- and $d$-coefficients consist of combinations of rate constants and reactants other than $X$. The $m$-coefficients can contain both positive and negative terms in the general case, but only positive terms under the *initial one-way condition*; the $d$-coefficients are strictly positive.

### 2.4.3. Essential versus non-essential reactants

Numerator terms in the initial rate law must all contain a forward cyclic pathway for the forward reaction. If some substrate or activator essential to the forward reaction were absent, it would not be possible to draw a complete forward cyclic pathway. Accordingly, although the numerator may include arrow-patterns devoid of non-essential ligands, it cannot include arrow-patterns devoid of any essential ligand:

*If X is non-essential:*

$$v_0 = \frac{m_z \mathrm{X}^z + m_{z-1}\mathrm{X}^z + \cdots + m_1\mathrm{X} + m_0}{d_z \mathrm{X}^z + d_{z-1}\mathrm{X}^{z-1} + \cdots + d_1\mathrm{X} + d_0}$$

*If X is essential:*

$$v_0 = \frac{m_z \mathrm{X}^z + m_{z-1}\mathrm{X}^{z-1} + \cdots + m_1\mathrm{X}}{d_0 \mathrm{X}^z + d_{z-1}\mathrm{X}^{z-1} + \cdots + d_1\mathrm{X} + d_0}$$

This is merely the formalization of a self-evident result; surely there can be no reaction if a supposedly essential ingredient is missing!

## 2.5. Generalized One-Substrate Mechanism

In the reversible Mechanism 2.I, the interconversion between $ES$ and $EP$ represents a purely intramolecular isomerization. There is little reason to stop short at a single isomerization, or for that matter isomerization of the enzyme-reactant complexes but not the free enzyme. Eliminating both of these restrictions gives a completely generalized model mechanism for one-substrate reactions:

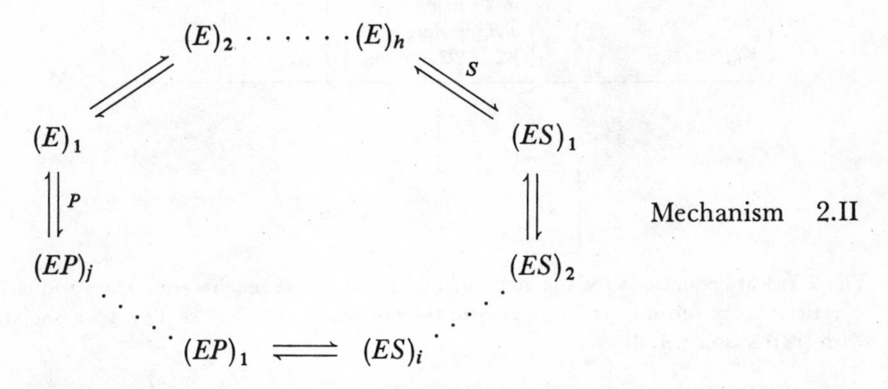


Mechanism 2.II

The form of the rate law for this mechanism is:

$$\frac{v}{(E)_0} = \frac{m_s\mathbf{S} - m_p\mathbf{P}}{d_{sp}\mathbf{SP} + d_s\mathbf{S} + d_p\mathbf{P} + d_0}. \tag{2.9}$$

The topological rationale for rate law (2.9) is straightforward. First, since the mechanism is an unbranched sequence, the numerator will consist of just one positive arrow-pattern composed of all the forward arrows, and one negative arrow-pattern composed of all the reverse arrows. The former is represented as $m_s\mathbf{S}$, and the latter as $m_p\mathbf{P}$. There are numerous denominator patterns, which cannot be higher than first-degree in either **S** or **P**, because only the one $(E)_h$ species reacts with $S$, and the one $(E)_1$ species reacts with $P$. Consequently the denominator is a first-degree polynomial with respect to both **S** and **P**.

A number of other useful equations can be derived from Eqn (2.9), and these derivations are summarized in Fig. 2.1. Equation (2.10) shows the initial forward velocity $v_f$, which is obtained when $P$ is absent from the system. Equation (2.11) shows the initial reverse velocity $v_r$, which is obtained when $S$ is absent from the system. Combination of these equations with the equilibrium constant $K_{eq}$ leads to Eqn (2.12) which establishes the relationship between the equilibrium and initial-rate parameters. First obtained by Haldane (1930), it is known as a *Haldane relation*. The exact

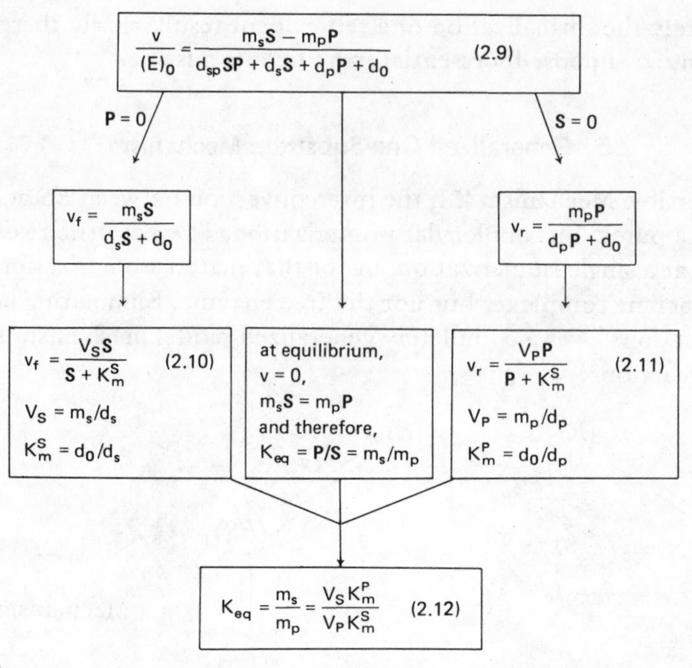


Fig. 2.1. Rate equations for the generalized one-substrate mechanism. The various coefficients are subscripted according to the concentration factors they are associated with (Pettersson, 1970).

compositions of the $m$ and $d$ coefficients in Eqn (2.9) are dependent on the many details of Mechanism 2.II, i.e. the exact number of $(E)$, $(ES)$ and $(EP)$ species, and the nature of their rate constants. However, the hyperbolic forms of $v_f$ and $v_r$, as well as the Haldane relation, are independent of such details. They are valid for Mechanism 2.II, as well as any other unbranched mechanism in which $S$ and $P$ each react with only one enzyme species.

The appearance of an **SP**-term in rate law (2.9) but not rate law (2.7) reflects the isomerization of free enzyme in the former but not the latter instance. Whenever the free enzyme does not isomerize, $S$ and $P$ would be reacting with the same enzyme species:

$$\cdots \rightleftharpoons EP \underset{P}{\rightleftharpoons} E \overset{S}{\rightleftharpoons} ES \rightleftharpoons \cdots$$

The appearance of **S** and **P** in the same arrow patterns would be disallowed by the structural rule, which limits any pattern to only one arrow from any enzyme species. On the other hand, whenever the free enzyme isomerizes, $S$ and $P$ would be reacting with different enzyme species

$$\cdots \underset{P}{\rightleftharpoons} E_1 \rightleftharpoons \cdots \rightleftharpoons E_h \overset{S}{\rightleftharpoons} \cdots$$

and **S** and **P** may appear jointly in the same denominator patterns. This explains the **SP**-term in rate law (2.9) for Mechanism 2.II, and in the rate laws for related mechanisms (Cleland, 1963a; Taraszka and Alberty, 1964).

Cennamo (1969) has suggested a method for detecting the SP-term. If substrate concentration is raised infinitely in the presence of a finite product concentration, rate law (2.9) is transformed into (2.9a):

$$\frac{v}{(E)_0} = \frac{m_s}{d_{sp}\mathbf{P} + d_s} \,. \tag{2.9a}$$

This substrate-saturated velocity would be decreased by **P**. Were the **SP**-term not present, $d_{sp}$ would be equal to zero, and this substrate-saturated velocity would be unaffected by **P**. An alternative that may be offered, which avoids the use of very high concentrations, is to employ a constant proportion of substrate and product, so that $\mathbf{P} = r\mathbf{S}$. This fixes the ratio between the substrate and product added to the system, and transforms rate law (2.9) into (2.9b):

$$\frac{v}{(E)_0} = + \frac{(m_s - m_p r)\,\mathbf{S}}{d_{sp} r \mathbf{S}^2 + (d_s + d_p r)\,\mathbf{S} + d_0} \,. \tag{2.9b}$$

Consequently the variation of the velocity with the substrate-product mixture would be nonhyperbolic, yielding nonlinear Lineweaver-Burk, Eadie, and Hanes plots of the data. Were the **SP**-term not present, Eqn (2.13) would be applicable instead of Eqn (2.9b):

$$\frac{v}{(E)_0} = \frac{(m_s - m_p r)\mathbf{S}}{(d_s + d_p r)\mathbf{S} + d_0} \,. \tag{2.13}$$

Consequently the variation of the velocity with the substrate-product mixture would be hyperbolic, yielding linear Lineweaver-Burk, Eadie, and Hanes plots of the data instead. In either (2.9b) or (2.13), the velocity could be positive, proceeding in the forward direction, or negative, proceeding in the reverse direction. The direction depends on the relative magnitudes of $m_s$ and $m_p r$, but does not affect the use of the method. Additional, more involved but equally valid, methods for detecting the **SP**-term have been described by Taraszka and Alberty (1964), Ray and Roscelli (1964), Darvey (1972), and Britton (1973).

## Problem 2.1

Mechanism 5.III from Chapter 5 brings about the catalysed reaction of $A + B = P + Q$. In the forward reaction, $A$ and $B$ add to the enzyme in random sequences. In the reverse reaction, $P$ and $Q$ add in the order of $P$ before $Q$.

$$
\begin{array}{ccccccccc}
 & & EB & & & & & & \\
 & \overset{k_1 B}{\nearrow}\underset{k_{-1}}{\swarrow} & & \overset{k_3 A}{\searrow}\underset{k_{-3}}{\nwarrow} & & & & & \\
E & & & & \text{Central-complex} & \underset{k_{-6}Q}{\overset{k_6}{\rightleftharpoons}} & EP & \underset{k_{-8}P}{\overset{k_8}{\rightleftharpoons}} & E \\
 & \overset{k_2 A}{\searrow}\underset{k_{-2}}{\nwarrow} & & \overset{k_4 B}{\nearrow}\underset{k_{-4}}{\swarrow} & & & & & \\
 & & EA & & & & & & 
\end{array}
$$

Mechanism 5.III

Use the structural rule to derive the steady-state rate law for this mechanism when $A$, $B$, $P$ and $Q$ are all present in the system. What happens to the rate law when $P$ is initially absent? when $Q$ is initially absent? or when $P$ and $Q$ are initially absent? Note the important simplification achieved in each case through the elimination of all arrow patterns containing the initially absent product.

# 3 Graphical Methods

The kinetic study of an enzyme mechanism is a two-way translation:

| Mechanism topology | ⟷ | Rate law | ⟷ | Rate behaviour |
|---|---|---|---|---|

In one direction, plausible model mechanisms are formulated, and the rate behaviour predicted by their rate laws analysed and compared to experiment. In the other direction, experimental rate observations are fitted to an appropriate form of rate law, with the objective of deducing the topology of the reaction-mechanism from the characteristics of this rate law. The translations between mechanism topology and rate law depend on the detailed derivation of rate laws or a system of topological reasoning; the translations between rate law and rate behaviour depend on the use of *graphical* and *statistical* methods.

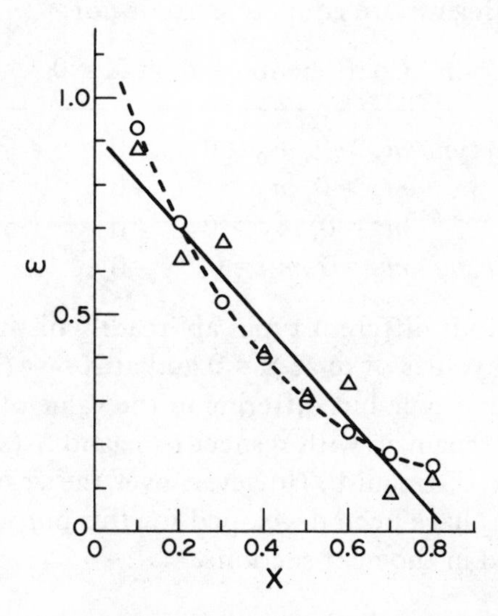


Fig. 3.1. Graphical versus statistical detection of nonlinearity. Graphically, the points in set (a), shown as open triangles, appear to be randomly scattered about the solid straight line; but the points in set (b), shown as open circles, appear to conform to the dashed nonlinear curve. Statistically, the total sum of squares of deviation from the solid straight line is the same for the two sets of points.

Sophisticated statistical methods are on the whole more powerful than graphical methods. However, graphical methods are less prone to overlook systematic deviations than unsophisticated statistical methods. For example, the sample points in Fig. 3.1 illustrate this. In set (a), the points are randomly scattered about a straight line, but in set (b) they fall on a smooth non-linear curve. Although the distinction between the two sets is graphically unmistakable, it is not statistically detectable solely on the basis of the sum of squares of deviation from linearity, which is the same in the two sets. Only a more sophisticated enquiry into the distribution of the deviations could establish the nonidentity of the two sets.

## 3.1. Types of Functions

An enzyme mechanism is known by the rate and binding functions it predicts. In general, an experimental measurement $\omega$ (reaction velocity in kinetic studies, or fractional saturation in binding studies) is related to the concentration of any ligand $X$ by a ratio between two polynomials in X:

$$\omega = \frac{m_z X^z + m_{z-1} X^{z-1} + \cdots + m_1 X + m_0}{d_z X^z + d_{z-1} X^{z-1} + \cdots + d_1 X + d_0} \,. \tag{3.1}$$

The coefficients $m_z$ and $m_0$ are sometimes equal to zero. Accordingly we have to distinguish between four major types of function, depending on whether these coefficients are equal to zero or not:

| Type of $\omega$(X) Function | Coefficients | $\omega$ at $X = 0$ | $\omega$ at $X = \infty$ |
|---|---|---|---|
| *Substrate (essential) type* | $m_z > 0, m_0 = 0$ | $=0$ | $>0$ |
| *Modifier type* | $m_z > 0, m_0 > 0$ | $>0$ | $>0$ |
| *Inhibitor type* | $m_z = 0, m_0 > 0$ | $>0$ | $=0$ |
| *Substrate-inhibitor type* | $m_z = 0, m_0 = 0$ | $=0$ | $=0$ |

Functions of these four different types are readily distinguishable on the basis of the limiting values of $\omega$ at $X = 0$ and at $X = \infty$ (Fig. 3.2). Functions of the same type but differing in the value of $z$, which defines the *degree* of the mechanism with respect to ligand $X$ (see Section 2.4.2), are more difficult to distinguish. However, over the years a variety of graphical procedures have been developed for this purpose. These procedures are surveyed in the next sections.

## 3.2. Substrate (Essential) Type

Rate functions fall into this type when $X$ is a reaction substrate or essential activator (see Section 2.4.3). Its first, second and third-degree cases are

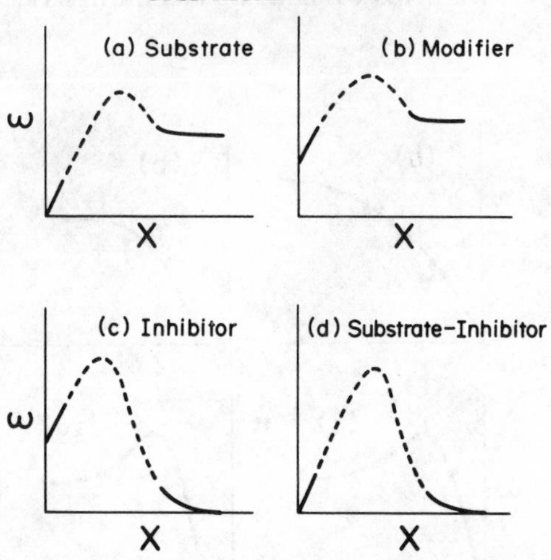


Fig. 3.2. Four fundamental types of rate functions. (a) Substrate (essential) type; (b) Modifier type; (c) Inhibitor type; and (d) Substrate-inhibitor type.

$$\omega = \frac{m_1\mathrm{X}}{d_1\mathrm{X} + d_0} \tag{3.2}$$

$$\omega = \frac{m_2\mathrm{X}^2 + m_1\mathrm{X}}{d_2\mathrm{X}^2 + d_1\mathrm{X} + d_0} \tag{3.3}$$

$$\omega = \frac{m_3\mathrm{X}^3 + m_2\mathrm{X}^2 + m_1\mathrm{X}}{d_3\mathrm{X}^3 + d_2\mathrm{X}^2 + d_1\mathrm{X} + d_0}\,. \tag{3.4}$$

Distinction between these different equations is aided by a number of graphical procedures.

(a) The plot of $\omega$ *versus* X yields a rectangular hyperbola in the first-degree case, but the higher-degree cases will yield nonhyperbolic curves, e.g. sigmoid curves or curves exhibiting a maximum. Botts (1958) established that curves for the second-degree function could exhibit up to one maximum and two inflection points (Fig. 3.3). More complex curve shapes may be exhibited by third- or higher-degree functions.

(b) *Linearized plots* of the rectangular hyperbola include the Lineweaver-Burk plot of $1/\omega$ versus 1/X, Eadie plot of $\omega$ versus $\omega$/X, Hanes plot of X/$\omega$ versus X. The Scatchard plot of $\omega$/X versus $\omega$, which is simply the inverse of the Eadie plot, is most widely used in binding studies. If an experimental system conforms to the first-degree Eqn (3.2), the data when plotted according to each of these plots would adhere to strict linearity. However, if the system conforms to the higher-degree (3.3) or (3.4), all these plots might become visibly nonlinear. Hofstee (1952, 1959) suggested that the different linearized plots are not equally effective in revealing

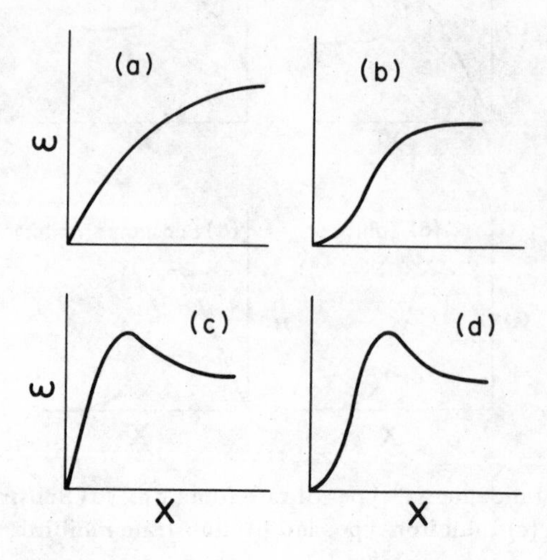


Fig. 3.3. Four possible curve shapes for the second-degree substrate type function (Eqn 3.3). Curve (d) has a sigmoidal rise and passes through a maximum; (c) lacks the sigmoidal rise; (b) lacks the maximum; (a) has neither of these traits. The relative magnitudes of the coefficients in Eqn (3.3) determine which of these shapes will prevail. (After Botts 1958.)

experimental deviations from linearity, the Eadie plot being graphically much more discerning than the others. This suggestion was substantiated by Dowd and Riggs (1965), but doubted by Dixon and Webb (1959). Figure 3.4 shows velocity measurements on *E. coli* phosphofructokinase, established by Atkinson and Walton (1965) and Blangy, Buc and Monod (1968) to be a nonhyperbolic, fourth-degree system. The nonlinear trend in these measurements is more evident in the Eadie plot than in either the Lineweaver-Burk or Hanes plot, a case which would support Hofstee's suggestion. In view of the considerable labour involved in accumulating accurate experimental data on enzymic rate functions, the prudent investigator may be well advised to subject the data to examination by all three graphical forms and to compare the results. This is desirable especially if the parameters $V_X$ and $K_m^X$ are to be determined by purely graphical means.

When linearity has been established, and statistical estimation of $V_X$ and $K_m^X$ is to be undertaken, the Hanes and the Eadie plots each offer distinct advantages, the former being generally preferred to the latter; the Lineweaver-Burk plot again proves to be less effective (see Section 11.5).

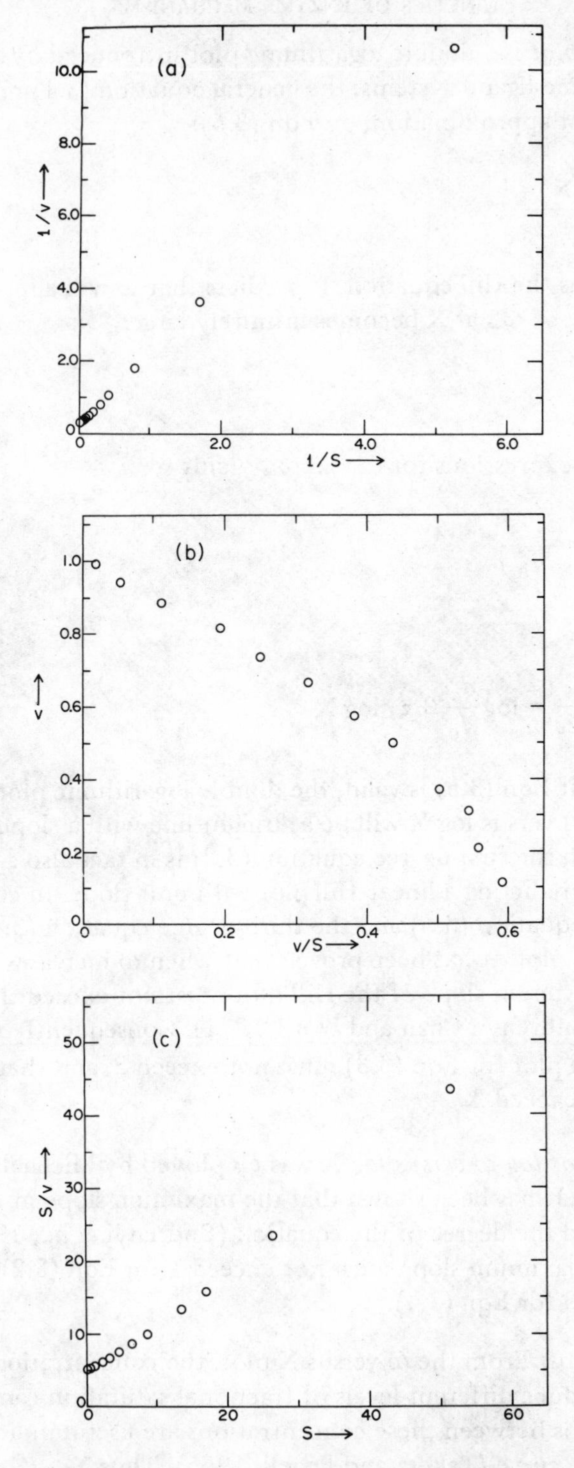


Fig. 3.4. Use of linear plots to detect nonhyperbolic behaviour. The same set of nonhyperbolic rate measurements on *E. coli* phosphofructokinase (from Blangy *et al.*, 1968) is plotted in the form of (a) Lineweaver-Burk plot; (b) Eadie plot; and (c) Hanes plot.

(c) *The Hill plot* is a double logarithmic plot introduced by Hill (1910). For some enzyme-ligand systems, the general equation (3.1) might be replaced, to first approximation, by Eqn (3.5):

$$\omega = \frac{m_z \mathbf{X}^z}{d_z \mathbf{X}^z + d_0} \ . \tag{3.5}$$

This is known as the Hill equation. It predicts that $\omega$ will approach a saturation value of $\omega_\infty$ as **X** becomes infinitely large:

$$\omega_\infty = \frac{m_z}{d_z} \ .$$

Combining the expressions for $\omega$ and $\omega_\infty$ yields

$$\frac{\omega}{\omega_\infty - \omega} = \frac{d_z \mathbf{X}^z}{d_0}$$

and finally,

$$\log \frac{\omega}{\omega_\infty - \omega} = \log \frac{d_z}{d_0} + z \cdot \log \mathbf{X}.$$

Consequently, if Eqn (3.5) is valid, the double logarithmic plot of $\log \omega/(\omega_\infty - \omega)$ versus $\log \mathbf{X}$ will be a straight line with a slope equal to $z$. In this regard, the first-degree equation (3.2) is in fact also a first-degree Hill equation, predicting a linear Hill plot with unit slope. In contrast, the second-degree equation (3.3) and the third-degree equation (3.4) predict a nonlinear Hill plot. It has been proven that when $\omega$ increases monotonically with **X**, the maximum slope of the Hill plot must not exceed the degree of the equation (Endrenyi, Chan and Wong 1971). Consequently the maximum slope in the Hill plot for Eqn (3.3) must not exceed 2, and that for Eqn (3.4) must not exceed 3.

(d) *The Plot of log* $\omega$ *versus log* **X** was employed by Michaelis and Menten (1913). It has been shown that the maximum slope in this plot must not exceed the degree of the equation (Endrenyi *et al.*, 1971). Consequently the maximum slope must not exceed 1 for Eqn (3.2), 2 for Eqn (3.3), and 3 for Eqn (3.4).

(e) $\mathbf{X}_i/\mathbf{X}_j$ *ratios.* From the $\omega$ versus **X** plot, the concentrations of $X$ required to produce different levels of fractional saturation can be determined, and ratios between these concentrations are useful indicators of the shape of the curve (Taketa and Pogell, 1965). Thus $\mathbf{X}_{0.9}/\mathbf{X}_{0.1}$, the ratio between the concentrations at 90 and 10% saturation, should be 81 if the curve is a rectangular hyperbola conforming to Eqn (3.2). Any

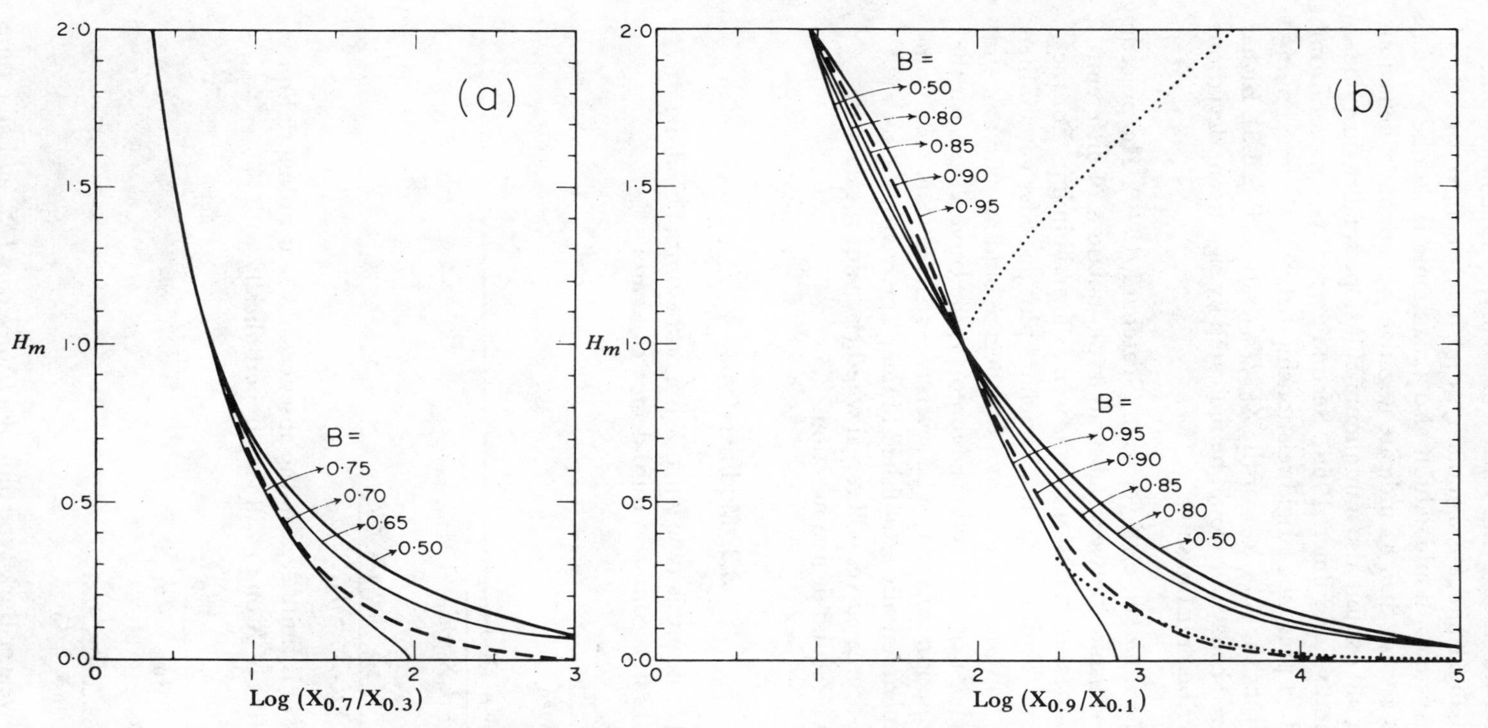


Fig. 3.5. Relationship between the maximum Hill slope, $H_m$, and (a) $\mathbf{X}_{0.7}/\mathbf{X}_{0.3}$ or (b) $\mathbf{X}_{0.9}/\mathbf{X}_{0.1}$ for a second-degree substrate-type function (Eqn 3.3). **B** is the level of $\omega$ at which the Hill slope attains a maximum value. Once **B** is determined, the second-degree function requires that the relationship between $H_m$ and either ratio should conform to the standard curves shown. The same curves hold for any **B** and (1 − **B**). The dotted curve in part (b) is a third-degree curve for **B** = 0.8 included for comparison. For example, if a maximum Hill slope of 1.5 occurs at 80% saturation, $H_m$ would be equal to 1.5, and **B** would be equal to 0.8. The second-degree function would require the $\mathbf{X}_{0.9}/\mathbf{X}_{0.1}$ ratio to be close to 1.3. A ratio close to 2.7 would rule out a second-degree function, but would be compatible with a third-degree function. (From Endrenyi *et al.*, 1971.)

deviation from 81 would be symptomatic of nonhyperbolic behaviour. Ratios smaller than 81 indicate sigmoidal, or positively cooperative, behaviour; such systems should exhibit also a maximum Hill slope greater than one. Ratios greater than 81 indicate negatively cooperative behaviour; such systems should exhibit also a minimum Hill slope smaller than one (Koshland, Némethy and Filmer, 1966). Second-degree systems conforming to Eqn (3.3) in fact prescribe a rigid relationship between the maximum Hill slope and either the $X_{0.7}/X_{0.3}$ or the $X_{0.9}/X_{0.1}$ ratios (Fig. 3.5). Higher than second-degree behaviour would be indicated by significant deviations from these prescribed relationships.

(f) *The Scatchard plot* of $\omega/X$ versus $\omega$ (Scatchard, 1949). The preceding criteria for distinguishing between substrate-type functions of different degrees apply to measurements of rate as well as ligand-binding. For the treatment of ligand-binding data, the Scatchard plot provides yet another useful criterion. This plot is linear for a first-degree binding function, and nonlinear for higher-degree functions. Moreover, the curve shapes for the second-degree function turn out to be severely restricted (Fig. 3.6). Comparison of an experimentally established binding curve with the set of standard second-degree curves will reveal whether or not the data are compatible with a second-degree function.

## 3.3. Modifier Type

Rate functions fall into this type when $X$ is a non-essential modifier of the reaction. Its first, second and third-degree cases are:

$$\omega = \frac{m_1 X + m_0}{d_1 X + d_0} \tag{3.6}$$

$$\omega = \frac{m_2 X^2 + m_1 X + m_0}{d_2 X^2 + d_1 X + d_0} \tag{3.7}$$

$$\omega = \frac{m_3 X^3 + m_2 X^2 + m_1 X + m_0}{d_3 X^3 + d_2 X^2 + d_1 X + d_0}\,. \tag{3.8}$$

In all three cases $\omega$ is finite even in the absence of $X$. According to Eqn (3.6), the net effect of X on $\omega$ will vary hyperbolically with X:

$$\Delta\omega = \frac{m_1 X + m_0}{d_1 X + d_0} - \frac{m_0}{d_0}$$

$$= \frac{(m_1 d_0 - m_0 d_1) X}{d_1 d_0 X + d_0^2}\,.$$

Therefore a Lineweaver-Burk type plot of $1/\Delta\omega$ versus $1/X$, an Eadie type

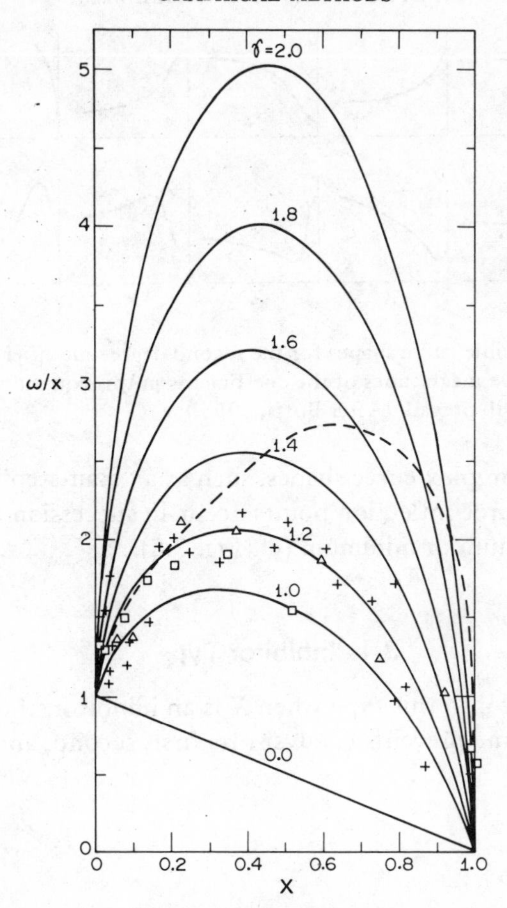


Fig. 3.6. Standard Scatchard plot for a second-degree ligand-binding function. The standard curves are calculated for different coefficients of cooperativity $\gamma$ (see Section 7.3.2). If the binding measurements obtained at different ligand concentrations follow the general contours of these standard curves, they can be accounted for by a second-degree function. Otherwise the second-degree function has to be rejected in favour of a higher-degree one. For example, the three sets of succinate-binding points (crosses, triangles and squares) obtained by Changeux, Gerhart and Schachman (1968) on aspartate transcarbamylase conform closely to a second-degree contour. In contrast, the data of Roughton (1963) on the binding of oxygen to hemoglobin (dashed line) are clearly inconsistent with a second-degree function. (From Endrenyi *et al.*, 1971.)

plot of $\Delta\omega$ versus $\Delta\omega/X$, and a Hanes type plot of $X/\Delta\omega$ versus X, are all predicted to be linear. These plots will be nonlinear for systems conforming to Eqn (3.7) or (3.8). Also, while $\omega$ will increase or decrease hyperbolically with X according to the first-degree Eqn (3.6), it is compatible with ten different curve shapes on the basis of the second-degree Eqn (3.7) (Botts, 1958). These curve shapes, shown in Fig. 3.7, contain up to one maximum and two inflection points. The third-degree Eqn (3.8) would be compatible

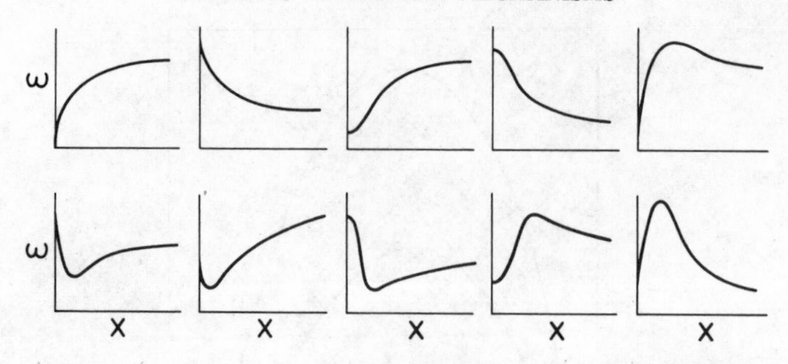


Fig. 3.7. Ten possible curve shapes for the second-degree modifier type function (Eqn 3.7). The relative magnitudes of the coefficients in this equation determine which of the curve shapes will prevail. (After Botts, 1958.)

with even more complex curve shapes, such as a "stair-step" type of curve in which two or three inflection points occur in succession without an intervening maximum or minimum (cf. Fig. 7.6).

## 3.4. Inhibitor Type

Rate functions fall into this type when $X$ is an inhibitor, the binding of which to the enzyme prevents catalysis. Its first, second, and third-degree cases are:

$$\omega = \frac{m_0}{d_1 X + d_0} \tag{3.9}$$

$$\omega = \frac{m_1 X + m_0}{d_2 X^2 + d_1 X + d_0} \tag{3.10}$$

$$\omega = \frac{m_2 X^2 + m_1 X + m_0}{d_3 X^3 + d_2 X^2 + d_1 X + d_0}\,. \tag{3.11}$$

The *Dixon plot* of $1/\omega$ versus X is a valuable method for analysing this type of function (Segal, Kachmar and Boyer, 1952; Dixon, 1963a). This plot will be linear according to Eqn (3.9), but nonlinear according to (3.10) or (3.11). However, in all three instances $1/\omega$ will increase linearly with X as X becomes very large. In contrast, modifier-type functions predict that $1/\omega$ will level off to a finite value as X becomes very large (Fig. 3.8).

In the above equations, the degree of the denominator polynomial exceeds by one the degree of the numerator polynomial. This difference between the two polynomials can be greater than one in rare instances. For example, Eqn (3.11a) shows the third-degree function with a discrepancy of two, and Eqn (3.11.b) shows the third-degree function with a discrepancy of three:

$$\omega = \frac{m_1 X + m_0}{d_3 X^3 + d_2 X^2 + d_1 X + d_0} \tag{3.11a}$$

$$\omega = \frac{m_0}{d_3 X^3 + d_2 X^2 + d_1 X + d_0}. \tag{3.11b}$$

Such instances might be analysed by plotting log $(1/\omega)$ versus log **X**. As **X** becomes very large, this plot will become linear with slope =1 for Eqn (3.11), linear with slope = 2 for Eqn (3.11a), and linear with slope = 3 for Eqn (3.11b).

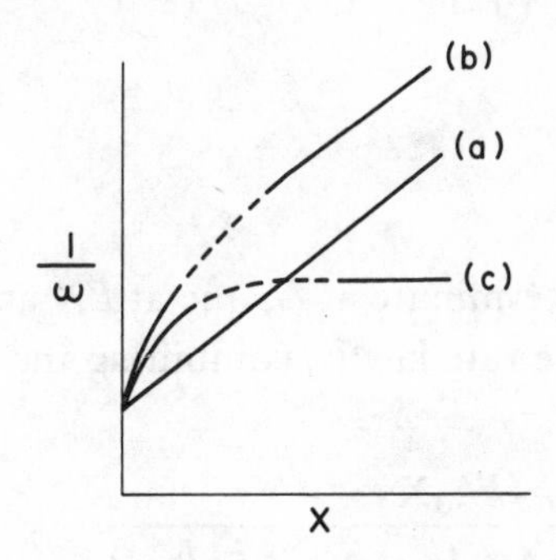


Fig. 3.8. Dixon plot for (a) First-degree inhibitor type function; (b) Higher-degree inhibitor type functions; and (c) Modifier type functions.

### 3.5. Substrate-Inhibitor Type

Inhibition by excess substrate occurs with many enzymes. Haldane (1930) treated a simple model mechanism giving rise to this type of behaviour:

$$E \underset{k_{-1}}{\overset{k_1 X}{\rightleftharpoons}} EX \xrightarrow{k_2} E + \text{Products} \qquad \text{Mechanism 3.I}$$

$$EX \underset{k_{-3}}{\overset{k_3 X}{\rightleftharpoons}} EX_2$$

Its initial rate law is easily obtained using the structural rule (Section 2.3). There are three enzyme species, viz. $E$, $EX$ and $EX_2$. Every numerator arrow-pattern must contain an arrow from each enzyme species, and must contain also a complete forward reaction cycle. The only admissible pattern is $k_1 X \cdot k_2 \cdot k_{-3}$ (with $k_1 X$ and $k_2$ providing the reaction cycle):

$$\boxed{E} \xrightarrow{k_1 X} \boxed{EX} \xrightarrow{k_2} \boxed{E}, \qquad \boxed{EX_2} \xrightarrow{k_{-3}} \boxed{EX}$$

The denominator consists of all possible noncyclic patterns, each containing two arrows leaving from two different enzyme species. There are altogether four admissible patterns:

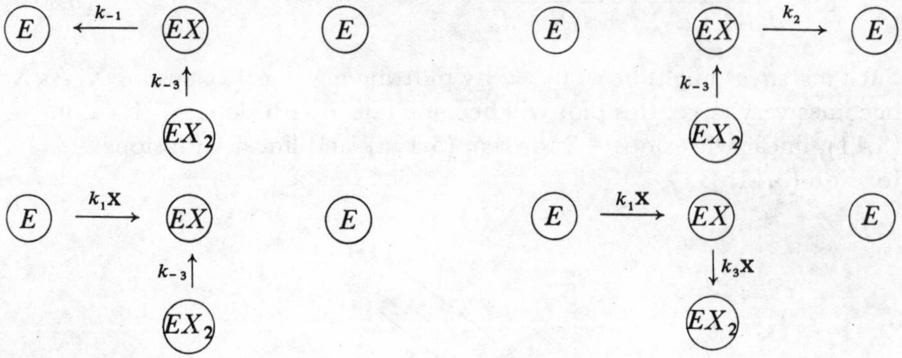

$$E \xleftarrow{k_{-1}} EX \qquad EX_2 \xrightarrow{k_{-3}} EX \qquad \qquad EX \xrightarrow{k_2} E \qquad EX_2 \xrightarrow{k_{-3}} EX$$

$$E \xrightarrow{k_1 \mathbf{X}} EX \qquad EX_2 \xrightarrow{k_{-3}} EX \qquad \qquad E \xrightarrow{k_1 \mathbf{X}} EX \qquad EX \xrightarrow{k_3 \mathbf{X}} EX_2$$

Two of the four patterns terminate at $E$, one at $EX$ and one at $EX_2$. The complete expression of the rate law is, combining the numerator and denominator patterns:

$$v_0 = \frac{k_1 k_2 k_{-3} \mathbf{X}}{k_1 k_3 \mathbf{X}^2 + k_1 k_{-3} \mathbf{X} + (k_{-1} k_{-3} + k_2 k_{-3})} \tag{3.12}$$

This second-degree equation belongs to the substrate-inhibitor type. Velocity is zero when $\mathbf{X}$ is zero, and again when $\mathbf{X}$ becomes infinitely large. The $k_3\mathbf{X}$ arrow leads into a dead-end. It cannot contribute in any way to the catalysed reaction, and is barred from the numerator of the rate-law. This explains the presence of an $\mathbf{X}^2$-term in the denominator but not in the numerator, causing the velocity to approach zero as $\mathbf{X}$ becomes very large. On the other hand, when $\mathbf{X}$ is very small, $k_1 k_3 \mathbf{X}^2$ would be negligibly small compared to the other terms in the denominator, and Eqn (3.12) is reduced to (3.12a):

$$v_0 = \frac{k_1 k_2 \mathbf{X}}{k_1 \mathbf{X} + k_{-1} + k_2} = \frac{V_X \cdot \mathbf{X}}{\mathbf{X} + K_m^X} . \tag{3.12a}$$

There is no significant formation of $EX_2$ under these conditions, and Eqn (3.12a) is precisely the rate law of Briggs and Haldane for the uninhibited one-substrate mechanism. Dixon and Webb (1958) have emphasized that, since the Lineweaver-Burk plot displays the low concentration points with special prominence, it would be most suitable for the evaluation of $V_X$ and $K_m^X$ under these conditions.

### Problem 3.1

What manner of a model mechanism would predict a rate law of the form of Eqn (3.11a)? or Eqn (3.11b)?

# 4 Inhibitors and Activators

Enzymes often interact reversibly with ligands other than their reaction substrates and products. These ligands under the various guises of inhibitors, modifiers, activators, and allosteric effectors can enter into the mechanism topology and influence the reaction velocity. Added exogenously, they can be versatile agents for the experimental and therapeutic control of biological systems. Generated endogenously, they are important cellular and hormonal regulators of metabolic function.

## 4.1. General Inhibition Mechanism

A general inhibition mechanism that describes the interaction of substrate $S$ and reversible inhibitor $I$ with the enzyme is:

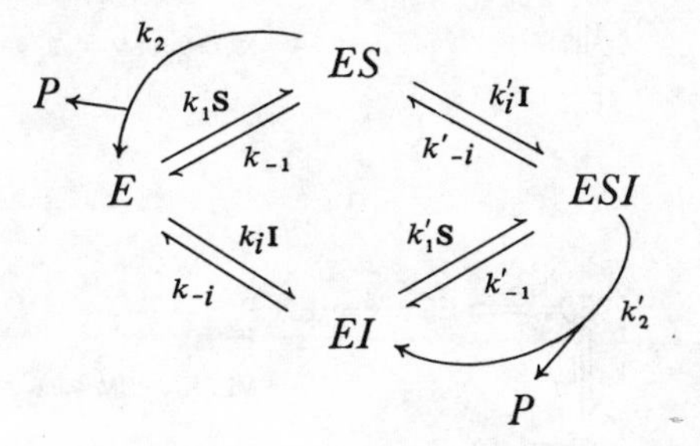


Mechanism 4.I

When the inhibitor is absent, only the enzyme species $E$ and $ES$ are formed; addition of the inhibitor gives rise to the two new species $EI$ and $ESI$. The $k_2$-arrow represents the catalytic formation of reaction product from $ES$, and the $k'_2$-arrow represents the catalytic formation of reaction product from $ESI$. The steady-state rate law for this mechanism was first derived by Botts and Morales (1953). It indicates that the $v_0(\mathrm{S})$ and $v_0(\mathrm{I})$ functions are both second-degree funtions:

$$v_0 = \frac{m_2\mathrm{S}^2 + m_1\mathrm{S}}{d_2\mathrm{S}^2 + d_1\mathrm{S} + d_0},$$

$$v_0 = \frac{m_2\mathrm{I}^2 + m_1\mathrm{I} + m_0}{d_2\mathrm{I}^2 + d_1\mathrm{I} + d_0}.$$

This is hardly surprising. There are two enzyme species to react with *S*, two to react with *I*, and no special topology to prevent both the **S**-arrows or both the **I**-arrows from appearing in the same arrow patterns.

## 4.2. Mechanisms Predicting a Hyperbolic $v_0(S)$

In the general Mechanism 4.I, substrate reacts only with *E* in the absence of inhibitor, but with both *E* and *EI* in the presence of inhibitor; correspondingly $v_0(S)$ changes from first-degree to second-degree. However, there are two kinds of mechanisms for which $v_0(S)$ remains first-degree even in the presence of the inhibitor, viz. (a) steady-state mechanisms which provide only one enzyme species to react with *S* even when inhibitor is added; and (b) mechanisms for which the quasi-equilibrium simplification is valid.

### 4.2.1. Steady-state mechanisms

If substrate reacts only with a single enzyme species even in the presence of the inhibitor, the steady-state rate law will remain first-degree hyperbolic even in the presence of the inhibitor. Such is the case with the mechanisms portrayed in Fig. 4.1:

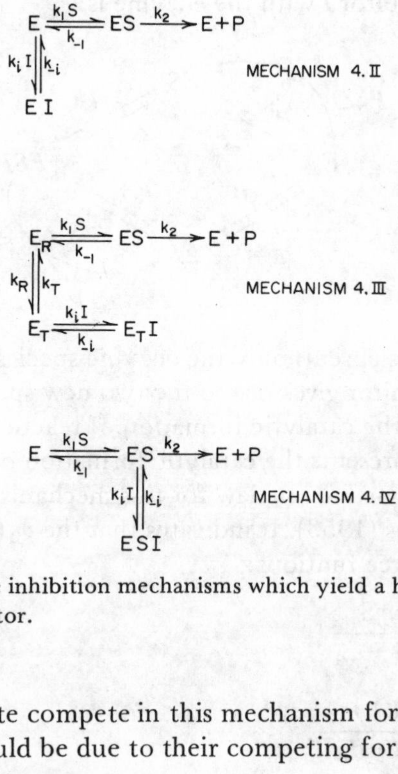

$$E \underset{k_{-1}}{\overset{k_1 S}{\rightleftharpoons}} ES \xrightarrow{k_2} E + P, \qquad E \underset{k_{-i}}{\overset{k_i I}{\rightleftharpoons}} EI$$

MECHANISM 4.II

$$E_R \underset{k_{-1}}{\overset{k_1 S}{\rightleftharpoons}} ES \xrightarrow{k_2} E + P, \qquad E_R \underset{k_T}{\overset{k_R}{\rightleftharpoons}} E_T, \qquad E_T \underset{k_{-i}}{\overset{k_i I}{\rightleftharpoons}} E_T I$$

MECHANISM 4.III

$$E \underset{k_{-1}}{\overset{k_1 S}{\rightleftharpoons}} ES \xrightarrow{k_2} E + P, \qquad ES \underset{k_{-i}}{\overset{k_i I}{\rightleftharpoons}} ESI$$

MECHANISM 4.IV

Fig. 4.1. Steady-state inhibition mechanisms which yield a hyperbolic $v_0(S)$ even in the presence of inhibitor.

*Mechanism 4.II*

Inhibitor and substrate compete in this mechanism for reaction with the free enzyme. This could be due to their competing for binding at the identical site on the enzyme; but it would also result if the binding of either prevented completely the binding of the other by whatever means. The steady-state rate law for this mechanism, to be derived in Problem 4.1 along with those for Mechanisms 4.III and 4.IV, has the algebraic form of

$$v_0 = \frac{\mathrm{S} \cdot V_S}{\mathrm{S} + K_m^S \left(1 + \dfrac{\mathrm{I}}{K_I}\right)} \tag{4.1}$$

Since the $k_1\mathrm{S}$ and $k_i\mathrm{I}$ arrows both leave from the same enzyme species $E$, they cannot appear together in the same arrow-patterns in the rate law; therefore no S · I-term is permitted. When S is raised to saturating levels, the terms in the rate law which do not carry S become negligible compared to the terms which do. Therefore $v_0$ will approach the same saturation value of $V_S$ irrespective of the presence of the inhibitor. This failure of I to affect the saturation velocity stems from the lack of an SI-term in the rate law, and the inhibition is said to be *competitive* in nature.

*Mechanism 4.III*

In this mechanism the enzyme exists in two distinct conformations. Conformation $E_R$ reacts exclusively with substrate, and conformation $E_T$ reacts exclusively with inhibitor. The transition between $E_R$ and $E_T$ is usually referred to as an *allosteric transition* governed by the *allosteric constant L:*

$$L = \frac{(E_T)}{(E_R)} = \frac{k_R}{k_T}\,. \tag{4.2}$$

The steady-state law has the algebraic form of:

$$v_0 = \frac{\mathrm{S} \cdot V_S}{\mathrm{S} + K_m^S \left(1 + \dfrac{\mathrm{I}}{K_I + K_I/L}\right)} \tag{4.3}$$

The topology of this mechanism incorporates an unusual restriction. The $k_i\mathrm{I}$ and $k_{-i}$ arrows form a local cycle and are not allowed to appear in the same arrow-patterns in the rate law. As a result $k_i\mathrm{I}$ appears only in denominator patterns which terminate at *EI*. Because the $E_R$- and $E_T$ reaction loops are sparsely interconnected only through $k_R$ and $k_T$, arrow-patterns which terminate at *EI* necessarily include $k_R$, and by the same token exclude $k_1\mathrm{S}$ which leaves from the same enzyme species as $k_R$. Consequently, $k_i\mathrm{I}$ and $k_1\mathrm{S}$ never appear together and no SI-term appears in the rate law. The inhibition by *I* is once again *competitive,* entirely without any effect on $V_S$.

*Mechanism 4.IV*

In this mechanism the inhibitor binds only to the $ES$ complex, and by so doing prevents the catalytic breakdown of substrate to product. Its steady-state rate law is:

$$v_0 = \frac{\mathrm{S} \cdot V_S}{\mathrm{S}(1 + \mathrm{I}/K_I') + K_m^S} = \frac{\mathrm{S} \cdot V_S/(1 + \mathrm{I}/K_I')}{\mathrm{S} + K_m^S/(1 + \mathrm{I}/K_I')}. \tag{4.4}$$

The appearance of an **SI**-term but no **I**-term in the denominator is again explicable in terms of mechanism topology. The $k_i\mathrm{I}$ arrow leads into the dead-end species $ESI$, and can appear only in arrow-patterns which terminate there. The other species $E$ and $ES$ each have to contribute an arrow to such patterns, and $E$ only has $k_1\mathrm{S}$ to contribute. This explains why $k_i\mathrm{I}$ must be accompanied by $k_1\mathrm{S}$, yielding an **SI**-term but excluding an **I**-term. Due to this absence of an **I**-term, inhibitor $I$ must effect $V_S$ and $K_m^S$ by precisely the same factor, $(1 + \mathrm{I}/K_I')$, and the inhibition is said to be *uncompetitive* in nature.

### 4.2.2. Quasi-equilibrium mechanisms

When the catalytic steps $k_2$ and $k_2'$ in the general mechanism 4.I are slow relative to the other steps, the binding of substrate and inhibitor to the enzyme would approach a state of quasi-equilibrium. Quasi-equilibrium represents a limiting condition of the steady-state, and the quasi-equilibrium rate law is obtainable from the steady-state rate law by setting $k_2$ and $k_2'$ to be very small. Or it may be derived from a direct consideration of the binding equilibria, as will be seen.

At quasi-equilibrium, the distribution of enzyme amongst the four enzyme species $E$, $ES$, $EI$ and $ESI$ is determined completely by the four dissociation constants $K_S$, $K_S'$, $K_I$ and $K_I'$:

$$\frac{(E) \cdot \mathrm{S}}{(ES)} = K_S \qquad \frac{(E) \cdot \mathrm{I}}{(EI)} = K_I$$

$$\frac{(EI) \cdot \mathrm{S}}{(ESI)} = K_S' \qquad \frac{(ES) \cdot \mathrm{I}}{(ESI)} = K_I' \tag{4.5}$$

These four constants are related. For instance, the two routes for calculating $(E)$ from $(ESI)$ must be equivalent in order not to violate the thermodynamic principle of *microscopic reversibility*:

$$\text{upper route:} \quad (E) = \frac{K_S}{\mathrm{S}} \cdot (ES) = \frac{K_S}{\mathrm{S}} \cdot \frac{K_I'}{\mathrm{I}} \cdot (ESI)$$

$$\text{lower route:} \quad (E) = \frac{K_I}{\mathrm{I}} \cdot (EI) = \frac{K_I}{\mathrm{I}} \cdot \frac{K_S'}{\mathrm{S}} \cdot (ESI)$$

Equivalence of the two routes means that

$$\frac{K'_S}{K_S} = \frac{K'_I}{K_I}. \tag{4.6}$$

Therefore the effect of inhibitor on substrate binding is precisely the same as that of substrate on inhibitor binding. Since product is formed in both the $k_2$ and $k'_2$ steps, the reaction velocity is expressible as

$$v_0 = \frac{k_2(ES) + k'_2(ESI)}{(ESI) + (ES) + (EI) + (E)}. \tag{4.7}$$

Substituting Eqns (4.5) and (4.6) into (4.7) yields the rate law in terms of substrate and inhibitor concentrations:

$$v_0 = \frac{\mathrm{S}\cdot(k_2 + k'_2\mathrm{I}/K'_I)/(1 + \mathrm{I}/K'_I)}{\mathrm{S} + K_S(1 + \mathrm{I}/K_I)/(1 + \mathrm{I}/K'_I)}. \tag{4.8}$$

Mechanistically, it is important to draw distinction between several different quasi-equilibrium cases, viz., Mechanisms 4. Va-e, which differ from one another with respect to the equality or inequality between $k_2$ and $k'_2$, and between $K_I$ and $K'_I$ (the combination of $k'_2 = k_2$ and $K'_I = K_I$ is not a valid case, because the inhibitor would be totally devoid of any observable effect on the reaction.) The rate law for each of these mechanisms is obtainable from Eqn (4.8) simply by inserting the special values assigned to $k'_2$ and $K'_I$:

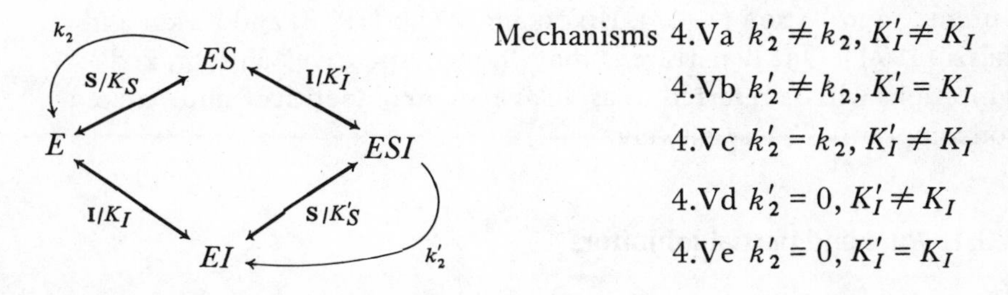


### 4.2.3. Conditional parameters: $V_{(S)}$ and $K^S_{(m)}$

The rate law for Mechanisms 4.II–4.V in the absence of inhibitor is:

$$v_0 = \frac{\mathrm{S}\cdot V_S}{\mathrm{S} + K^S_m}$$

$$V_S = k_2 \tag{4.9}$$

$$K^S_m = \frac{(k_2 + k_{-1})}{k_1} \text{ at steady-state, or } k_{-1}/k_1 \text{ at quasi-equilibrium.}$$

Although for this group of mechanisms the rate law remains hyperbolic when inhibitor is added, the saturation velocity $V_S$ or Michaelis constant $K_m^S$ may be affected by the inhibitor; consequently it is necessary to rewrite the rate law in the form of:

$$v_0 = \frac{S \cdot V_{(S)}}{S + K_{(m)}^S} \tag{4.10}$$

$V_{(S)}$ and $K_{(m)}^S$ are respectively the *conditional saturation velocity* and the *conditional Michaelis constant:* their values are conditional upon the inhibitor concentration employed. By writing Eqns (4.1), (4.3), (4.4) and (4.8) in the form of (4.10), the expressions for $V_{(S)}/V_S$ and $K_{(m)}^S/K_m^S$ appropriate to the different inhibition mechanisms are readily obtained and are entered in Table 4.1. For instance, in the case of Mechanism 4.II and its Eqn (4.1), we have

$$V_{(S)} = V_S, \quad \text{and therefore} \quad V_{(S)}/V_S = 1$$
$$K_{(m)}^S = K_m^S\,(1 + I/K_I), \quad \text{and therefore} \quad K_{(m)}^S/K_m^S = 1 + I/K_I.$$

## 4.3. Nomenclature and Differentiation

The foregoing survey of inhibition mechanisms predicting a hyperbolic $v_0$(S) is by no means exhaustive, and the idiosyncrasies of many enzymes may demand consideration of additional model mechanisms. Many nomenclatures have been proposed for inhibition mechanisms. The nomenclature to be adopted in this section is largely that developed by Ebersole, Gutentag and Wilson (1943), Dixon and Webb (1958) and Keleti and Fajszi (1971). One departure is that "noncompetitive" and "mixed" inhibitions will be referred to as "classic-noncompetitive" and "mixed-noncompetitive", respectively.

### 4.3.1. Pure and partial inhibitors

When enzyme species containing the inhibitor are catalytically inactive, $v_0$(I) will be an inhibitor-type function. Reaction velocity at any fixed substrate concentration will fall to zero with high enough concentrations of inhibitor, i.e. $V_I = 0$, and the inhibition is *pure.* On the other hand, when catalysis can proceed within the enzyme-substrate-inhibitor complex, $v_0$(I) will be a modifier-type function. Reaction velocity will approach a finite asymptote at high inhibitor concentrations, i.e. $V_I \neq 0$, and the inhibition is *partial.*

Mechanisms 4.II-4.IV, as well as 4.Vd and 4.Ve, lead to pure inhibition. Mechanisms 4.Va, 4.Vb and 4.Vc, where *ESI* is deemed to be catalytically active, lead to partial inhibition.

### 4.3.2. Competitive, uncompetitive, and noncompetitive inhibitions

The varying effects of inhibitor on the saturation velocity and the Michaelis constant suffice to distinguish between four patterns of inhibitions:

(a) *Competitive inhibition:* the saturation velocity is unaltered by inhibitor, i.e. $V_{(S)} = V_S$.

(b) *Uncompetitive inhibition:* the ratio between the Michaelis constant and the saturation velocity is unaltered by inhibitor, i.e.,

$$\frac{K^S_{(m)}}{V_{(S)}} = \frac{K^S_m}{V_S}, \quad \text{which also means} \quad \frac{K^S_{(m)}}{K^S_m} = \frac{V_{(S)}}{V_S}.$$

(c) *Classic-noncompetitive inhibition:* the Michaelis constant is unaltered by inhibitor, i.e. $K^S_{(m)} = K^S_m$.

(d) *Mixed-noncompetitive inhibition:* the Michaelis constant, the saturation velocity, and the ratio between them, are all altered by the inhibitor.

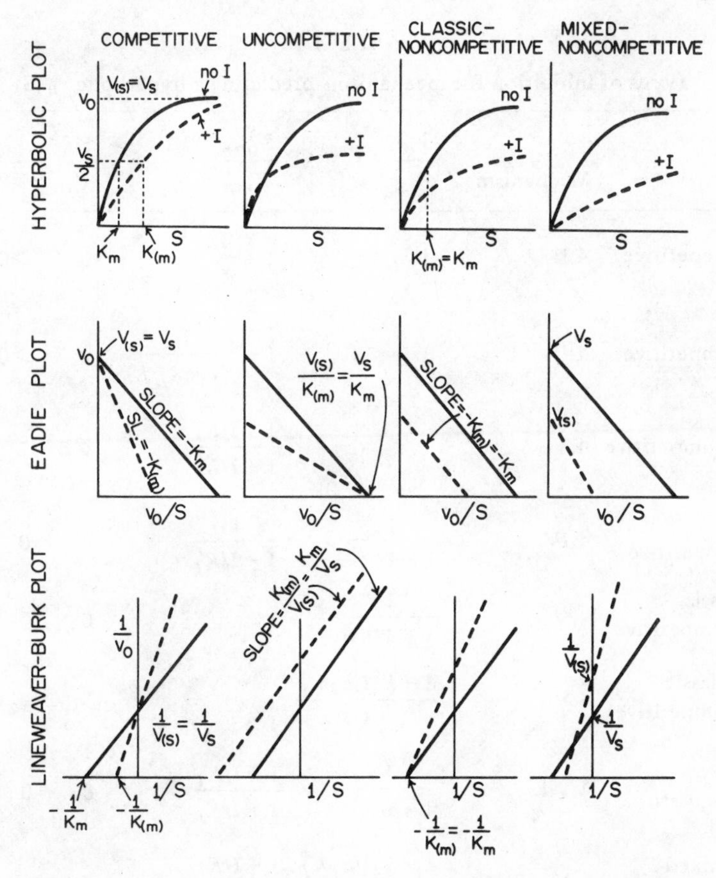


Fig. 4.2. Four types of inhibition patterns as revealed by the hyperbolic, Eadie, or Lineweaver-Burk plot.

Ebersole, Gutentag and Wilson called type (c) "noncompetitive", and Dixon and Webb called type (d) "mixed". However, Cleland (1963a) called type (d) "noncompetitive". Thus the term "noncompetitive" has been used in two dissimilar ways. The nomenclature adopted above, recognizing the distinction between the *classic* and *mixed* varieties of noncompetitive behaviour, eliminates possible confusion because of this double usage.

These four different types of inhibition give rise to four different patterns in the hyperbolic or linearized plots of the dependence of velocity on substrate concentration (Fig. 4.2). Together with the distinction between pure ($V_I = 0$) and partial ($V_I \neq 0$) inhibitions, this yields eight categories of inhibition patterns. Mechanisms 4.II–4.V are categorized on this basis in Table 4.1.

The competitive inhibition of Mechanism 4.II and the allosteric inhibition of Mechanism 4.III cannot be differentiated solely by their inhibition patterns, but they can be differentiated by experimental criteria developed for characterizing allosteric mechanisms (Chapter 8). For example, a direct

TABLE 4.1

Types of inhibition for mechanisms predicting a hyperbolic $v_0$ (S)

| Type of inhibition | Mechanism | $\frac{V_{(S)}}{V_S}$ | $\frac{K^S_{(m)}}{K^S_m}$ | $V_I$ | $\frac{V_{(S+I)}}{V_S}$ |
|---|---|---|---|---|---|
| Pure, competitive | 4.II | 1 | $1 + \frac{I}{K_I}$ | 0 | $>0, <1$ |
| Pure, competitive | 4.III | 1 | $1 + \frac{I}{K_I + K_I/L}$ | 0 | $>0, <1$ |
| Partial, competitive | 4.Vc | 1 | $\frac{1 + I/K_I}{1 + I/K'_I}$ | $>0$ | 1 |
| Pure, uncompetitive | 4.IV | $\frac{1}{1 + I/K'_I}$ | $\frac{1}{1 + I/K'_I}$ | 0 | 0 |
| Pure, classic-noncompetitive | 4.Ve | $\frac{1}{1 + I/K_I}$ | 1 | 0 | 0 |
| Partial, classic-noncompetitive | 4.Vb | $\frac{1 + k'_2 I/k_2 K_I}{1 + I/K_I}$ | 1 | $>0$ | $k'_2/k_2$ |
| Pure, mixed-noncompetitive | 4.Vd | $\frac{1}{1 + I/K'_I}$ | $\frac{1 + I/K_I}{1 + I/K'_I}$ | 0 | 0 |
| Partial, mixed-noncompetitive | 4.Va | $\frac{1 + k'_2 I/k_2 K'_I}{1 + I/K'_I}$ | $\frac{1 + I/K_I}{1 + I/K'_I}$ | $>0$ | $k'_2/k_2$ |

physical measurement of enzyme conformation may be performed. According to Mechanism 4.III ($E_T$) the inhibitor-binding conformation should vary with inhibitor concentration as follows:

$$\frac{(E_T)}{(E)_0} = \frac{m_1\mathbf{I} + m_0}{d_1\mathbf{I} + d_0} . \tag{4.11}$$

On the other hand, according to Mechanism 4.II, binding of inhibitor may or may not induce a measurable change in enzyme conformation. If it does, the induced conformation should vary with inhibitor concentration in the form of

$$\frac{(EI)}{(E)_0} = \frac{m_1\mathbf{I}}{d_1\mathbf{I} + d_0} . \tag{4.12}$$

Equation (4.12) predicts a curve which goes through the origin, but Eqn (4.11) does not. This testable difference is due to the pre-existence of a conformational transition in Mechanism 4.III but not in Mechanism 4.II.

The quasi-equilibrium Mechanisms 4.Vb and 4.Ve predict classic-non-competition. If the additional suppositions are made that inhibitor binds equally to both $E$ and $ES$, but $EI$ cannot bind substrate to form $ESI$, classic-noncompetition would be predicted under steady-state conditions as well (Botts and Morales, 1953).

## 4.4. Method of Joint Saturations

In the partial-inhibition Mechanisms 4.Va and 4.Vb the ratio $k_2'/k_2$ reflects the effect of inhibitor on the efficiency of catalytic transformation. Experimentally, it may be evaluated by the following method:

(a) At some substrate concentration $\mathbf{S}'$, determine the inhibitor concentration $\mathbf{I}'$ which will reduce the velocity by about half. This will ensure that the rate effects due to $\mathbf{I}'$ are not overwhelmed by those due to $\mathbf{S}'$, and *vice versa*.

(b) Add to the system increasing concentrations of substrate and inhibitor at the fixed ratio of $\mathbf{S}':\mathbf{I}'$, and determine the asymptotic velocity approached at very high concentrations. This velocity, jointly saturated by substrate and inhibitor, may be represented as $V_{(S+I)}$.

(c) The ratio between $k_2'$ and $k_2$ is equal to the ratio between $V_{(S+I)}$ and $V_S$:

$$\frac{k_2'}{k_2} = \frac{V_{(S+I)}}{V_S} .$$

The basis for this proposed method is straightforward. In the absence of inhibitor, enzyme accumulates in the form of *ES* when substrate is saturating, and $V_S = k_2$. However, it accumulates in the form of *ESI* when substrate and inhibitor are jointly saturating, and $V_{(S+I)} = k'_2$. Therefore the ratio between the two saturation velocities is equal to $k'_2/k_2$.

It is evident from Table 4.1 that observations on $V_{(S+I)}$ also will contribute to the differentiation of other inhibition mechanisms. For example, in Mechanisms 4.IV, 4.Vd and 4.Ve, joint saturation by substrate and inhibitor will entrap all the enzyme into the unproductive complex *ESI*, and $V_{(S+I)}$ will fall to zero. In contrast, *ESI* is not formed in Mechanisms 4.II and 4.III, and $V_{(S+I)}$ will be finite, albeit smaller than $V_S$ on account of the counterproductive competition between substrate and inhibitor. Finally, the competitive partial-inhibition Mechanism 4.Vc predicts that $V_{(S+I)} = V_S$, because in this case $k'_2 = k_2$.

## 4.5. Determination of Inhibition Constants

The dissociation constants $K_I$ and $K'_I$ describe the dissociation of inhibitor from the *EI* and the *ESI* complexes, respectively. For the various inhibition mechanisms, these constants can be calculated from the ratios $V_{(S)}/V_S$ and $K^S_{(m)}/K^S_m$ expressed in Table 4.1, as follows:

Mechanism 4.II:

$$K_I = \frac{\mathrm{I}}{\dfrac{K^S_{(m)}}{K^S_m} - 1}$$

Mechanisms 4.III:

$$\text{apparent } K_I = K_I(1 + 1/L) = \frac{\mathrm{I}}{\dfrac{K^S_{(m)}}{K_m} - 1}$$

(i.e. the true $K_I$ can be determined only if the allosteric constant $L$ is known)

Mechanism 4.IV, 4.Vd:

$$K'_I = \frac{\mathrm{I}}{\dfrac{V_S}{V_{(S)}} - 1}$$

Mechanism 4.Ve:

$$K_I = \frac{\mathbf{I}}{\dfrac{V_S}{V_{(S)}} - 1}$$

Mechanisms 4.Va, 4.Vc, and 4.Vd:

$$\frac{K^S_{(m)}}{K^S_m} = \frac{1 + \dfrac{\mathbf{I}}{K_I}}{1 + \dfrac{\mathbf{I}}{K'_I}}$$

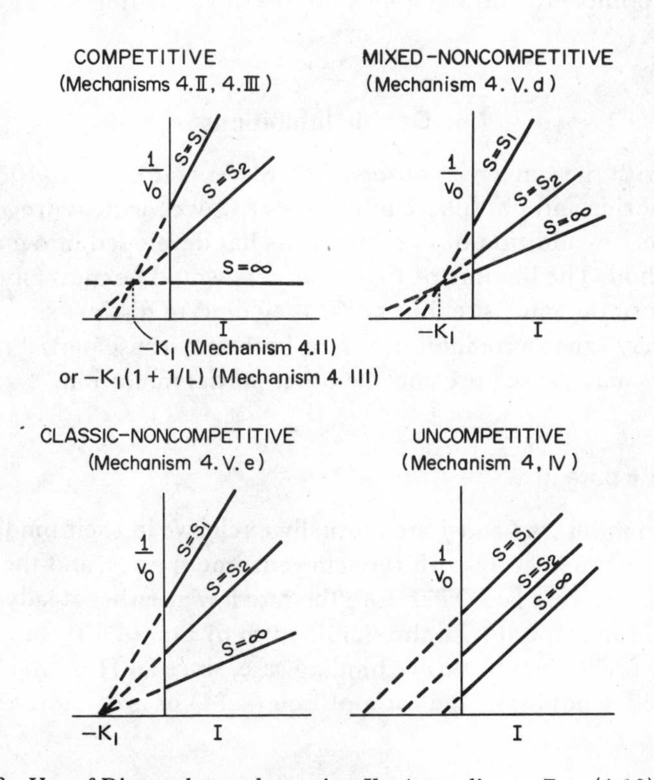


Fig. 4.3. Use of Dixon plot to determine $K_I$. According to Eqn (4.10), $1/v_0$ would be independent of **S** under the condition of $K^S_{(m)}/V_{(S)} = 0$. This condition is fulfilled at $\mathbf{I} = -K_I$ for Mechanisms 4.II, 4.Vd and 4.Ve, or at $\mathbf{I} = -K_I(1 + 1/L)$ for Mechanism 4.III (see Table 4.1). The Dixon plots obtained at various substrate concentrations converge when this condition is fulfilled. Fulfilment of this condition is not possible in the case of Mechanism 4.IV, which accordingly predicts parallel Dixon plots. On the other hand, fulfilment of this condition would cause the velocity to become infinite in the case of Mechanism 4.Ve, which accordingly predicts that the Dixon plots converge on the **I**-axis. Mechanism 4.Vd differs from 4.II and 4.III insofar that the saturation velocity obtained at infinite substrate concentration will vary with **I** in the former but not the latter cases. (Extended from Dixon, (1953a.)

and the two unknowns, $K_I$ and $K_I'$, can be evaluated from the two $K^S_{(m)}/K^S_m$ values estimated at any two different levels of **I**.

Graphically, the pure-inhibition mechanisms all predict a linear Dixon plot of $1/v_0$ versus **I** (see Section 3.4). As Fig. 4.3 shows, such straight-line plots obtained at various substrate levels give different patterns of convergence for the different mechanisms. They run parallel to one another for the uncompetitive Mechanism 4.IV. However, they converge at $\mathbf{I} = -K_I$, either on the **I**-axis for the classic noncompetitive Mechanism 4.Ve, or off the *I*-axis for the competitive Mechanism 4.II and the mixed-noncompetitive Mechanism 4.Vd. They also converge, at $\mathbf{I} = -K_I(1 + 1/L)$, off the **I**-axis for the competitive Mechanism 4.III. Therefore the various convergent points provide a graphical means of estimating $K_I$, or $K_I(1 + 1/L)$.

## 4.6. Double Inhibitions

Beginning with the synergism observed by Slater and Bonner (1952) between fluoride and phosphate inhibitions of succinic dehydrogenase, the simultaneous addition of two inhibitors has developed into a useful kinetic method. The binding of two inhibitors would be mutually exclusive if they bind to the same site, but not if they bind to distinct sites. Both inhibitors may cause pure inhibitions, or both may cause partial inhibitions, or one may cause pure and the other partial inhibition.

### 4.6.1. Pure + pure

If two pure inhibitors *I* and *J* are mutually exclusive in their binding, they would react competitively with the same enzyme species, and there can be no **IJ**-term in the rate law. Therefore the rate law at either steady-state or quasi-equilibrium would take the simple form of Eqn (4.13). In contrast, the rate law for the nonexclusive binding case, since an **IJ**-term would not be prohibited, would take the form of Eqn (4.14) or even more complex forms:

$$\text{exclusive binding:} \qquad v_0 = \frac{m_0}{d_i\mathbf{I} + d_j\mathbf{J} + d_0} \tag{4.13}$$

$$\text{nonexclusive binding:} \qquad v_0 = \frac{m_0}{d_{ij}\mathbf{IJ} + d_i\mathbf{I} + d_j\mathbf{J} + d_0} \tag{4.14}$$

Experimental distinction between exclusive and nonexclusive binding depends on a differentiation between these two rate laws, which may be achieved through one of the three available methods (Fig. 4.4):

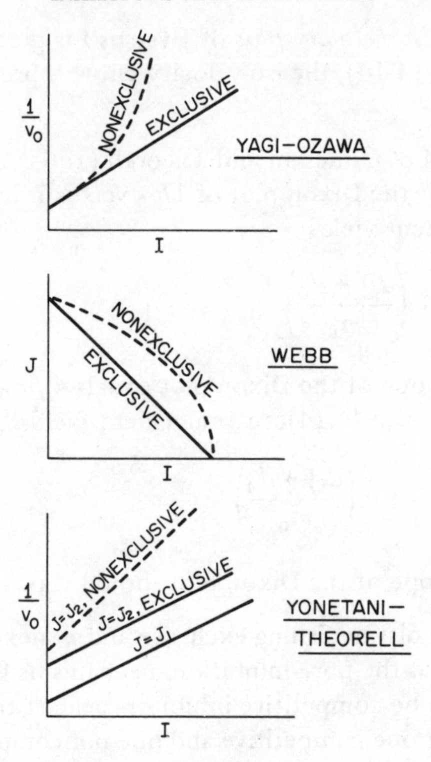


Fig. 4.4. Methods for distinguishing between *exclusive* and *nonexclusive* binding of two pure inhibitors, *I* and *J*.

(a) In the method of Yagi and Ozawa (1960), the two inhibitors *I* and *J* are added to the reaction system at a fixed ratio *r*, so that $\mathbf{J} = r\mathbf{I}$. This converts Eqn (4.13) into a first-degree inhibitor-function in **I**:

$$v_0 = \frac{m_0}{(d_i + rd_j)\mathbf{I} + d_0} \tag{4.13a}$$

and the Dixon plot of $1/v_0$ versus **I** is predicted to be linear. In contrast, this places a second-degree **I**-term into Eqn (4.14):

$$v_0 = \frac{m_0}{rd_{ij}\mathbf{I}^2 + (d_i + rd_j)\mathbf{I} + d_0} \tag{4.14a}$$

and the Dixon plot is predicted to be exponential.

(b) In the method of Webb (1963) a constant reaction velocity is maintained by counterbalancing changes in **I** by appropriate changes in **J**. In the case of Eqn (4.13), velocity will stay constant if

$$\Delta\mathbf{I} = -\frac{d_j}{d_i}\,\Delta\mathbf{J}$$

Consequently *the iso-velocity plot* of J versus I is predicted to be linear. In the case of Eqn (4.14), the iso-velocity curve is predicted to be non-linear.

(c) The method of Yonetani and Theorell (1964) calls for an examination of the slope of the Dixon plot of $1/v_0$ versus I. In the case of Eqn (4.13), rearrangement yields

$$\frac{1}{v_0} = \left(\frac{d_i}{m_0}\right) \mathrm{I} + \left(\frac{d_j\mathrm{J} + d_0}{m_0}\right) \tag{4.13b}$$

Accordingly, the slope of the Dixon plot equals $d_i/m_0$, and is independent of J. In the case of Eqn (4.14), rearrangement yields

$$\frac{1}{v_0} = \left(\frac{d_{ij}\mathrm{J} + d_i}{m_0}\right) \mathrm{I} + \left(\frac{d_j\mathrm{J} + d_0}{m_0}\right) \tag{4.14b}$$

Accordingly the slope of the Dixon plot should vary with J.

These methods for distinguishing exclusive and nonexclusive binding are valid whether or not the pure-inhibition mechanisms for *I* and *J* are alike, e.g. they may both be competitive inhibitors against the substrate, both noncompetitive, or one competitive and one noncompetitive.

### 4.6.2. Pure + partial

Rate laws for the pure + partial and partial + partial combinations have been derived and analysed by Keleti and Fajszi (1971). If *J* is a partial inhibitor and *I* a pure inhibitor, the rate-law numerator will contain J so that $v_0$ will not vanish at infinite J. However, it will not contain I so that $v_0$ will vanish at infinite I. Further, there will be an IJ-term in the denominator in the case of nonexclusive binding, but not in the case of exclusive binding. Therefore the predicted rate laws are:

exclusive binding:
$$v_0 = \frac{m_j\mathrm{J} + m_0}{d_i\mathrm{I} + d_j\mathrm{J} + d_0} \tag{4.15}$$

nonexclusive binding:
$$v_0 = \frac{m_j\mathrm{J} + m_0}{d_{ij}\mathrm{IJ} + d_i\mathrm{I} + d_j\mathrm{J} + d_0} \tag{4.16}$$

The method of Yagi and Ozawa is again applicable. Using $\mathrm{J} = r\mathrm{I}$, Eqns (4.15) and (4.16) become respectively (4.15a) and (4.16a):

$$v_0 = \frac{rm_j\mathrm{I} + m_0}{(d_i + rd_j)\mathrm{I} + d_0} \tag{4.15a}$$

$$v_0 = \frac{rm_j\text{I} + m_0}{rd_{ij}\text{I}^2 + (\text{d}_i + rd_j)\text{I} + d_0} \tag{4.16a}$$

Equation (4.15a) is a first-degree modifier function, for which the Eadie-type plot of $\Delta v_0$ versus $\Delta v_0/\text{I}$ (Section 3.3) is predicted to be linear. Equation (4.16a) is a second-degree inhibitor type function, and the plot is predicted to be nonlinear. Also, the Dixon plot of $1/v_0$ versus **I** should approach a finite asymptote for Eqn (4.15a), but increase indefinitely for Eqn (4.16a).

### 4.6.3. Partial + partial

If *I* and *J* are both partial inhibitors, both concentration factors would appear in the rate-law numerator, and the appropriate rate laws would be:

exclusive binding: $$v_0 = \frac{m_i\text{I} + m_j\text{J} + m_0}{d_i\text{I} + d_j\text{J} + d_0} \tag{4.17}$$

nonexclusive binding: $$v_0 = \frac{m_{ij}\text{IJ} + m_i\text{I} + m_j\text{J} + m_0}{d_{ij}\text{IJ} + d_i\text{I} + d_j\text{J} + d_0} \tag{4.18}$$

Using $\text{J} = r\text{I}$ in the method of Yagi and Ozawa, these equations become respectively (4.17a) and (4.18a):

$$v_0 = \frac{(m_i + rm_j)\text{I} + m_0}{(d_i + rd_j)\text{I} + d_0} \tag{4.17a}$$

$$v_0 = \frac{rm_{ij}\text{I}^2 + (m_i + rm_j)\text{I} + m_0}{rd_{ij}\text{I}^2 + (d_i + rd_j)\text{I} + d_0} \tag{4.18a}$$

Equation (4.17a) is a first-degree modifier function, and Eqn (4.18a) is a second-degree modifier function. The former predicts a linear Eadie-type plot of $\Delta v_0$ versus $\Delta v_0/I$, but the latter predicts a nonlinear plot.

## 4.7. Generalized Treatments of Inhibition

So far we have been concerned with the basic inhibition models for one-substrate reactions. The range of plausible models is far wider for multi-substrate reactions, but the kinetic methods examined above can be combined with those to be examined in Chapters 5–9 in dealing with multi-substrate inhibition and activation. The studies of Henderson (1968) and Clark (1970) have amply demonstrated the feasibility of a general treatment for the inhibition kinetics of multisubstrate systems. It is also useful here to take note of some observations which are valid for inhibition phenomena in multisubstrate as well as one-substrate systems.

### 4.7.1. Inhibition rules

For any linear mechanism topology of the form

$$\cdots \rightleftharpoons EX_i \overset{S}{\rightleftharpoons} \cdots \rightleftharpoons EX_j \rightleftharpoons \cdots \xrightarrow[\text{irreversible step}]{} $$
$$EX_j \underset{}{\overset{I}{\rightleftharpoons}} EX_jI$$

Hearon (1952) has formulated a series of rigorous inhibition rules that include the following:

(a) If $EX_i$ precedes $EX_j$ in the linear sequence, $K^S_{(m)}/V_{(S)} = K^S_m/V_S$, but $1/V_S$ increases linearly with **I**. This represents a *generalized uncompetitive inhibition.*

(b) If $EX_i$ is the same species as $EX_j$, $V_{(S)} = V_S$, but $K^S_{(m)}/V_{(S)}$ increases linearly with **I**. This represents a *generalized competitive inhibition.*

(c) If $EX_j$ precedes $EX_i$ in the linear sequence, both $1/V_{(S)}$ and $K^S_{(m)}/V_{(S)}$ increase linearly with **I**. This represents a *generalized mixed-noncompetitive inhibition.*

Inhibition rules for linear sequences also have been formulated by Cleland (1963b). These are not rigorous and are more useful for the treatment of product inhibitions in linear multisubstrate mechanisms (Section 6.6).

### 4.7.2. Multiple dead-ends

If two ligands react with different enzyme species in a mechanism, their concentrations can usually appear jointly in the same arrow-patterns in the rate law. However, if their reaction steps are located in regions of the mechanism that are sparsely connected, their joint appearance may be prevented by topological constraints. Mechanism 4.III is such an example: the ligands $S$ and $I$ react with different enzyme species, but no **SI**-term is permitted. The tetrameric allosteric Mechanism of Monod, Wyman and Changeux (Fig. 8.6) is another example: $S$ reacts with eight enzyme species, but the steady-state $v_0$(S) function is only fourth-degree. Still another example is encountered when two ligands react with separate enzyme species to form a dead-end, but their concentrations cannot appear jointly in the rate law. Take for instance the following topology:

$$E_m \underset{k_{-m}}{\overset{k_m\mathbf{I}}{\rightleftharpoons}} E_mI$$
$$E_nI \underset{k_{-n}}{\overset{k_n\mathbf{I}}{\rightleftharpoons}} E_n \rightleftharpoons \cdots\cdots\cdots \rightleftharpoons E_j \underset{k_{-j}}{\overset{k_j\mathbf{J}}{\rightleftharpoons}} E_jJ \qquad \text{Mechanism 4.VI}$$

In this diagram the $k_m$**I** and $k_{-m}$ arrows form a closed cycle, and are therefore forbidden by the structural rule to appear together. Therefore the $k_m$ I arrow appears only in those denominator patterns which terminate at $E_mI$, and nowhere in the numerator. By the same token, the $k_n$I arrow appears only in those denominator patterns which terminate at $E_nI$, and the $k_j$J arrow appears only in those denominator patterns which terminate at $E_jJ$. These three arrows thus occur only in separate denominator terms and not at all in numerator terms. The rate law carries neither an $I^2$-term nor an IJ-term. The implication is that separate dead-end reactions can never be synergistic with respect to their detrimental effects on the reaction.

### 4.7.3. Substrate inhibition

Any reaction step that leads to a dead-end enzyme species is barred from the rate-law numerator. If a substrate reacts to form a dead-end in the mechanism, the rate law may or may not become a substrate-inhibitor type. Mechanisms 4.VII and 4.VIII give two simple examples:

$$SE \underset{k_{-3}}{\overset{k_3\mathrm{S}}{\leftleftarrows}} E \underset{k_{-1}}{\overset{k_1\mathrm{S}}{\rightleftarrows}} ES \xrightarrow{k_2} E + P \qquad \text{Mechanism 4.VII}$$

$$E \underset{k_{-1}}{\overset{k_1\mathrm{S}}{\rightleftarrows}} ES \xrightarrow{k_2} E + P, \quad ES \underset{k_{-3}}{\overset{k_3\mathrm{S}}{\rightleftarrows}} ES_2 \qquad \text{Mechanism 4.VIII}$$

In Mechanism 4.VII, $E$ reacts with $S$ in alternative ways. The $k_1$S and $k_3$S arrows both leave from the same enzyme species, and so will never appear together in the same terms. Accordingly the rate law will be first-degree substrate type,

$$v_0 = \frac{m_1\mathrm{S}}{d_1\mathrm{S} + d_0}$$

and no inhibition by excess substrate is expected. In Mechanism 4.VIII, the $k_1$S arrow (from $E$), and the $k_3$S arrow (from $ES$) will appear together in the denominator of the rate law to create an $\mathrm{S}^2$-term. However, there is no $\mathrm{S}^2$-term in the numerator, because $k_3$S leads into a dead-end, and is barred from the numerator. This results in a second-degree inhibitor function (see Eqn 3.12):

$$v_0 = \frac{m_1\mathrm{S}}{d_2\mathrm{S}^2 + d_1\mathrm{S} + d_0}$$

Reaction velocity will be totally inhibited as S becomes infinitely large.

Inhibition by excess substrate is not a rare phenomenon. One plausible cause was suggested many years ago by Murray (1930). If the active site of an enzyme should consist of two or more *subsites* with affinities for different parts of the substrate molecule, the unproductive crowding of two molecules of the substrate on to the same active site might occur. The mechanism proposed by Wilson and Cabib (1956) to explain the inhibition of acetylcholinesterase by excess acetylcholine is a classic illustration of this possibility ($e$ = esteratic subsite on enzyme, $a$ = anionic subsite, $R_1COO$ = acetyl moiety, $R_2$ = choline moiety):

$$E\begin{smallmatrix} e \\ a \end{smallmatrix} \underset{}{\overset{R_1COOR_2}{\rightleftharpoons}} E\begin{smallmatrix} e\cdots \\ a\cdots \end{smallmatrix} R_1COOR_2 \xrightarrow[R_2OH]{} E\begin{smallmatrix} e\text{-}OOCR_1 \\ a \end{smallmatrix} \xrightarrow[R_1COOH]{} E\begin{smallmatrix} e \\ a \end{smallmatrix}$$

$$E\begin{smallmatrix} e\text{-}OOCR_1 \\ a \end{smallmatrix} \underset{}{\overset{R_1COOR_2}{\rightleftharpoons}} E\begin{smallmatrix} e\text{-}OOCR_1 \\ a\ldots R_2OOCR_1 \end{smallmatrix} \quad \text{(Dead-end Complex)}$$

In this mechanism, the second molecule of the substrate $R_1COOR_2$ binds to the acetylenzyme to form a dead-end complex, and the rate law will be second-degree substrate-inhibitor type with respect to $R_1COOR_2$.

In multisubstrate reactions, a substrate also may cause a dead-end by binding in place of a cosubstrate or a reaction product. The mechanism proposed by Raval and Wolfe (1963) for the oxaloacetate inhibition of malate dehydrogenase illustrates such a misdemeanor (OAA = oxaloacetate, MAL = malate):

$$E/\text{—}/OAA \overset{OAA}{\rightleftharpoons} E/\text{—}/\text{—} \overset{NADH}{\rightleftharpoons} E/NADH/\text{—} \overset{OAA}{\rightleftharpoons} E/NADH/OAA$$

$$E/\text{—}/\text{—} \overset{NAD}{\rightleftharpoons} E/NAD/\text{—} \overset{MAL}{\rightleftharpoons} E/NAD/MAL \rightleftharpoons E/NADH/OAA$$

$$E/NAD/\text{—} \overset{OAA}{\rightleftharpoons} E/NAD/OAA$$

In this mechanism oxaloacetate normally combines with the $E$/NADH/— to form the productive ternary complex $E$/NADH/OAA. However, it also may react with the free enzyme $E$/—/— in competition against $NAD^+$ and NADH, or react with the $E$/NAD/— complex in competition against malate; in either case an unproductive dead-end is formed. The rate law will be a second-degree substrate-inhibitor function in terms of OAA:

$$v_0 = \frac{m_1(\text{OAA})}{d_2(\text{OAA})^2 + d_1(\text{OAA}) + d_0}.$$

Neither of the dead-end OAA-arrows can appear in the numerator, which is accordingly first-degree. The two dead-end OAA-arrows cannot appear in the same arrow patterns in the denominator (cf. Mechanism 4.VI), which is accordingly second-degree even though OAA reacts with three different enzyme species in the mechanism.

## 4.8. Essential Metal Activators

Metallo-enzymes contain metal atoms firmly bound to their structures, but metal-activated enzymes depend on the binding of added ions for activity. For metal-activated enzymes which form a 1:1:1 ternary complex between enzyme, metal and substrate, Mildvan (1970) distinguished between four possible coordination schemes within the ternary complex:

*enzyme bridge complex:* $M\text{—}E\text{—}S$

*substrate bridge complex:* $E\text{—}S\text{—}M$

*metal bridge complex:* $E\text{—}M\text{—}S$

*cyclic complex:* $E\langle{}^{M}_{S}$ (E bonded to both M and S, with M—S bonded)

From the kinetic viewpoint, an important question regarding any metal-activated system is the identity of the pathways generating the productive ternary complex.

### 4.8.1. M—E—S coordination

If enzyme is to form a bridge between metal and substrate, these two ligands are expected to bind separately to the enzyme. Mechanism 4.IX is a simple model consistent with this expectation, providing pathways for the random additions of metal and substrate:

$$E \underset{M}{\overset{S}{\rightleftharpoons}} \begin{matrix} ES \\ EM \end{matrix} \underset{S}{\overset{M}{\rightleftharpoons}} EC \rightarrow E + M + P \qquad \text{Mechanism 4.IX}$$

Since $S$ and $M$ are both necessary for the reaction, and both react with two enzyme species in the mechanism, $v_0(\text{S})$ and $v_0(\text{M})$ are second-degree substrate (essential) type functions:

$$v_0 = \frac{m_2 S^2 + m_1 S}{d_2 S^2 + d_1 S + d_0}$$

$$v_0 = \frac{m_2 M^2 + m_1 M}{d_2 M^2 + d_1 M + d_0}. \tag{4.19}$$

If M is raised to saturating levels, all the $E$ would react with M, and only EM is left to react with $S$. Similarly, if S is raised to saturating levels, all the $E$ would react with $S$, and only $ES$ is left to react with $M$. Accordingly $V_S$(M) and $V_M$(S) become first-degree functions:

$$V_{(S)} = \frac{m_1 M}{d_1 M + d_0}$$

$$V_{(M)} = \frac{m_1 S}{d_1 S + d_0} \tag{4.20}$$

Mechanism 4.IX, like Mechanism 4.I, also simplifies to a quasi-equilibrium condition if the catalytic formation of product proves to be rate limiting. At quasi-equilibrium, $v_0$(M) and $v_0$(S) no less than $V_S$(M) and $V_M$(S) are all hyperbolic. Furthermore, if the binding of $M$ and $S$ are mutually independent, Mechanism 4.IX would behave analogously to Mechanism 4.Vb. The resulting relationship between $M$ and $S$ will be classic-non-competitive, i.e. the Michaelis constant for the substrate will be identical to the substrate-enzyme dissociation constant, and invariant with metal concentration:

$$K^S_{(m)} = K_S.$$

Similarly, the Michaelis constant for the metal will be identical to the metal-enzyme dissociation constant, and invariant with substrate concentration:

$$K^M_{(m)} = K_M.$$

The $Ca^{++}$ activation of guinea pig liver transglutaminase for *p*-nitrophenyl acetate hydrolysis, analysed by Folk, Cole and Mullooly (1967) conformed to this mode of behaviour. As Fig. 4.5 demonstrates, the $K_{(m)}$ for *p*-nitrophenyl acetate (given by the intercept on the abscissa of the Lineweaver-Burk plot) remained the same when $Ca^{++}$ concentration was varied from 0.16 mM up to 4.91 mM.

Mechanism 4.IX postulates that metal and substrate do not react with one another to form a metal-substrate complex. If in addition a 1 : 1 metal-substrate complex $C$ is reversibly formed, the result would be Mechanism

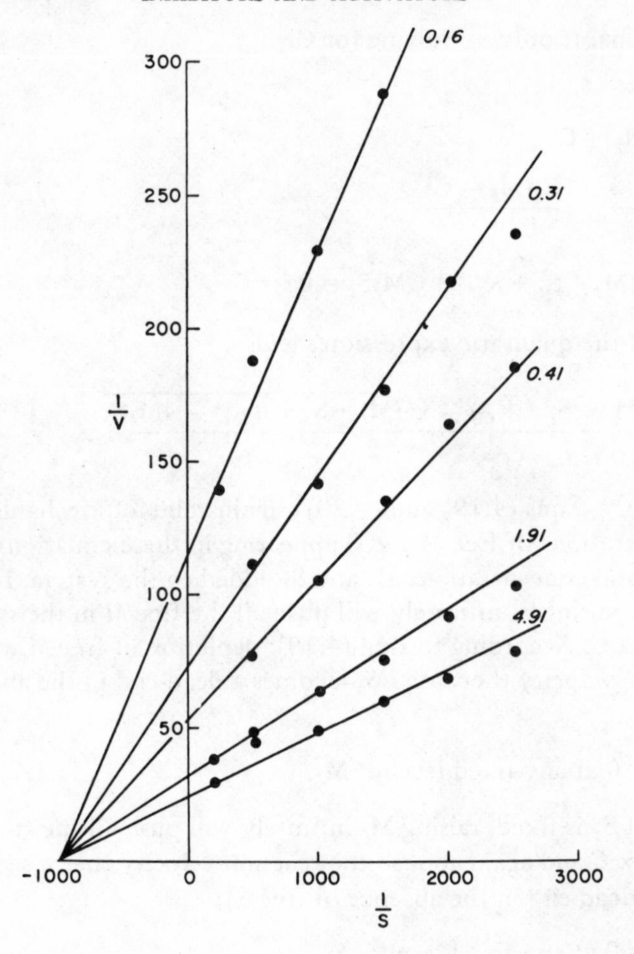


Fig. 4.5. Effects of $Ca^{++}$ concentration (indicated in mM for each line) on the velocity of *p*-nitrophenyl acetate hydrolysis by Guinea pig liver transglutaminase. (From Folk, Cole and Mullooly, 1967.)

4.X, the rate behaviour of which differs sharply from that of 4.IX:

$$E \underset{M}{\overset{S}{\rightleftharpoons}} ES \overset{M}{\rightleftharpoons} EC \rightarrow E + M + P \qquad E \rightleftharpoons EM \overset{S}{\rightleftharpoons} EC \qquad \text{Mechanism 4.X}$$

$$S + M \rightleftharpoons C$$

Due to the formation of the complex $C$, the amount of free substrate S present is no longer the same as the amount of total substrate $S_t$ (even neglecting as usual the much smaller amounts of substrate complexed in various enzyme species). Equally, the free metal M is no longer the same as the total metal $M_t$. Instead, the concentrations of the two free ligands are determinable only by solving for **C**:

$$\mathbf{S} = \mathbf{S}_t - \mathbf{C}$$

$$\mathbf{M} = \mathbf{M}_t - \mathbf{C}$$

$$K_C = \frac{(\mathbf{S}_t - \mathbf{C})(\mathbf{M}_t - \mathbf{C})}{\mathbf{C}} \tag{4.21}$$

$$\mathbf{C}^2 - (\mathbf{M}_t + \mathbf{S}_t + K_C)\mathbf{C} + \mathbf{M}_t\mathbf{S}_t = 0$$

Solution of the quadratic expression yields:†

$$\mathbf{C} = \frac{(\mathbf{M}_t + \mathbf{S}_t + K_C) - \sqrt{(\mathbf{M}_t + \mathbf{S}_t + K_C)^2 - 4\mathbf{M}_t\mathbf{S}_t}}{2}\,. \tag{4.22}$$

Consequently, Eqns (4.19) and (4.20) remain valid for Mechanism 4.X, but the concentrations of free **M** and **S** appearing in these equations deviate from the total concentrations $\mathbf{M}_t$ and $\mathbf{S}_t$ added to the system. In fact, if $\mathbf{M}_t$ is fixed, raising $\mathbf{S}_t$ infinitely will push all the free *M* in the system into the complex *C*. According to Eqn (4.19), depletion of free *M* will abolish the reaction velocity (because *ES* becomes a dead-end in the absence of free *M*):

$$V_{(S)} = 0 \text{ at any fixed level of } \mathbf{M}_t.$$

Similarly, if $\mathbf{S}_t$ is fixed, raising $\mathbf{M}_t$ infinitely will push all the free *S* into the complex *C* and again abolish the reaction velocity (because *EM* becomes a dead-end in the absence of free *S*):

$$V_{(M)} = 0 \text{ at any fixed level of } \mathbf{S}_t.$$

### 4.8.2. E–S–M coordination

If substrate forms a bridge between enzyme and metal, plausibly *S* or *C* but not *M* may bind to the free enzyme:

$$E \underset{S}{\rightleftharpoons} ES \underset{M}{\rightleftharpoons} EC \rightarrow E + M + P; \quad E \underset{C}{\rightleftharpoons} EC \qquad \text{Mechanism 4.XI}$$

$$S + M \rightleftharpoons C$$

For this mechanism, if $\mathbf{M}_t$ is fixed, raising $\mathbf{S}_t$ infinitely will deplete the free *M* and so block the conversion of *ES* to *EC*, effectively turning *ES* into a

† The quadratic in Eqn (4.21) has two roots, but only the root given in (4.22) is meaningful. The other root impossibly requires **C** to be greater than either $\mathbf{M}_t$ or $\mathbf{S}_t$.

dead-end complex:

$$E \underset{C}{\rightleftharpoons} EC \rightarrow E + M + P, \quad E \overset{S}{\rightleftharpoons} ES \dashrightarrow EC$$

The metal-substrate complex formed must not exceed the total substrate or total metal added; **C** cannot increase indefinitely as long as $\mathbf{M}_t$ is fixed. Therefore increases in **S** sooner or later will overwhelm **C** in their competition for $E$, and push all the enzyme into the dead-end $ES$, i.e.

$V_{(S)} = 0$ at any fixed level of $\mathbf{M}_t$.

On the other hand, if $\mathbf{S}_t$ is fixed, raising $\mathbf{M}_t$ infinitely will deplete the free $S$ and result in complete dissociation of $ES$, leaving the lower pathway operative:

$V_{(M)} \neq 0$ at any fixed level of $\mathbf{S}_t$.

The dissimilarity of the two saturation velocities clearly reflects the asymmetry in the mechanism topology.

If the substrate bridge has to be in place on the enzyme before metal can be added, the sequential Mechanism 4.XII becomes important:

$$E \overset{S}{\rightleftharpoons} ES \overset{M}{\rightleftharpoons} EC \rightarrow E + M + P \qquad \text{Mechanism 4.XII}$$

This mechanism predicts that the saturation velocities with respect to both substrate and metal will be finite; all of $v_0(\mathrm{S})$, $v_0(\mathrm{M})$, $V_S(\mathrm{M})$ and $V_M(\mathrm{S})$ will be first-degree, hyperbolic functions.

### 4.8.3. E—M—S coordination

If the metal is to form a bridge between enzyme and substrate, plausibly $M$ or $C$ but not $S$ may bind to the free enzyme:

$$E \underset{C}{\rightleftharpoons} EC \rightarrow E + M + P \qquad \text{Mechanism 4.XIII}$$

$$E \overset{M}{\rightleftharpoons} EM \overset{S}{\rightleftharpoons} EC$$

$$S + M \rightleftharpoons C$$

Since this mechanism is the converse of Mechanism 4.XI, we can expect that $V_{(S)}$ will be finite but $V_{(M)}$ will be zero: depletion of free **M** by excess $\mathbf{S}_t$ will create no dead-end, but depletion of **S** by excess $\mathbf{M}_t$ will convert $EM$ into a dead-end.

If the metal bridge must be in place on the enzyme before the substrate can add, the sequential Mechanisms 4.XIV and 4.XV become important:

$$E \underset{}{\overset{M}{\rightleftharpoons}} EM \underset{}{\overset{S}{\rightleftharpoons}} EC \rightarrow E + M + P \qquad \text{Mechanism 4.XIV}$$

$$E \underset{k_{-1}}{\overset{k_1\mathbf{M}}{\rightleftharpoons}} EM \underset{k_{-2}}{\overset{k_2\mathbf{S}}{\rightleftharpoons}} EC \xrightarrow{k_3} EM + P \qquad \text{Mechanism 4.XV}$$

In these sequential models, neither excess substrate nor excess metal gives rise to an inhibitory dead-end in the mechanism. The saturation velocities with respect to both substrate and metal will be finite.

Mechanism 4.XV is an unusual mechanism. It differs from 4.XIV in that release of the reaction product from the ternary complex leaves behind *EM* rather than *E*. Its steady-state rate law was derived by Dixon and Webb (1958) to be:

$$v_0 = \frac{k_1\mathbf{M}k_2\mathbf{S}k_3}{k_1\mathbf{M}k_2\mathbf{S} + k_1\mathbf{M}k_{-2} + k_1\mathbf{M}k_3 + k_{-1}k_{-2} + k_{-1}k_3}.$$

On the one hand, if **S** is held constant,

$$v_0 = \frac{V_{(M)}\mathbf{M}}{\mathbf{M} + K^M_{(m)}}$$

$$V_{(M)} = \frac{k_2k_3\mathbf{S}}{k_2\mathbf{S} + k_{-2} + k_3}, \qquad K^M_{(m)} = \frac{k_{-1}(k_{-2} + k_3)}{k_2\mathbf{S} + k_{-2} + k_3}.$$

On the other hand, if **M** is held constant.

$$v_0 = \frac{V_{(S)}\mathbf{S}}{\mathbf{S} + K^S_{(m)}}$$

$$V_{(S)} = k_3 \qquad K^S_{(m)} = \frac{(k_1\mathbf{M} + k_{-1})(k_{-2} + k_3)}{k_1k_2\mathbf{M}}.$$

Accordingly, $V_{(M)}$ increases hyperbolically with **S**, but $V_{(S)}$ does not vary with **M** at all! Topologically, when *S* is saturating, all of the *EM* reacts with *S* to form *EC* and none of it goes back to *E*. Since *E* is not formed via any other route, the flow from *E* to *EM* becomes one-way traffic, and *E* will be completely depleted by the time steady-state is established in the reaction system. Consequently, as long as there is sufficient metal to complex with all the enzyme molecules, $V_{(S)}$ will not be affected at all by the actual level of **M**. Experimentally, this mode of behaviour was suggested by Folk, Mullooly and Cole (1967) for the $Ca^{++}$ activation of transglutamination between carbobenzoxyglutaminylglycine and hydroxylamine catalysed by guinea pig liver transglutaminase.

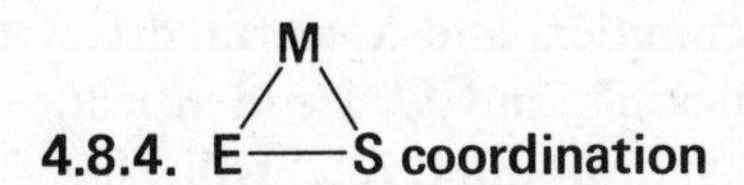

### 4.8.4. E—S coordination (cyclic M–E–S)

The cyclic coordination scheme imposes no special restriction on the sequence in which substrate and metal must bind to the enzyme. It is compatible with all the forementioned mechanisms, as well as 4.XVI, where *S*, *M* and *C* can bind randomly to the free enzyme:

$$\begin{array}{ccccc} & & ES & & \\ & {\scriptstyle S}\nearrow\!\!\swarrow & & \searrow\!\!\nwarrow{\scriptstyle M} & \\ E & & \overset{\longrightarrow}{\underset{C}{\longleftarrow}} & & EC \rightarrow E + M + P \\ & {\scriptstyle M}\searrow\!\!\nwarrow & & \nearrow\!\!\swarrow{\scriptstyle S} & \\ & & EM & & \end{array} \qquad \text{Mechanism 4.XVI}$$

$$S + M \rightleftharpoons C$$

This model mechanism has been widely applied, e.g. by Morrison, O'Sullivan and Ogston (1961) to creatine kinase, Gaffney and O'Sullivan (1964) to lombricine kinase, and Mildvan and Cohn (1966) to pyruvate kinase. It predicts that, if $\mathbf{M}_t$ is fixed, raising $\mathbf{S}_t$ infinitely will deplete all the free **M** and turn *ES* into a dead-end. No less, if $\mathbf{S}_t$ is fixed, raising $\mathbf{M}_t$ infinitely will deplete all the free **S** and turn *EM* into a dead-end. Therefore both $V_{(S)}$ and $V_{(M)}$ should be zero.

In contrast to Mechanism 4.XVI which is consistent only with the cyclic coordination scheme, the very simple Mechanism 4.XVII is consistent will all three coordination schemes of *E–S–M*, *E–M–S*, and cyclic M–E–S (triangle of *E*, *S*, *M*):

$$E \overset{C}{\rightleftharpoons} EC \rightarrow E + M + P \qquad \text{Mechanism 4.XVII}$$

$$S + M \rightleftharpoons C$$

This mechanism generates no dead-end if either $\mathbf{S}_t$ or $\mathbf{M}_t$ is raised infinitely, and both $V_{(S)}$ and $V_{(M)}$ will be finite.

### 4.8.5. Apparent equilibrium constant

Substrate and metal react to form a complex in some model mechanisms but not in others. Whenever substrate-metal and/or product-metal complexes are formed, $K_{\text{app}}$, the apparent equilibrium constant of the reaction as measured by total substrate and product concentrations, is related to the metal concentration as follows:

$$K_{\text{app}} = \frac{\mathbf{P}_{\text{total}}}{\mathbf{S}_{\text{total}}} = \frac{\mathbf{P} + (P\text{–}M\ \text{Complex})}{\mathbf{S} + (S\text{–}M\ \text{Complex})} = \frac{\mathbf{P}(1 + \mathbf{M}/K_c')}{\mathbf{S}(1 + \mathbf{M}/K_c)}$$

$K_c$ is the dissociation constant for the $S$–$M$ complex, and $K'_c$ is the dissociation constant for the $P$–$M$ complex. Unless $K_c$ and $K'_c$ are identical, i.e. metal has equal affinity toward substrate and product, $K_{app}$ will vary with metal concentration. In general then, observation of such a variation would be evidence for the significant formation of *substrate-metal* and/or *product-metal* complexes in the reaction system. For example, the apparent equilibrium constant for ATP-creatine transphosphorylase varied strikingly with $Mg^{++}$ concentration (Fig. 4.6), and this was interpreted by Noda, Kuby and Lardy (1954) in terms of the formation of $Mg^{++}$-ATP and $Mg^{++}$-ADP complexes in the system.

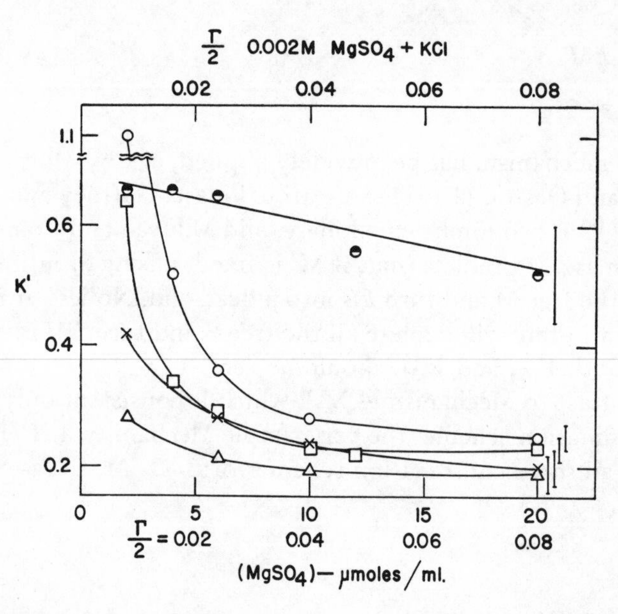


Fig. 4.6. Variation of the apparent equilibrium constant, namely (creatine phosphate)(ADP)/(ATP)(creatine), for ATP-creatine transphosphorylase with $Mg^{++}$ concentration. The different open symbols, obtained with different initial mixtures of creatine and ATP, indicated a striking variation when increasing amounts of $MgSO_4$ were added. In contrast, the half-filled symbols indicated only a much smaller variation when $Mg^{++}$ was held constant, and increasing amounts of KCl were added to change the ionic strength. (From Noda, Kuby and Lardy, 1954.)

## 4.8.6. Interchangeability criterion

In Mechanism 4.XVII only the complex $C$, and neither free $S$ nor free $M$, reacts with any enzyme species in the mechanism. Correspondingly, only **C**, but neither **S** nor **M**, appears in its rate law:

$$v_0 = \frac{m_1\mathbf{C}}{d_1\mathbf{C} + d_0}\,.$$

The expression for C given by Eqn (4.22) is symmetrical with respect to $\mathbf{M}_t$ and $\mathbf{S}_t$; the amount of $C$ formed from e.g., 1 mM $\mathbf{M}_t$ and 10 mM $\mathbf{S}_t$ is precisely the same as that formed from 10 mM $\mathbf{M}_t$ and 1 mM $\mathbf{S}_t$. *Therefore interchanging* $\mathbf{M}_t$ *and* $\mathbf{S}_t$ *will leave the reaction velocity unchanged.* This prediction, first recognized by Segal, Kachmar and Boyer (1952), holds for not only Mechanism 4.XVII, but also all other mechanisms where only $C$ reacts with any enzyme species, e.g. Mechanism 4.XVIII:

$$\begin{array}{ccl} E & \underset{}{\overset{C}{\rightleftharpoons}} & EC \rightarrow E + \text{product} \\ & & {\scriptstyle C}\updownarrow \\ & & EC_2 \end{array} \qquad \text{Mechanism 4.XVIII}$$

In this mechanism, excess $C$ entraps all the enzyme into the $EC_2$ dead-end. Its rate law would be a second-degree substrate-inhibitor function with respect to C:

$$v_0 = \frac{m_1\mathrm{C}}{d_2\mathrm{C}^2 + d_1\mathrm{C} + d_0}.$$

Since free $S$ and free $M$ do not take part in any of the arrows in the mechanism, they are excluded from the rate law, and interchanging $\mathbf{S}_t$ and $\mathbf{M}_t$ again will leave the reaction velocity unchanged.

### 4.8.7. Joint saturations

Instead of analysing $v_0(\mathbf{M}_t)$ at fixed $\mathbf{S}_t$, and $v_0(\mathbf{S}_t)$ at fixed $\mathbf{M}_t$, Kuby, Noda and Lardy (1954) and Keech and Barritt (1967) suggested the alternate experimental approach of jointly varying substrate and metal in the form of an equimolar mixture. The usefulness of this approach may be illustrated by a comparison of Mechanisms 4.XIX and 4.XX:

$$\begin{array}{ccl} ME & \overset{C}{\rightleftharpoons} & MEC \\ {\scriptstyle M}\updownarrow & & {\scriptstyle M}\updownarrow \\ E & \overset{C}{\rightleftharpoons} & EC \rightarrow E + \text{products} \end{array} \qquad \text{Mechanism 4.XIX}$$

$$\begin{array}{ccl} ME & & \\ {\scriptstyle M}\updownarrow & & \\ E & \overset{C}{\rightleftharpoons} & EC \rightarrow E + \text{products} \end{array} \qquad \text{Mechanism 4.XX}$$

If $\mathbf{S}_t$ is held constant, raising $\mathbf{M}_t$ infinitely will lead to a zero saturation velocity $V_{(M)}$ in both mechanisms: all the enzyme will be trapped in ME and $MEC$ in Mechanism 4.XIX, or $ME$ in Mechanism 4.XX. Therefore the two model mechanisms are not separable on this basis. On the other hand, if an equimolar mixture of substrate and metal is raised to infinitely high concentrations, the jointly-saturated velocity $V_{(S+M)}$ will be zero for Mechanism 4.XIX, where all the enzyme comes to be trapped in *MEC*, but not for Mechanism 4.XX. The latter prediction stems from the difference between **C** and **M** calculated from Eqns (4.21) and (4.22) under the equimolar condition of $\mathbf{M}_t = \mathbf{S}_t$:

$$\begin{aligned} \mathbf{C} - \mathbf{M} = 2\mathbf{C} - \mathbf{M}_t &= \mathbf{S}_t + K_c - \sqrt{(2\mathbf{S}_t + K_c)^2 - 4\mathbf{S}_t^2} \\ &= \sqrt{(\mathbf{S}_t + K_c)^2} - \sqrt{(\mathbf{S}_t + K_c)^2 + \mathbf{S}_t(2K_c - \mathbf{S}_t)} \\ &< 0 \quad \text{if} \quad \mathbf{S}_t \text{ is less than } 2K_c \\ &> 0 \quad \text{if} \quad \mathbf{S}_t \text{ exceeds } 2K_c \end{aligned}$$

Accordingly, **C** remains smaller than **M** only as long as $\mathbf{S}_t$ is less than $2K_c$ but will become greater than **M** as soon as $\mathbf{S}_t$ exceeds $2K_c$. As more and more of the equimolar mixture is added to the system then, **C** will accumulate faster than **M**, and in Mechanism 4.XX the formation of *EC* will be increasingly favoured over the formation of *ME*. Therefore the enzyme will escape entrapment into the otherwise dead-end species *ME*, and the jointly saturated velocity $V_{(S+M)}$ will be finite.

Morrison *et al.* (1961) have proposed yet another experimental approach for the study of metal activation systems, based on establishing the form of dependence of reaction velocity on the substrate-metal complex *C*. Unlike $\mathbf{S}_t$ and $\mathbf{M}_t$, which can be varied directly, **C** can be varied only indirectly by adjusting $\mathbf{S}_t$ and $\mathbf{M}_t$. If $K_c$ is known, the amounts of $\mathbf{S}_t$ and $\mathbf{M}_t$ added can be chosen so as to bring about a variation in **C** while either free **S** or free **M** remains fixed. If $K_c$ is not accurately known, trial computations would have to be repeated for different plausible values of $K_c$. Kuby and Noltman (1962) and Josse (1966) have given instructive treatments of this problem, and the usefulness of this approach involving the variation of **C** will be illustrated by the case of yeast pyrophosphatase in Section 4.8.9.

### 4.8.8. Differentiation between mechanisms

The fore-mentioned criteria, summarized in Table 4.2, suffice to achieve extensive resolution between Mechanisms 4.IX–4.XX. However, these mechanisms are fairly elementary examples, and more complex model mechanisms are readily visualized. A general model analysed by London and Steck (1969) contains the eight enzyme species *E, EM, ES, EC, ME, MEM, MES* and *MEC,* and it is not difficult to expand the model even further. The possibility of additional models suggests that caution should be exercised in interpreting some of the criteria. In particular, $V_{(M)}$ falling to zero might indicate an *EM* dead-end in the mechanism created by the exhaustion of free substrate, but *it might indicate alternatively that an*

TABLE 4.2

Differentiation between metal-activation mechanisms

In this table:

$V_{(S)}$ = saturation velocity with respect to $S_t$ at any fixed level of $M_t$
$V_{(M)}$ = saturation velocity with respect to $M_t$ at any fixed level of $S_t$
$V_{(S+M)}$ = joint-saturation velocity with respect to an equimolar mixture of $S_t$ and $M_t$
$K_{app}$ = apparent equilibrium constant

| Mechanism | Is $v_0$ ($S_t$) hyperbolic? | Is $v_0$ ($M_t$) hyperbolic? | $V_{(S)}$ | $V_{(M)}$ | $V_{(S+M)}$ | Does $K_{app}$ vary with $M_t$? | Are $S_t$ and $M_t$ inter-changeable? |
|---|---|---|---|---|---|---|---|
| 4.IX | No†‡ | No†§ | Finite | Finite | Finite | No | No |
| 4.X | No | No | 0 | 0 | Finite | Yes | No |
| 4.XI | No‡ | No | 0 | Finite | Finite | Yes | No |
| 4.XII | Yes | Yes | Finite | Finite | Finite | No | No |
| 4.XIII | No | No§ | Finite | 0 | Finite | Yes | No |
| 4.XIV | Yes | Yes | Finite | Finite | Finite | No | No |
| 4.XV | Yes | Yes | Finite | Finite | Finite | No | No |
| 4.XVI | No | No | 0 | 0 | Finite | Yes | No |
| 4.XVII | No‡ | No§ | Finite | Finite | Finite | Yes | Yes |
| 4.XVIII | No | No | Finite | Finite | 0 | Yes | Yes |
| 4.XIX | No | No | Finite | 0 | 0 | Yes | No |
| 4.XX | No | No | Finite | 0 | Finite | Yes | No |

† This becomes hyperbolic if quasi-equilibrium condition holds,
‡ This becomes hyperbolic as $M_t$ becomes saturating.
§ This becomes hyperbolic as $S_t$ becomes saturating.

*inhibitory 2:1 metal-substrate complex is formed.* The latter possibility was discussed in detail by Noat, Ricard, Borel and Got (1970) in connection with the inhibition of yeast hexokinase by excess $Mg^{++}$. Similarly, $V_{(S)}$ falling to zero might indicate an *ES* dead-end in the mechanism created by the disappearance of free metal, or alternatively that an inhibitory 1:2 metal-substrate complex is formed. Nevertheless, the topological criteria summarized in Table 4.2 illustrate how useful experimental approaches may be devised to clarify many features of any metal-activation mechanism. There are also other experimental approaches capable of providing mechanistic insight.

First, binding studies can reveal the types of association complexes which are formed between the enzyme and the substrate and metal ligands. Thus an enzyme that acts through an *M–E–S* or cyclic coordination may bind substrate in the absence of metal, or metal in the absence of substrate. An enzyme that acts through an *E–S–M* coordination may bind substrate in the absence of metal, but likely not metal in the absence of substrate. An enzyme that acts through an *E–M–S* coordination may bind metal in the absence of substrate, but likely not substrate in the absence of metal.

Furthermore, with some favourable systems, physical changes in the enzyme, metal or substrate within these complexes might be detected via changes in their optical, electronic or magnetic properties. By way of such detection the coordination schemes for pyruvate kinase (Mildvan and Cohn, 1966) and creatine kinase (O'Sullivan and Cohn, 1966) could be identified as:

$$\text{pyruvate kinase} \begin{cases} \text{Mn}^{++} \\ \text{ADP} \\ \text{pyruvate} \end{cases} \quad (\text{Mn}^{++}\text{–ADP linked})$$

$$\text{creatine kinase} \begin{cases} \text{ADP–Mn}^{++} \\ \text{creatine} \end{cases}$$

Secondly, metal-activation mechanisms such as 4.IX, 4.XII and 4.XIV are formally similar to some of the two-substrate mechanisms analysed in Chapter 5 and 6, and kinetic criteria developed for the latter mechanisms would be equally appropriate here. Finally, computer curve-fitting techniques have been developed to discriminate between enzyme models, and have found important applications in metal-activation systems. Josse (1966), Kowalik and Morrison (1968) and London and Steck (1969) have all made effective use of curve-fitting techniques to analyse such systems. The identification of the yeast pyrophosphatase mechanism to be examined in the next section also relied heavily on this approach.

### 4.8.9. Yeast Pyrophosphatase

The inorganic pyrophosphatase from yeast requires a divalent cation such as $Mg^{++}$ for catalysis. Since $Mg^{++}$ combines with pyrophosphate to form a 1:1 complex $C_1$, Rapoport, Hohne, Reich, Heitmann and Rapoport (1972) varied the concentration of the complex through appropriate adjustments of total magnesium and pyrophosphate concentrations. Fig. 4.7a shows that the relationship between reaction velocity and the 1:1 complex conformed to a family of linear Lineweaver-Burk plots at different fixed concentrations of free $Mg^{++}$, i.e., *a hyperbolic relationship.* Furthermore, the intercept of these Lineweaver-Burk plots, namely $1/V_{(C)}$, varied linearly with $1/Mg^{++}$, suggesting that the relationship between $V_{(C)}$ and free $Mg^{++}$ was also hyperbolic (Fig. 4.7b). These rate properties are consistent with Mechanism 4.XXI under quasi-equilibrium conditions:

$$E \underset{}{\overset{M}{\rightleftharpoons}} ME \overset{C_1}{\rightleftharpoons} MEC_1; \quad E \overset{C_1}{\rightleftharpoons} EC_1 \overset{M}{\rightleftharpoons} MEC_1; \quad MEC_1 \rightarrow \text{reaction products} \qquad \text{Mechanism 4.XXI}$$

Since only $MEC_1$, but not $EC_1$, can break down to form reaction products, both $M$ and $C_1$ are essential for reaction. Accordingly, under quasi-equilibrium conditions, reaction velocity would vary hyperbolically with both ligands. Figure 4.7c confirms the good agreement between the proposed mechanism

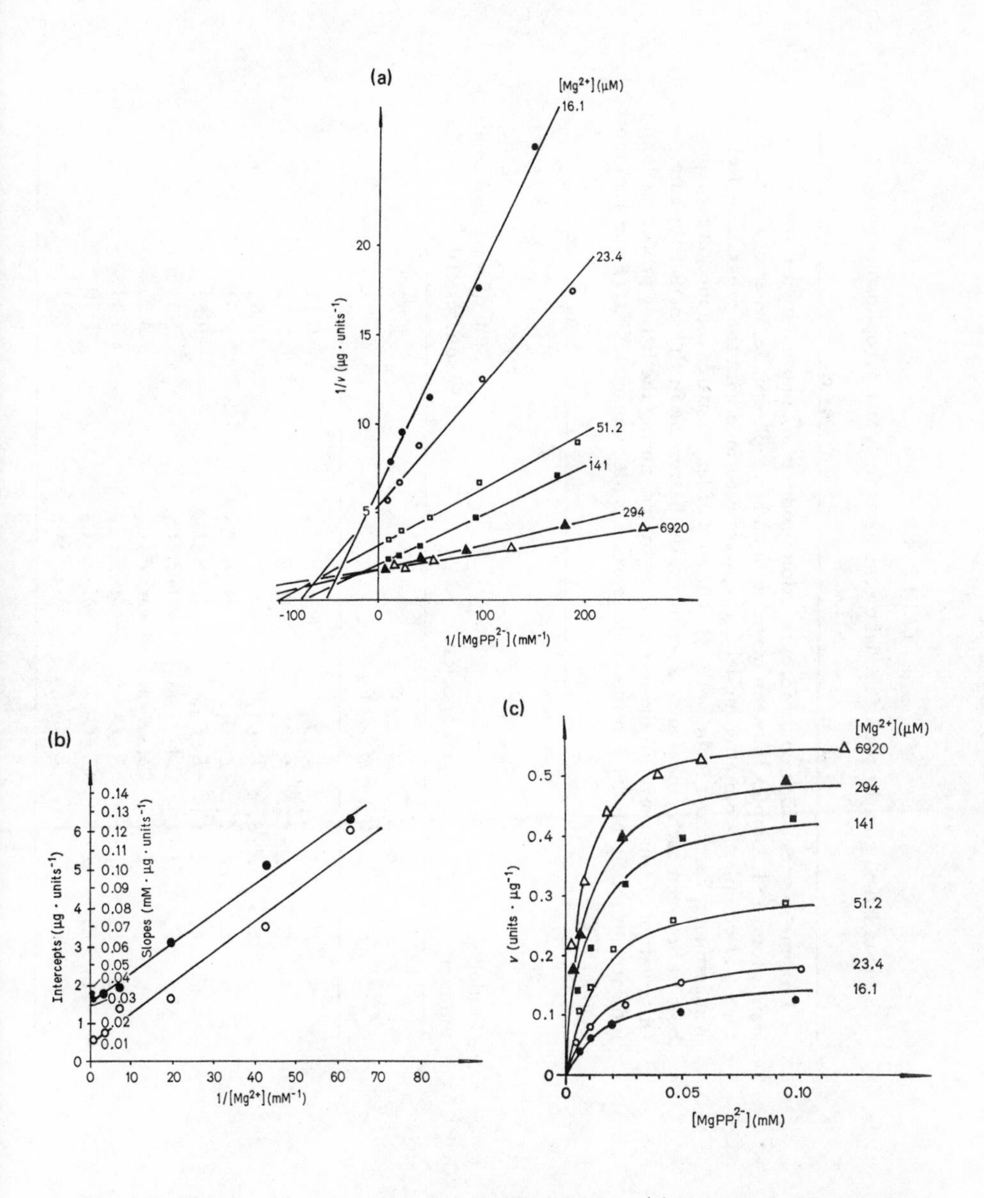


Fig. 4.7. Kinetics of yeast inorganic pyrophosphatase. (a) Variation of reaction velocity with 1 : 1 Mg-pyrophosphate complex at different fixed concentrations of free $Mg^{++}$. (b) Plot of the intercepts (open circles) and slopes (filled circles) from part (a) versus $1/Mg^{++}$. (c) Agreement between the data points and the theoretical curves calculated on the basis of the proposed mechanism. (From Rapoport *et al.*, 1972.)

TABLE 4.3

Statistical comparison of model mechanisms for yeast pyrophosphatase

Enzyme species which could form reaction products are marked by a + sign. $C_1$ stands for a 1:1 complex between magnesium and pyrophosphate, whereas $C_2$ stands for a 2:1 complex. Computer model fitting was performed with the program of Reich, Wangermann, Falck and Rohde (1972). The sum of least squares is a measure for the best fit between each model and experiment, the better the fit the smaller the sum. Observation of singularity in the information matrix during the fitting procedure is an indicator of misfit. Model 3 in this table represents Mechanism 4.XXI. (From Rapoport *et al.*, 1972.)

| Model number | Enzyme species | | | | Sum of least squares × $10^{-4}$ | Singularity of the information matrix |
|---|---|---|---|---|---|---|
| 1 | $E$, | $E \cdot C_1$, | $+M \cdot E \cdot C_1$ | | 2.73 | |
| 2 | $E$, | $E \cdot C_2$, | $+M \cdot E \cdot C_2$ | | 3.31 | |
| 3 | $E$, | $E \cdot C_1$, | $M \cdot E$, | $+M \cdot E \cdot C_1$ | 0.11 | |
| 4 | $E$, | $E \cdot C_2$, | $M \cdot E$, | $+M \cdot E \cdot C_2$ | 2.72 | + |
| 5 | $E$, | $E \cdot S$, | $E \cdot C_1$, | $+E \cdot C_2$ | 2.62 | + |
| 6 | $E$, | $E \cdot C_1$, | $+E \cdot C_2$, | $M \cdot E \cdot C_2$ | 1.51 | |
| 7 | $E$, | $E \cdot C_1$, | $+E \cdot C_2$, | $M \cdot E \cdot M$ | 0.51 | |
| 8 | $E$, | $E \cdot C_2$, | $M \cdot E \cdot M_2$, | $+M \cdot E \cdot C_2$ | 1.53 | |
| 9 | $E$, | $E \cdot C_2$, | $M \cdot E \cdot M$ | $+M \cdot E \cdot C_2$ | 0.23 | |
| 10 | $E$, | $E \cdot S$, | $E \cdot C_2$, | $+M \cdot E \cdot C_2$ | 3.31 | + |
| 11 | $E$, | $E \cdot S$, | $+E \cdot C_2$, | $M \cdot E \cdot C_2$ | 1.50 | |
| 12 | $E$, | $E \cdot C_1$, | $E \cdot C_2$, | $+M \cdot E \cdot C_2$ | 3.31 | + |

| | | | | | | | |
|---|---|---|---|---|---|---|---|
| 13 | $E$, | $E \cdot S$, | $E \cdot C_1$, | $M \cdot E$, | $+M \cdot E \cdot C_1$ | 0.09 | |
| 14 | $E$, | $E \cdot C_1$, | $E \cdot C_2$, | $M \cdot E \cdot M$, | $+M \cdot E \cdot C_2$ | 0.22 | + |
| 15 | $E$, | $E \cdot C_1$, | $E \cdot C_2$, | $M \cdot E$, | $+M \cdot E \cdot C_2$ | 2.62 | + |
| 16 | $E$, | $E \cdot C_1$, | $M \cdot E$, | $M \cdot E \cdot M$, | $+M \cdot E \cdot C_2$ | 0.16 | + |
| 17 | $E$, | $M . E$, | $M : E \cdot M$, | $M \cdot E \cdot C_1$, | $+M \cdot E \cdot C_2$ | 0.10 | |
| 18 | $E$, | $E \cdot S$, | $E \cdot C_1$, | $+E \cdot C_2$, | $M \cdot E \cdot M$ | 0.48 | |
| 19 | $E$, | $E \cdot S$, | $+E \cdot C_2$, | $M \cdot E \cdot M$, | $M \cdot E \cdot C_2$ | 2.82 | + |
| 20 | $E$, | $E \cdot S$, | $+E \cdot C_2$, | $M \cdot E$, | $M \cdot E \cdot C_2$ | 0.15 (obvious trends) | |
| 21 | $E$, | $E \cdot C_1$, | $E \cdot C_2$, | $M \cdot E \cdot M_2$, | $+M \cdot E \cdot C_2$ | 1.53 | + |
| 22 | $E$, | $E \cdot C_1$, | $M \cdot E$, | $M \cdot E \cdot M_2$, | $+M \cdot E \cdot C_2$ | 1.54 | + |
| 23 | $E$, | $E \cdot C_2$, | $M \cdot E$, | $M \cdot E \cdot M_2$, | $+M \cdot E \cdot C_2$ | 0.53 | + |
| 24 | $E$, | $E \cdot S$, | $E \cdot M$, | $M \cdot E \cdot M$, | $+M \cdot E \cdot C_2$ | 0.25 | + |
| 25 | $E$, | $E \cdot S$, | $E \cdot C_2$, | $M \cdot E \cdot M$, | $+M \cdot E \cdot C_2$ | 0.23 | + |
| 26 | $E$, | $E \cdot S$, | $E \cdot C_1$, | $M \cdot E \cdot M$, | $+M \cdot E \cdot C_2$ | 0.78 | + |
| 27 | $E$, | $E \cdot S$, | $E \cdot C_1$, | $M \cdot E$, | $+M \cdot E \cdot C_2$ | 2.60 | + |
| 28 | $E$, | $E \cdot S$, | $E \cdot C_2$, | $M \cdot E$, | $+M \cdot E \cdot C_2$ | 2.62 | + |
| 29 | $E$, | $E \cdot S$, | $E \cdot C_1$, | $M \cdot E$, $M \cdot E \cdot S$, | $+M \cdot E \cdot C_1$ | 0.35 | + |
| 30 | $E$, | $+E \cdot C_1$, | $+E \cdot C_2$ | | | 3.12 | + |
| 31 | $E$, | $+E \cdot C_2$, | $+M \cdot E \cdot C_2$ | | | 1.59 | + |
| 32 | $E$, | $+E \cdot C_1$, | $M \cdot E$, | $+M \cdot E \cdot C_1$ | | 0.11 ($V_1$ estimated to $10^{-12}$) | + |
| 33 | $E$, | $+E \cdot C_2$, | $M \cdot E$ | $+M \cdot E \cdot C_2$ | | 3.99 | |
| 34 | $E$, | $+E \cdot C_2$, | $M \cdot E \cdot M$, | $+M \cdot E \cdot C_2$ | | 0.19 | |
| 35 | $E$, | $E \cdot S$, | $+E \cdot C_1$, | $+E \cdot C_2$ | | 0.29 | |
| 36 | $E$, | $E \cdot C_1$, | $+E \cdot C_2$, | $+M \cdot E \cdot C_2$ | | 2.49 | + |
| 37 | $E$, | $+E \cdot C_1$, | $E \cdot C_2$, | $+M \cdot E \cdot C_2$ | | 3.22 | + |
| 38 | $E$, | $+E \cdot C_1$, | $+E \cdot C_2$, | $+M \cdot E \cdot C_2$ | | 2.50 | + |

and experiment. In Table 4.3, the adequacy of Mechanism 4.XXI (namely Model 3 in the table) was statistically compared with that of a wide range of alternative model mechanisms. Models 1 and 2 each consisted of only three enzyme species; their poor fit to experiment was reflected by a large sum of squares of residuals. Amongst the other models, only Models 13, 17 and 32 gave as satisfactory a fit as Model 3. Model 32 was less stringent than Model 3 in that reaction products could be formed from either $EC_1$ or $MEC_1$. However, the calculated rate of product formation from $EC_1$ ($V_1$ in the table) was negligible, so that Model 32 in effect approached Model 3. For Model 17, the calculated concentration of free $E$ proved to be negligible over the entire experimental range of magnesium levels, so that it was also in effect the same as Model 3, the only difference being the presence of an additional metal ion on all four significant enzyme species. Only Model 13, which included the additional enzyme species $ES$, constituted a real alternative to Model 3. It is common to both Models 3 and 13 that only $MEC_1$ could form product. Consequently, before catalysis could proceed, two molecules of the metal must be liganded to the enzyme, one to a site for free-metal, and one to a site for the metal-substrate complex. In confirmation, it was observed that whereas $Mg^{++}$, $Co^{++}$ and $Mn^{++}$ could bind to both sites, $Ca^{++}$ could bind (when complexed with $PP_i$) to the site for the metal-substrate complex, but not to the site for free-metal (Rapoport, Hohne, Heitmann and Rapoport, 1973). Clearly the two sites were dissimilar.

## Problem 4.1

Use the structural rule to derive the rate laws for Mechanisms 4.II, 4.III and 4.IV,

# 5 Branching Mechanisms

In the biological world, the reactions of photosynthesis and chemosynthesis create organic covalent bonds, and reactions like those of the hydrolases destroy these bonds. In between, group-transfer reactions rearrange and reconstruct the wide spectrum of organic molecules of monomeric, oligomeric and polymeric dimensions which constitute the material basis of life. Some enzymic group transfers involve three substrates, but more typically they involve two substrates and two products:

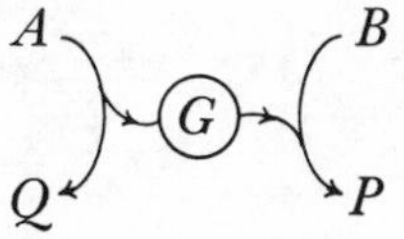

$$A \quad Q \quad (G) \quad B \quad P$$

Group $(G)$, which may be a pair of hydrogens, a phosphoryl, a carboxyl-, or some other moiety, is transferred from $A$ to $B$ in the forward reaction, and from $P$ to $Q$ in the reverse reaction. The transfer may occur directly between the two substrates or, as Doudoroff, Barker and Hassid (1947) first discovered from isotope-exchange experiments on sucrose phosphorylase, the enzyme in some cases may act as an intermediate covalent acceptor in a double transfer. The list of enzymes which exhibit such a mode of *enzyme substitution*, or *covalent catalysis*, is now a very long one. Given two substrates, two products, and the possibility of enzyme substitution, the fundamental enzyme species to consider are the free enzyme, four binary complexes, two ternary complexes, and the group-substituted enzyme $E$–$G$. The general model comprising these enzyme species is Mechanism 5.I, and most two-substrate mechanisms are derivatives of this general case. Some of these derivatives provide branching pathways for the addition of substrates and/or release of products (Fig. 5.1), and some only an ordered pathway in which the substrates and products interact with the enzyme according to a predetermined sequence. The rate behaviour of the branching ones, which have been analysed by Wong and Hanes (1962), are examined in this chapter, and the linear ones in the next chapter. It might be noted that, in Mechanisms 5.I, 5.V, 5.VI and 5.VII, the substituted enzyme $E$–$G$ exists side by side with the ternary complexes $EAB$ and $EPQ$. Fundamentally, these two

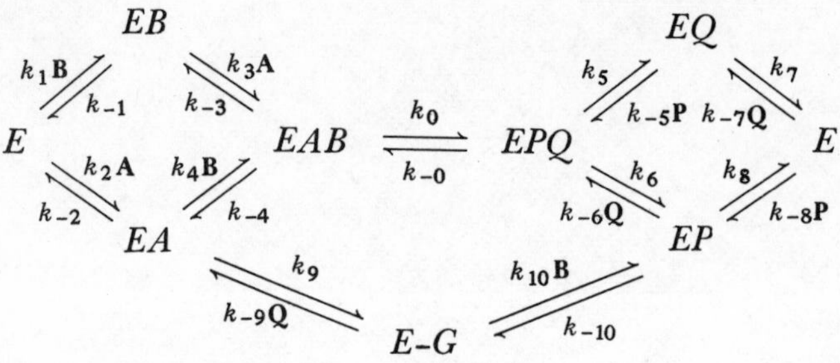

$EB$ $EQ$
$k_1\mathbf{B}$ $k_{-1}$ $k_3\mathbf{A}$ $k_{-3}$ $k_5$ $k_{-5}\mathbf{P}$ $k_7$ $k_{-7}\mathbf{Q}$
$E$ $EAB \overset{k_0}{\underset{k_{-0}}{\rightleftharpoons}} EPQ$ $E$
$k_2\mathbf{A}$ $k_{-2}$ $k_4\mathbf{B}$ $k_{-4}$ $k_6$ $k_{-6}\mathbf{Q}$ $k_8$ $k_{-8}\mathbf{P}$
$EA$ $EP$
$k_9$ $k_{-9}\mathbf{Q}$ $k_{10}\mathbf{B}$ $k_{-10}$
$E\text{–}G$

Mechanism 5.I

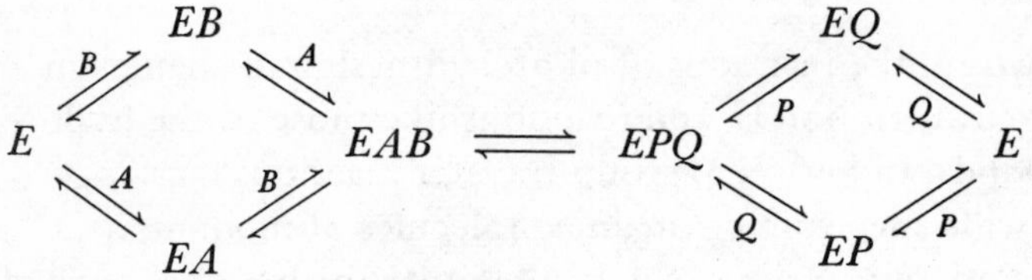

$EB$ $EQ$
$B$ $A$ $P$ $Q$
$E$ $EAB \rightleftharpoons EPQ$ $E$
$A$ $B$ $Q$ $P$
$EA$ $EP$

Mechanism 5.II

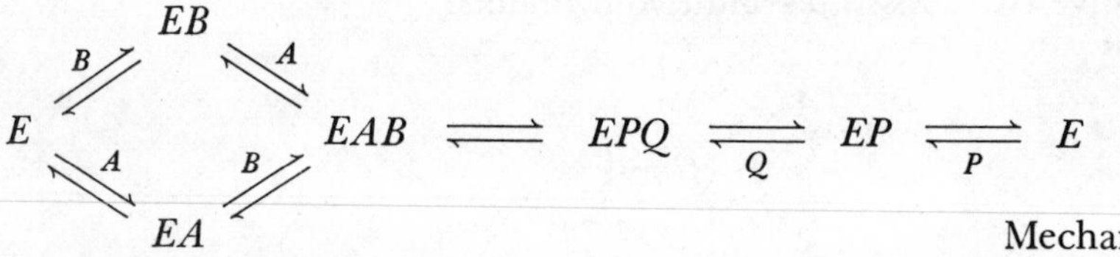

$EB$
$B$ $A$
$E$ $EAB \rightleftharpoons EPQ \underset{Q}{\rightleftharpoons} EP \underset{P}{\rightleftharpoons} E$
$A$ $B$
$EA$

Mechanism 5.III

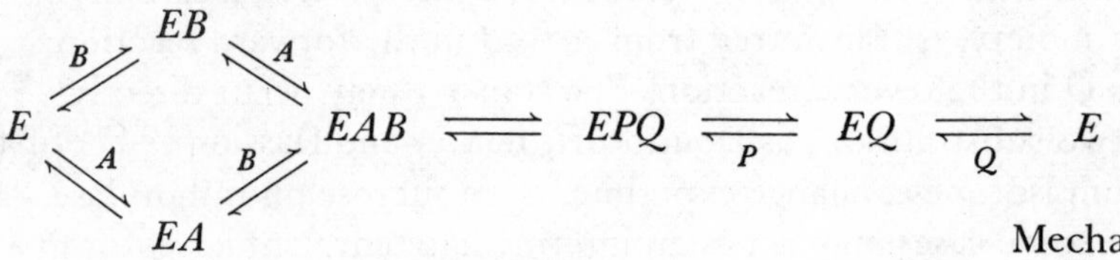

$EB$
$B$ $A$
$E$ $EAB \rightleftharpoons EPQ \underset{P}{\rightleftharpoons} EQ \underset{Q}{\rightleftharpoons} E$
$A$ $B$
$EA$

Mechanism 5.IV

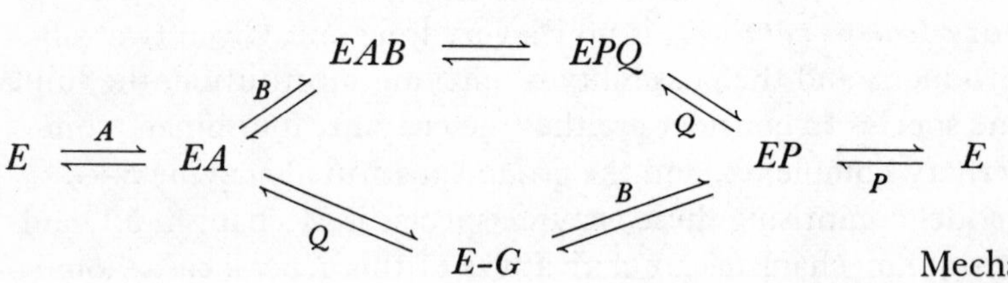

$EAB \rightleftharpoons EPQ$
$B$ $Q$
$E \overset{A}{\rightleftharpoons} EA$ $EP \underset{P}{\rightleftharpoons} E$
$Q$ $B$
$E\text{–}G$

Mechanism 5.V

$EB \overset{A}{\rightleftharpoons} EAB \rightleftharpoons EPQ \underset{P}{\rightleftharpoons} EQ$
$B$ $Q$
$E$ $E$
$A$ $P$
$EA \underset{Q}{\rightleftharpoons} E\text{–}G \overset{B}{\rightleftharpoons} EP$

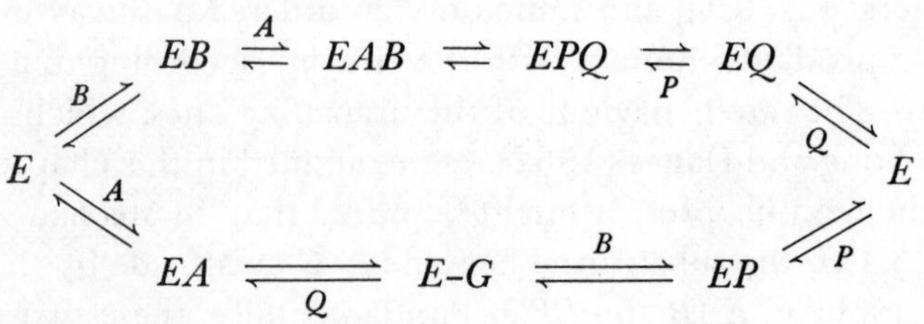

Mechanism 5.VI

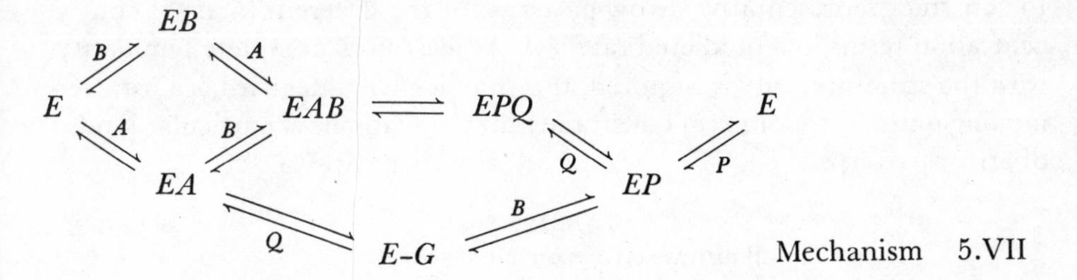


Fig. 5.1. Branching model mechanisms for group-transfer reactions between two substrates and two products.

kinds of enzyme species are not mutually exclusive of one another: the formation of a covalent linkage between the transferred group and the enzyme, e.g., an acyl enzyme, phosphoryl enzyme, etc., may give rise to a freely-existing $E$–$G$, or may occur within the ternary complexes, *or both.*

## 5.1. Effects of Substrate Concentrations

The steady-state rate behaviour of any branching mechanism is determined by its rate law, which in turn is determined by the *structural rule*. To define the most prominent features of its rate behaviour requires little more than identifying the form of its rate law by inspecting the topological arrangement of its enzyme species.

### 5.1.1. Initial velocity

In Mechanism 5.I, substrate $A$ reacts with two enzyme species, and $\mathbf{A}^2$-arrow patterns acceptable to the structural rule are readily constructed. The dependence of the initial forward velocity $v_f$ on $\mathbf{A}$ is therefore second-degree, substrate-type. Substrate $B$ reacts with three enzyme species, $\mathbf{B}^3$-arrow patterns are readily constructed, and the dependence is third-degree:

$$\text{at fixed } \mathbf{B},\ v_f = \frac{m_2\mathbf{A}^2 + m_1\mathbf{A}}{d_2\mathbf{A}^2 + d_1\mathbf{A} + d_0}, \tag{5.1}$$

$$\text{at fixed } \mathbf{A},\ v_f = \frac{m_3\mathbf{B}^3 + m_2\mathbf{B} + m_1\mathbf{B}}{d_3\mathbf{B}^3 + d_2\mathbf{B}^2 + d_1\mathbf{B} + d_0}. \tag{5.2}$$

The velocity dependences for both directions of the reaction likewise may be determined by topological reasoning for each of Mechanisms 5.I–5.VII. The results are summarized in Table 5.1: the degree of the $v_f(\mathbf{A})$ function in every instance depends on the number of enzyme species $A$ reacts with in the mechanism, and a similar consideration holds for $v_f(\mathbf{B})$, $v_r(\mathbf{P})$ and $v_r(\mathbf{Q})$ as well. Still, it is always advisable, in applying topological reasoning, to construct representative arrow-patterns for the different kinds of concentration terms in a predicted rate law. At least until complete familiarity with the structural rule is acquired, this practice will safeguard against overlooking unusual topological constraints that may disallow particular kinds of arrow-patterns.

TABLE 5.1
Substrate effects on initial velocities

| | Forward reaction | | Reverse reaction | |
|---|---|---|---|---|
| Mechanism | $v_f$(A) | $v_f$(B) | $v_r$(P) | $v_r$(Q) |
| 5.I | 2° | 3° | 2° | 3° |
| 5.II | 2° | 2° | 2° | 2° |
| 5.III | 2° | 2° | 1° | 1° |
| 5.IV | 2° | 2° | 1° | 1° |
| 5.V | 1° | 2° | 1° | 2° |
| 5.VI | 2° | 2° | 2° | 2° |
| 5.VII | 2° | 3° | 1° | 2° |

*A* and *B* are the substrates for the initial forward velocity $v_f$ and *P* and *Q* are the substrates for the initial reverse velocity $v_r$. All the velocity functions are substrate-type. 1° = first-degree (hyperbolic); 2° = second-degree; and 3° = third-degree.

Whenever a substrate function predicted by a steady-state branching mechanism is second or third degree, deviations from the hyperbolic relationship are expected, and a nonlinear Eadie, Lineweaver–Burk or Hanes plot is predicted. The rate profile also may show a maximum or a sigmoidal region. However, such nonhyperbolic trends may be either pronounced or subdued, *depending on the numerical relationships between the rate constants* (Botts and Morales, 1953; Dalziel, 1958; Walter, 1962; London, 1968; Pettersson, 1969). When they are subdued, they can be detected only if the data are examined critically, e.g. with the more discriminating Eadie plot rather than the less discriminating Lineweaver–Burk plot (see Fig. 3.4). When they are pronounced, the steady-state branching mechanisms capable of predicting them would constitute important models to be considered, along with allosteric mechanisms involving homotropic interactions, for the explanation of sigmoidal kinetics (Botts, 1958; Ferdinand, 1966; Sweeny and Fisher, 1968).

### 5.1.2. Saturation velocity

Equation (5.1) describes the $v_f$(A) function for Mechanism 5.I. When the concentration of *A* is raised, the $d_0$ term in Eqn (5.1) does not change, $m_1$A and $d_1$A increase linearly, whereas $m_2$A$^2$ and $d_2$A$^2$ increase exponentially. At very high levels of A, the latter terms become predominant, and all others become negligible by comparison:

$$\text{as } A \to \infty \qquad v_f \to \frac{m_2 A^2}{d_2 A^2} = \frac{m_2}{d_2} = V_{(A)}. \tag{5.3}$$

The term $m_2A^2$ includes all the numerator arrow patterns, and $d_2A^2$ all the denominator patterns, that contain both the $k_2$A and $k_3$A arrows. Such patterns have two common properties:

(a) Since they include $k_2$A, they must not include $k_{-2}$, which forms a closed cycle with $k_2$A. Also they must not include $k_1$**B**, $k_{-7}$Q or $k_{-8}$**P**, all of which competitively leave from the same enzyme species as $k_2$A.

(b) Since they include $k_3$A, they must not include $k_{-3}$, which forms a closed cycle with $k_3$A. Also they must not include $k_{-1}$, which competitively leaves from the same enzyme species as $k_3$A.

Consequently, $k_{-2}$, $k_1$**B**, $k_{-7}$Q, $k_{-8}$**P**, $k_{-3}$ and $k_{-1}$ all cannot appear in the rate expression for $V_{(A)}$. The exclusion of $k_1$**B** is most important. It means that, of the three **B**-arrows in the mechanism, only $k_4$**B** and $k_{10}$**B** will appear in the rate expression for $V_{(A)}$, which is accordingly a second-degree function with respect to **B**:

$$V_{(A)} = \frac{m_{22}B^2 + m_{21}B}{d_{22}B^2 + d_{21}B + d_{20}}. \tag{5.4}$$

By the same token, if **B** is raised to saturation, the arrow-patterns in $m_3B^3$ and $d_3B^3$ in Eqn (5.2) will define the saturation velocity $V_{(B)}$. These patterns perforce include the three **B**-arrows, and exclude $k_2$A which competes with $k_1$**B**. Accordingly, of the two A-arrows in the mechanism only

TABLE 5.2
Cosubstrate effects on saturation velocities

| Mechanism | Forward reaction $V_A$(B) | Forward reaction $V_B$(A) | Reverse reaction $V_P$(Q) | Reverse reaction $V_Q$(P) |
|---|---|---|---|---|
| 5.I | 2° | 1° | 2° | 1° |
| 5.II | 1° | 1° | 1° | 1° |
| 5.III | 1° | 1° | 1° | 1° |
| 5.IV | 1° | 1° | 1° | 1° |
| 5.V | 2° | 1° | 2° | 1° |
| 5.VI | 1° | 1° | 1° | 1° |
| 5.VII | 2° | 1° | 2° | 1° |

All the functions are substrate type. 1° = first-degree (hyperbolic); 2° = second-degree.

$k_3\mathbf{A}$ is allowed in the rate expression for $V_{(B)}$, which is therefore a first-degree function with respect to **A**:

$$V_{(B)} = \frac{m_{31}\mathbf{A}}{d_{31}\mathbf{A} + d_{30}}. \tag{5.5}$$

All forms of saturation velocity for Mechanisms 5.I–5.VII in both directions of the reaction likewise may be determined by topological reasoning. The results are summarized in Table 5.2: the degree of the $V_A(\mathbf{B})$ function depends on the number of enzyme species which react with **B** but not with **A**, and a similar consideration holds for the other substrate-cosubstrate pairs.

## 5.2. Effects of Product Concentrations

Addition of $P$ or $Q$ to the forward reaction system introduces extra arrow-patterns to the rate law for the forward initial velocity. However, as long as one of the two products is absent, none of the extra patterns will be negative (see Section 2.4.1).

When $P$ is added to Mechanism 5.I, $k_{-5}\mathbf{P}$ and $k_{-8}\mathbf{P}$ are no longer zero, and extra arrow-patterns containing one or both of these arrows can be constructed. Both numerator and denominator patterns will contain $k_{-5}\mathbf{P}$. However, free $E$ being an essential junction point in the mechanism, all forward pathways in the numerator patterns must include either $k_1\mathbf{B}$ or $k_2\mathbf{A}$, and exclude $k_{-8}\mathbf{P}$ on account of its leaving from the same enzyme species as $k_1\mathbf{B}$ and $k_2\mathbf{A}$. Therefore only denominator patterns will contain $k_{-8}\mathbf{P}$, and the function $v_f(\mathbf{P})$ will be second-degree inhibitory:

$$v_f = \frac{m_1\mathbf{P} + m_0}{d_2\mathbf{P}^2 + d_1\mathbf{P} + d_0}. \tag{5.6}$$

If substrate $A$ is now raised to saturating concentrations, only those terms which contain both **A**-arrows remain significant in the rate law. All others are eliminated. Since $k_{-8}\mathbf{P}$ leaves from the same enzyme species as $k_2\mathbf{A}$, it is disallowed even in the denominator, and only $k_{-5}\mathbf{P}$ appears in the rate expression for $V_{(A)}$. The saturation rate function $V_A(\mathbf{P})$, namely the dependence of $V_{(A)}$ on **P**, is accordingly a first-degree modifier function:

$$V_{(A)} = \frac{m_1\mathbf{P} + m_0}{d_1\mathbf{P} + d_0}. \tag{5.7}$$

Similarly, the saturation function $V_B(\mathbf{P})$ also will be first-degree modifier-type. In effect, at fixed non-saturating concentrations of $A$ and $B$, the forward reaction is totally inhibited by high concentrations of product $P$ blocking the entry of free $E$ into forward reaction. However, if either $A$ or $B$ is raised to saturating levels, this blockade is overcome, and the inhibitory-type $v_f(\mathbf{P})$ is transformed into a modifier-type $V_A(\mathbf{P})$ or $V_B(\mathbf{P})$. Similarly, the predicted effects of single added products on all the branching mechanisms may be determined by inspecting the mechanism topologies. These predictions are summarized in Tables 5.3 and 5.4. Consider Mechanism 5.V,

TABLE 5.3
Product effects on initial velocities

| | Forward reaction | | Reverse reaction | |
|---|---|---|---|---|
| Mechanism | $v_f$(**P**) | $v_f$(**Q**) | $v_r$(**A**) | $v_r$(**B**) |
| 5.I | 2° *i* | 3° *i* | 2° *i* | 3° *i* |
| 5.II | 2° *i* | 2° *i* | 2° *i* | 2° *i* |
| 5.III | 1° *i* | 1° *i* | 2° *i* | 2° *i* |
| 5.IV | 1° *i* | 1° *i* | 2° *i* | 2° *i* |
| 5.V | 1° *i* | 2° *i* | 1° *i* | 2° *i* |
| 5.VI | 2° *i* | 2° *i* | 2° *i* | 2° *i* |
| 5.VII | 1° *i* | 2° *i* | 2° *i* | 3° *i* |

All the functions are inhibitory type. 1° = first-degree; 2° = second-degree; 3° = third-degree. (*i* stands for inhibitory)

TABLE 5.4
Product effects on saturation velocities

| Mechanism | $V_A$(**P**) | $V_A$(**Q**) | $V_B$(**P**) | $V_B$(**Q**) | $V_P$(**A**) | $V_P$(**B**) | $V_Q$(**A**) | $V_Q$(**B**) |
|---|---|---|---|---|---|---|---|---|
| 5.I | 1° *m* | 2° *m* | 1° *m* | 1° *m* | 1° *m* | 2° *m* | 1° *m* | 1° *m* |
| 5.II | 1° *m* | 1° *m* | 1° *m* | 1° *m* | 1° *m* | 1° *m* | 1° *m* | 1° *m* |
| 5.III | 0° | 1° *i* | 0° | 1° *i* | 1° *m* | 1° *m* | 2° *i* | 2° *i* |
| 5.IV | 1° *i* | 0° | 1° *i* | 0° | 2° *i* | 2° *i* | 1° *m* | 1° *m* |
| 5.V | 0° | 2° *i* | 1° *i* | 1° *i* | 0° | 2° *i* | 1° *i* | 1° *i* |
| 5.VI | 0° | 1° *i* | 1° *i* | 0° | 0° | 1° *i* | 1° *i* | 0° |
| 5.VII | 0° | 2° *i* | 0° | 1° *i* | 1° *m* | 2° *m* | 2° *i* | 2° *i* |

The inhibitory-type functions are marked *i*, and the modifier-type functions are marked *m*. 0° = zero-degree (no observable product effects); 1° = first-degree; 2° = second-degree. (See Chapter 3 regarding methods for distinguishing between the different forms of function.)

for example. There are one **P**-arrow and two **Q**-arrows; $v_f$(**P**) is first-degree inhibitory, and $v_f$(**Q**) is second-degree inhibitory. *If A is raised to saturation*, $V_A$(**P**) becomes zero degree, i.e. no observable product effects (the **P**-arrow leaves from free $E$ in competition against the saturating **A**-arrow: it cannot appear in the rate expression for $V_A$). In contrast, $Q$ reacts with $E$–$G$ and $EP$, neither of which reacts with $A$, and $V_A$(**Q**) remains second-degree inhibitory. *If B is raised to saturation instead*, the **P**-arrow can enter into the denominator of the rate law, but not into the numerator, because the A-arrow it competes against is essential to all forward reaction cycles. Accordingly, $V_B$(P) is first-degree inhibitory. As for $V_B$(Q), the Q-arrow from $E$–$G$ is excluded when B is saturating because it is competing against a saturating B-arrow. The Q-arrow from $EP$ can appear in the denominator, but not in the numerator because it is competing against the essential forward conversion of $EP$ to $E$. Accordingly, $V_B$(Q) is also first-degree inhibitory.

Taken together, the predictions presented in Tables 5.1–5.4 provide a sufficient basis for a complete resolution between Mechanisms 5.I–5.VII, each of which is distinguished by its own unique combination of velocity functions. In all instances the form of a velocity function is indicative of some pertinent feature of the mechanism topology in terms of the arrangement of various reaction steps within the mechanism. The strength of identification of any branching reaction mechanism ultimately depends on the extent to which the various topological features so indicated could be integrated into a consistent whole. The kinetic analysis of horse liver alcohol dehydrogenase illustrates this process of integration.

## 5.3. Horse Liver Alcohol Dehydrogenase

The alcohol dehydrogenase from horse liver catalyses the reversible oxidation of ethanol by NAD$^+$ to yield acetaldehyde and NADH. Theorell and Chance (1951) established that there are at least three significant enzyme species in the mechanism, viz. the free enzyme $E$, the $EO$ complex ($O$ stands for the oxidized coenzyme NAD$^+$), and the $ER$ complex ($R$ stands for the reduced coenzyme NADH). However, the existence of greater complexity was indicated by the finding of excess-ethanol inhibition by Theorell, Nygaard and Bonnichsen (1955), and Dalziel and Dickinson (1966). The

*($E$/$R$/eth)

$k_8$R $k_9$ $k_{-9}$R

*($E$/–/eth)

$k_{10}$O $k_{-10}$ $k_{-7}$ $k_7$eth

$E$/–/eth *($E$/$O$/eth)

$k_1$eth $k_{-1}$ $k_3$O $k_{-3}$ $k_{11}$

$E$/–/– [$E$/$O$/eth ⇌ $E$/$R$/ald] $\underset{k_{-5}\text{ald}}{\overset{k_5}{\rightleftharpoons}}$ $E$/$R$/– $\underset{k_{-6}\text{R}}{\overset{k_6}{\rightleftharpoons}}$ $E$/–/–

$k_2$O $k_{-2}$ $k_4$ eth $k_{-4}$

$E$/$O$/–

Mechanism 5.VIII

identification of the branching Mechanism 5.VIII for this enzyme by Hanes, Bronskill, Gurr and Wong (1972) was founded on the method of topological reasoning, and it illustrates the experimental application of this method.

### 5.3.1. Low-substrate pathways

The sets of kinetic observations by Hanes *et al.* (1972) are presented in Figs. 5.2–5.6. Each set was designed to establish the form of dependence of an initial or saturation velocity on the concentration of a substrate or product. Sets 2 and 4 in Fig. 5.2 show that the $v_f$(ethanol) profile for the forward reaction of ethanol oxidation was divisible into (a) a low to moderate

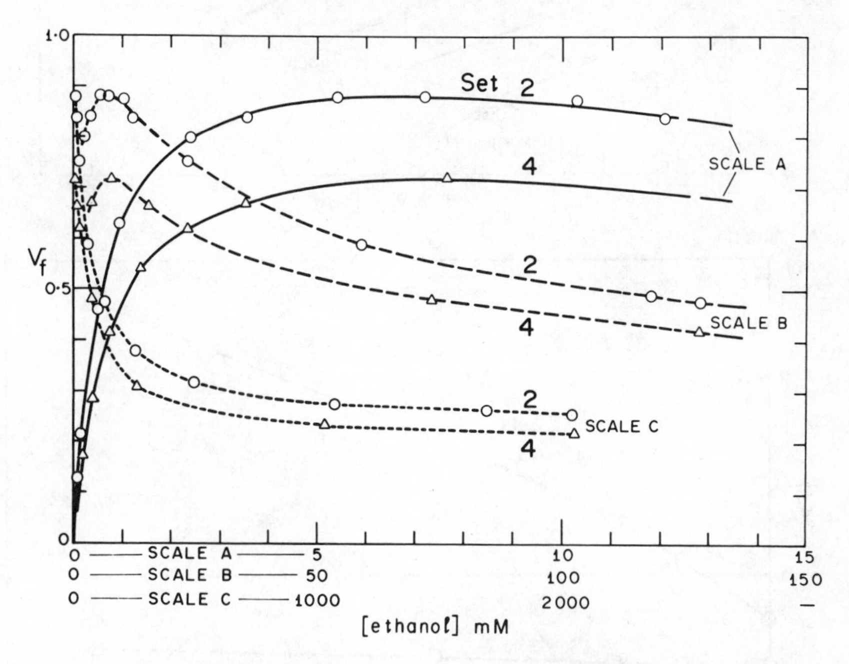


Fig. 5.2. The rate function $v_f$(ethanol), at different $NAD^+$ concentrations in Sets 2 and 4, showed a maximum at about 7–10 mM ethanol. Excess ethanol (1–2 M) caused the velocity to approach a depressed but finite asymptote. (From Hanes *et al.*, 1972.)

ethanol range between 0–7 mM., within which the velocity rose to a maximum, and (b) an excess ethanol range between 7 mM–2 M, within which the velocity slowly declined toward a finite asymtote. Throughout the lower ethanol range, the nonlinear Eadie plots for Sets 20–24 (Fig. 5.3) show the $v_f(NAD^+)$ relationship to be higher-degree for a series of different ethanol concentrations. However, when ethanol was saturating, the linear curve B indicates $V_{eth}(NAD^+)$ to be first-degree. This behaviour points to the random additions of ethanol and $NAD^+$ represented in Mechanism 5.VIII. Two

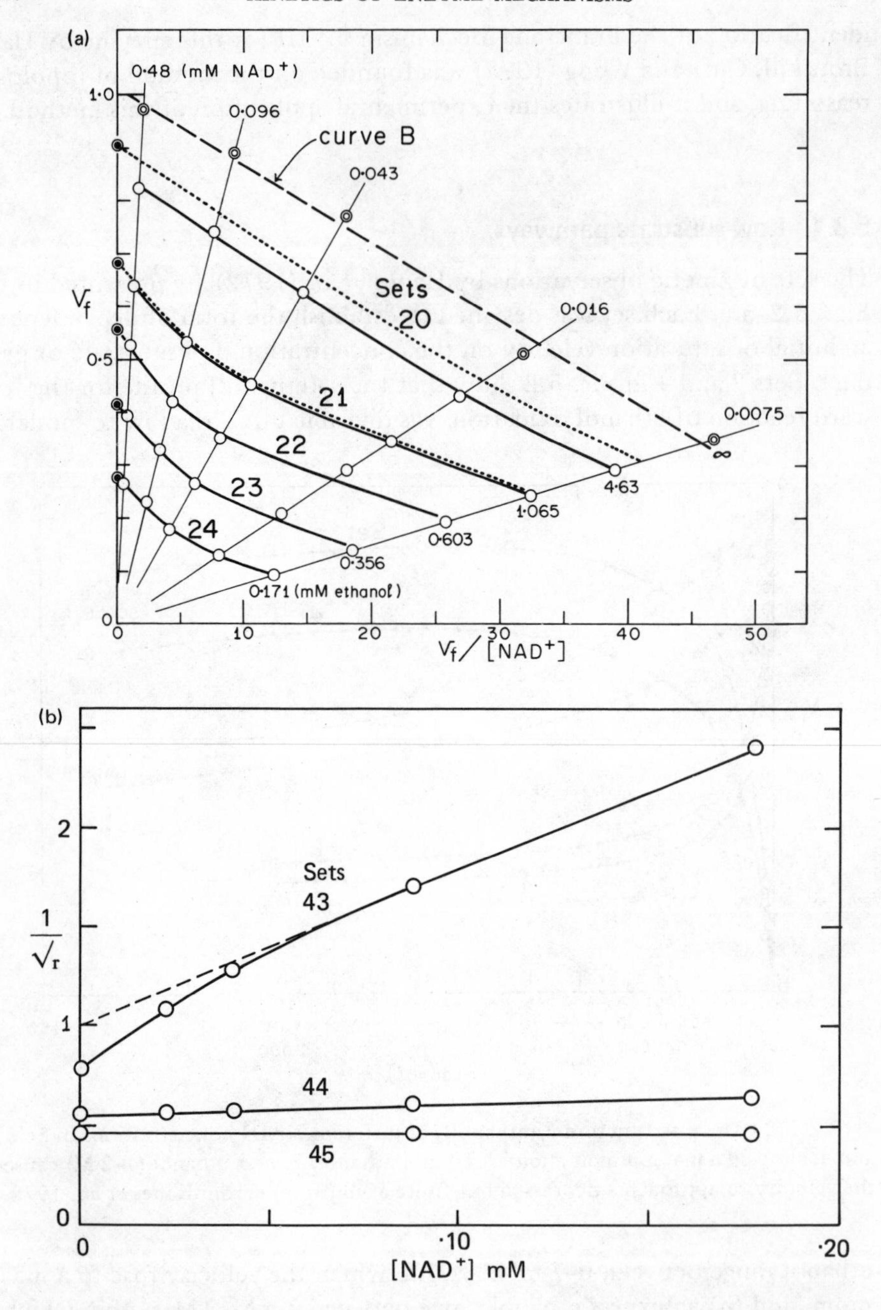


Fig. 5.3. Rate effects of $NAD^+$. Sets 20–24, obtained at nonsaturating ethanol levels, gave nonlinear Eadie plots that indicate a higher-degree $v_f(NAD^+)$ substrate dependence. Curve B, applicable to saturating ethanol levels within the non-excess range, gave a linear plot that indicates a first-degree $V_{eth}(NAD^+)$. The product effects by $NAD^+$ on the reverse reaction of acetaldehyde reduction were higher-degree inhibitory (i.e. nonlinear Dixon plot) in Set 43 at non-saturating NADH, but increasingly approached zero degree (i.e. no product effects) in Sets 44 and 45 as NADH was raised to saturation. (From Hanes *et al.*, 1972.)

enzyme species, namely the free enzyme $E/—/—$ and the enzyme-ethanol binary complex $E/—/$eth, reacted with $NAD^+$ at non-saturating ethanol levels, and brought about a second-degree $v_f(NAD^+)$ dependence. At saturating ethanol levels, $NAD^+$ could not compete against ethanol and all the free enzyme reacted with ethanol. Only $E/—/$eth reacted with $NAD^+$, giving rise to a first-degree $V_{eth}(NAD^+)$ dependence. The conclusion that $NAD^+$ reacted with more than one species was corroborated by Set 43, which shows $v_r(NAD^+)$ to be a higher-degree function when $NAD^+$ was added as a product to the reverse reaction mixture. As for the reduced co-enzyme (Fig. 5.4), the nonlinear Eadie plots of Sets 25 and 26 suggest a higher-degree $v_r$(NADH), and the nonlinear Dixon plot of Set 35 suggests a higher-degree $v_f$(NADH). Therefore NADH reacted with more than one enzyme species in the mechanism. Moreover, these enzyme species must be the same ones $NAD^+$ reacted with, namely $E/—/—$ and $E/—/$eth, because Set 45 indicates a zero-degree $V_{NADH}(NAD^+)$, and Sets 36 and 38 indicate a zero-degree $V_{NAD}$(NADH). Further in confirmation, when ethanol was raised to saturation, NADH could not compete against ethanol for $E/—/—$, and the linear Dixon plot of Set 37 points to a first-degree $V_{eth}$(NADH) relationship.

In Fig. 5.5, the linear Eadie plots for Sets 27 and 28 indicate a first-degree $v_r$(acetaldehyde) function, and the linear Dixon plots for Sets 31–33 indicate

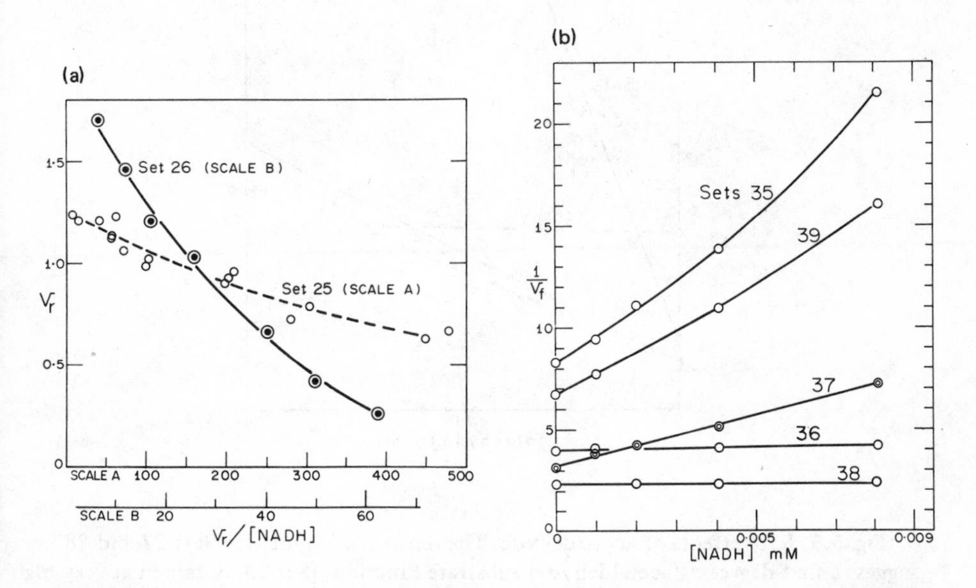


Fig. 5.4. Rate effects of NADH. In Sets 25 and 26, NADH was varied as a substrate, and the nonlinearity of the Eadie plots points to a higher-degree $v_r$(NADH). In Sets 35–39, NADH was added as reaction product to the forward reaction system. Set 35: low $NAD^+$, low ethanol. Set 36: saturating $NAD^+$, low ethanol. Set 37: low $NAD^+$, saturating but inexcessive ethanol. Set 38: saturating $NAD^+$, saturating but inexcessive ethanol. Set 39: low $NAD^+$, excess ethanol. (From Hanes *et al.*, 1972.)

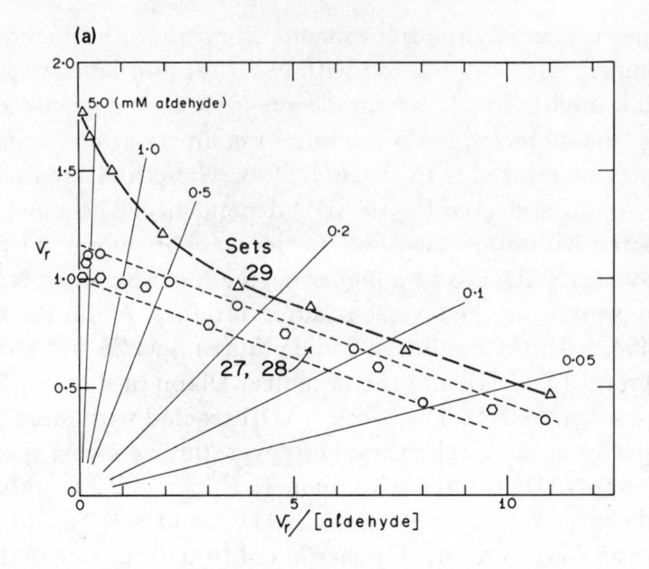


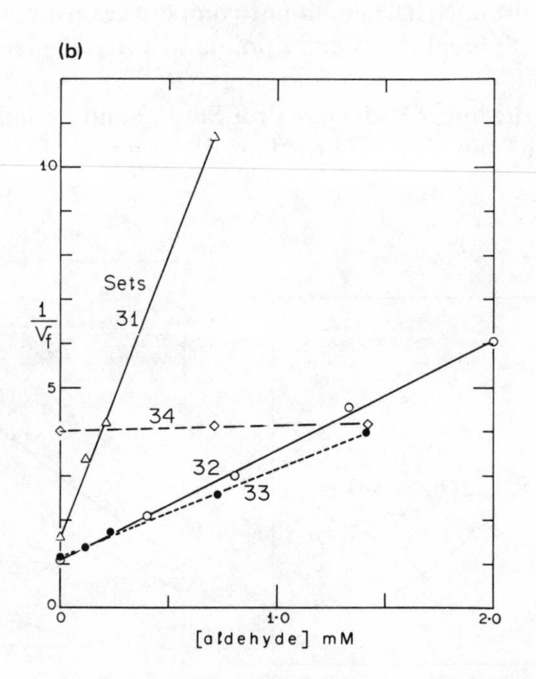


Fig. 5.5 Rate effects of acetaldehyde. The linear Eadie plots for Sets 27 and 28 suggest a first-degree $v_r$(acetaldehyde) substrate function. (Set 29, obtained at very high NADH, shows a nonhyperbolic activation traceable to impurities formed in stored acetaldehyde.) The linear Dixon plots for Sets 31 and 32 (low ethanol), 33 (saturating but non-excess ethanol) suggest a first-degree $v_f$(acetaldehyde) inhibitor function. The lack of an acetaldehyde effect in Set 34 (at excess ethanol concentration) suggests that acetaldehyde competes with ethanol when the excess-ethanol pathway is operating. (From Hanes *et al.*, 1972.)

a first-degree $v_f$(acetaldehyde) function. Therefore acetaldehyde reacted with a single enzyme species, which would be the enzyme-NADH binary complex *E*/*R*/— in Mechanism 5.VIII.

Thus the rate functions for both directions of the reaction are altogether explained by Mechanism 5.VIII with its random pathways for $NAD^+$ and ethanol additions, and an ordered pathway for NADH and acetaldehyde additions. Since no simple alternative can offer the same combination of topological features as defined by the individual rate functions, the explanation appears to be a unique one.

### 5.3.2. Excess-substrate pathways

As ethanol is increased above about 7 mM, it reacts to a noticeable extent with the enzyme-NADH binary complex *E*/*R*/—, in competition with acetaldehyde and bringing about a progressive lowering of the reaction rate. This competition is indicated in Sets 2, and 46–49, by the altered $v_f$(ethanol) profile in the presence of aldehyde, and by the diminishing aldehyde effects as the asymptotic forward velocity is approached at very high ethanol concentrations (Fig. 5.6). The *(*E*/*R*/eth) ternary complex so formed is formed also by reaction of NADH with *E*/—/eth. On release of NADH, it yields a *(*E*/—/eth) complex which is conformationally distinct from the normal *E*/—/eth complex produced by reaction of ethanol with free enzyme. The reasons for this conclusion are threefold.

First, if *(*E*/—/eth) and *E*/—/eth were the same, the rate law would require $V_{(NAD)}/K_{(m)}^{NAD}$ to remain constant throughout the excess ethanol range. In fact, this ratio decreased continually with ethanol concentration (Fig. 5.6). Secondly, according to Mechanism 5.VIII, NADH would react with *E*/—/— and *E*/—/eth at low ethanol concentrations, with only *E*/—/eth at saturating but non-excess ethanol concentrations, and finally with *E*/—/eth and *(*E*/—/eth) within the excess-ethanol range. In keeping with this expectation, $v_f$(NADH) was found to change from higher-degree (Set 35) to first-degree (Set 37) and again back to higher-degree (Sets 39, 53, 54) as the ethanol concentration was raised. The reversion to higher degree would not happen if *E*/—/eth and *(*E*/—/eth) were the same enzyme species.

Thirdly, the $v_f(NAD^+)$ function predicted for Mechanism 5.VIII operating under excess-ethanol concentrations presents an unusual diagnostic feature. Within the excess-ethanol range, the three ethanol-arrows $k_1$-(ethanol), $k_4$(ethanol) and $k_7$(ethanol) predominate over other arrows, so that only arrow patterns which contain all three of these arrows remain significant in the rate law. Since $k_{-1}$ forms a local cycle with $k_1$(ethanol), it is excluded. Consequently, the species *E*/—/eth can only contribute the $k_3$O-arrow to any arrow-pattern. Also, in the absence of any added NADH, the $k_{-9}$R-arrow equals zero, and the species *(*E*/—/eth) can only contribute the $k_{10}$O-arrow to any arrow-pattern. Since the structural rule requires all

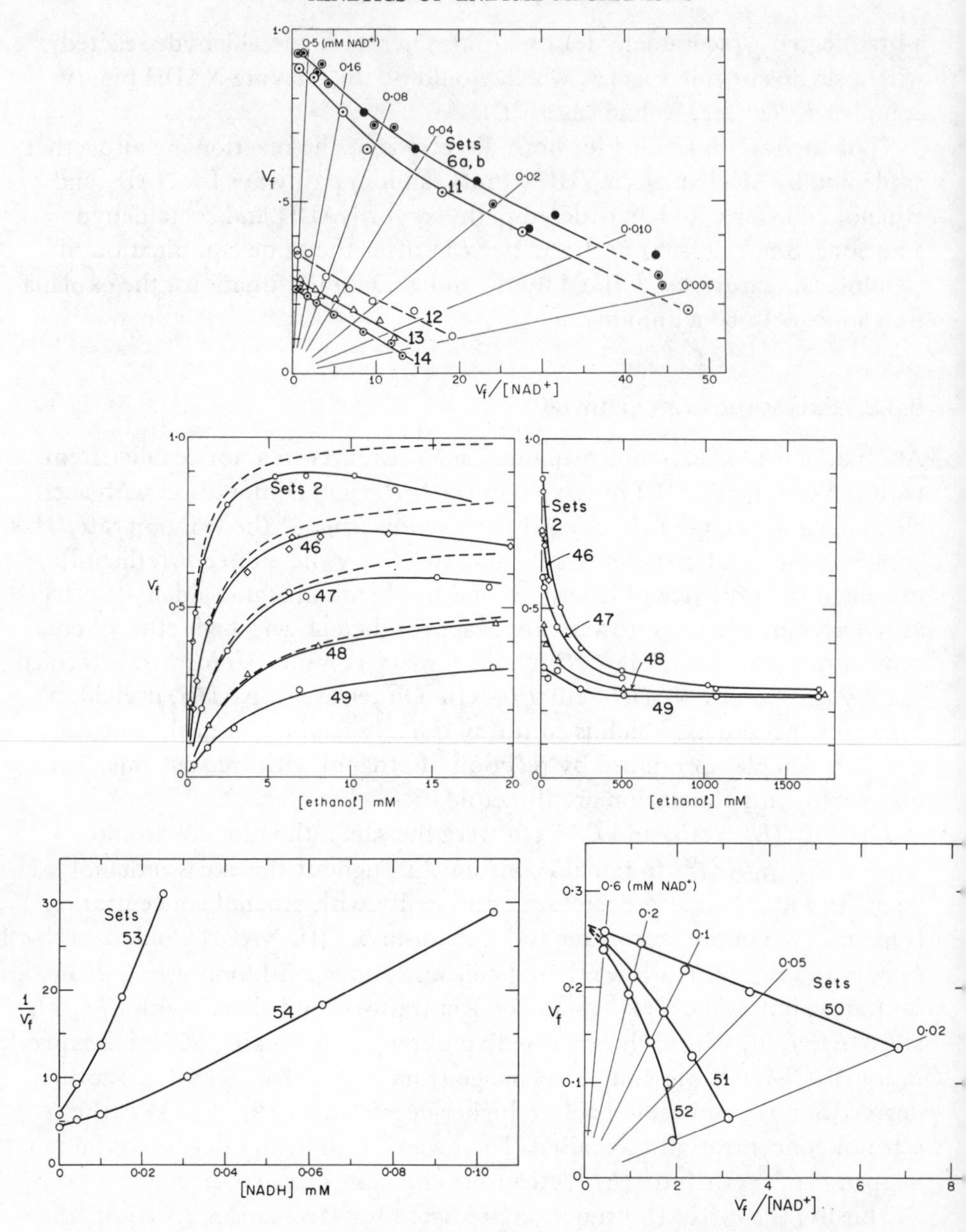


Fig. 5.6. Rate behaviour within the excess-ethanol range. Sets 6 and 11 were obtained at low, and 12–14 at excess, ethanol concentrations. Excess ethanol depressed $V_{(NAD)}$ (vertical intercept) without greatly affecting $K^{NAD}_{(m)}$ (slope). Therefore the ratio $V_{(NAD)}/K^{NAD}_{(m)}$ decreased with increasing ethanol. Sets 2 (with no added acetaldehyde), and 46–49 (with increasing acetaldehyde) pointed to the competition between acetaldehyde and excess ethanol: aldehyde abolished the maximum in the $v_f$(ethanol) profile, and at the same time excess ethanol (1 M–2 M) abolished the product effects due to aldehyde. The nonlinear Dixon plots for Sets 53 and 54, obtained at different NAD⁺ concentrations but both in excess ethanol, suggest that $v_f$(NADH) was higher-degree within the excess ethanol range. This would be the case only if $E$/—/eth and *($E$/—/eth) in Mechan-

enzyme species to contribute an arrow to all numerator patterns, the latter must contain both $k_3$O and $k_{10}$O. Since it also requires all but one enzyme species to contribute an arrow to all denominator patterns, the latter must contain $k_3$O, or $k_{10}$O, or both. Therefore the $v_f$(NAD$^+$) relationship is a second-degree function, but one which is devoid of the $m_1$O and $d_0$ terms usually present in a second-degree function. As a result the relationship is reducible to hyperbolic form:

$$v_f = \frac{m_2\mathrm{O}^2}{d_2\mathrm{O}^2 + d_1\mathrm{O}} = \frac{m_2\mathrm{O}}{d_2\mathrm{O} + d_1} \text{ (\textit{hyperbolic form})}.$$

If NADH were added initially to the reaction system (i.e. as a single product), *(*E*/—/eth) would have the extra arrow $k_{-9}$R to contribute. The above topological constraint would be relieved, and arrow-patterns for the $m_1$O and $d_0$ terms could then be constructed. Thus the $v_f$(NAD$^+$) relationship in the presence of NADH is predicted to be nonhyperbolic:

$$v_f = \frac{m_2\mathrm{O}^2 + m_1\mathrm{O}}{d_2\mathrm{O}^2 + d_1\mathrm{O} + d_0} \text{ (\textit{nonhyperbolic form})}.$$

This unusual prediction of a switch from a hyperbolic $v_f$(NAD$^+$) to a nonhyperbolic one was verified by Sets 50–52. In Set 50, no NADH was added, and the $v_f$(NAD$^+$) relationship was linear in the Eadie plot. In Sets 51 and 52, NADH was added, and the relationship in both cases was nonlinear in the Eadie plot. Consequently, not only were there two distinct enzyme species, namely *E*/—/eth and *(*E*/—/eth), to react with NAD$^+$ under excess ethanol concentrations, but their placement in the mechanism also exactly conformed to the unusual topological constraint being tested.

## Problem 5.1

Figure 5.7 shows the rate observations on an enzyme-catalysed reaction of the type $A + B = Q + P$, in which a group $G$ is transferred from $A$ to $B$ in the forward direction, or from $P$ to $Q$ in the reverse direction. Of the symbols employed, $v_f$ stands for the initial forward velocity, $v_r$ the initial reverse velocity, $V_A$ the forward velocity saturated with respect to $A$, $V_B$ the forward velocity saturated with respect to $B$, and $V_P$ the reverse velocity saturated with respect to $P$. Identify the kinetic mechanism on the basis of these observations.

---

ism 5.VIII represented distinct enzyme species. Finally, in Set 50 no NADH was added, and $v_f$(NAD$^+$) was apparently first-degree; in Sets 51 and 52, different amounts of NADH were added and the dependence became higher-degree. This unusual switch induced by NADH is a consequence of the topological constraints embodied in Mechanism 5.VIII. (From Hanes *et al.*, 1972.)

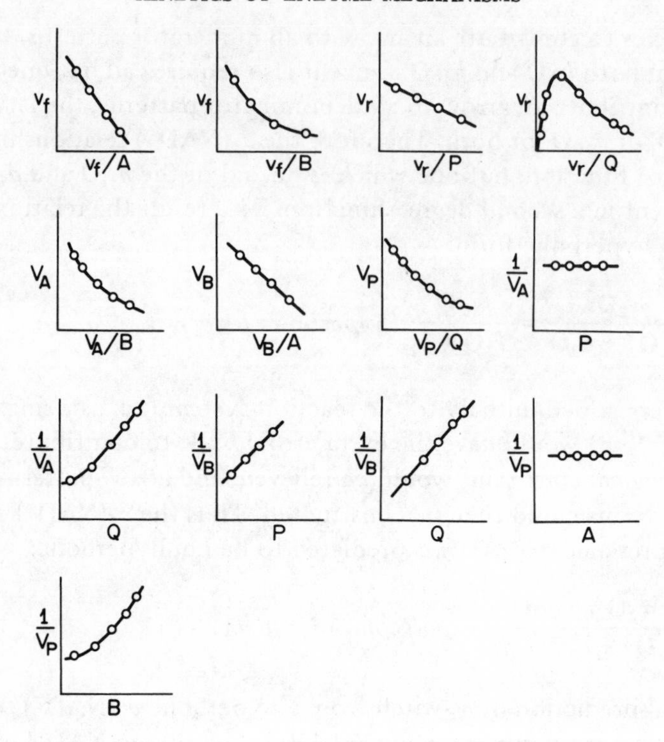


Fig. 5.7 Rate observations on a two-substrate group-transfer reaction.

# 6 Linear Mechanisms

In a multisubstrate reaction, if the enzyme binds the substrates and releases the products in a fixed order, the mechanism would be a linear, unbranching chain of reaction steps. As a rule, in such mechanisms each substrate or product reacts with only one enzyme species. Accordingly, their rate functions are uniformly first-degree, and kinetic criteria rather different from those for treating branching mechanisms are required for their differentiation.

## 6.1. Two-Substrate Mechanisms

Any linear mechanism in which substrates $A$ and $B$, and products $P$ and $Q$, each react with one enzyme species can give rise to only one positive arrow-pattern and one negative arrow-pattern in the numerator of the rate-law. The former consists of all the forward arrows in the mechanism, and the latter all the reverse arrows†:

$$\frac{v}{(E)_0} = \frac{m_{11}\mathrm{AB} - m'_{11}\mathrm{PQ}}{\mathrm{DENOMINATOR}}$$
$$m_{11} = k_1 k_2 k_3 \ldots k_n \tag{6.1}$$
$$m'_{11} = k_{-1} k_{-2} k_{-3} \ldots k_{-n}$$

The DENOMINATOR also cannot exceed first-degree with respect to each of the four reactants. Therefore, the forward initial rate-law in the absence of reaction products has the general form of

$$v_f = \frac{\mathrm{AB}}{V_f^{-1}\mathrm{AB} + \phi_B \mathrm{A} + \phi_A \mathrm{B} + \phi_{AB}} \tag{6.2}$$

and the reverse initial rate-law has the general form of

$$v_r = \frac{\mathrm{PQ}}{V_r^{-1}\mathrm{PQ} + \phi_Q \mathrm{P} + \phi_P \mathrm{Q} + \phi_{PQ}}. \tag{6.3}$$

† The reason is that the structural rule requires that the positive pattern contains a complete pathway for the forward reaction, and the negative pattern contains a complete pathway for the negative reaction (Section 2.3).

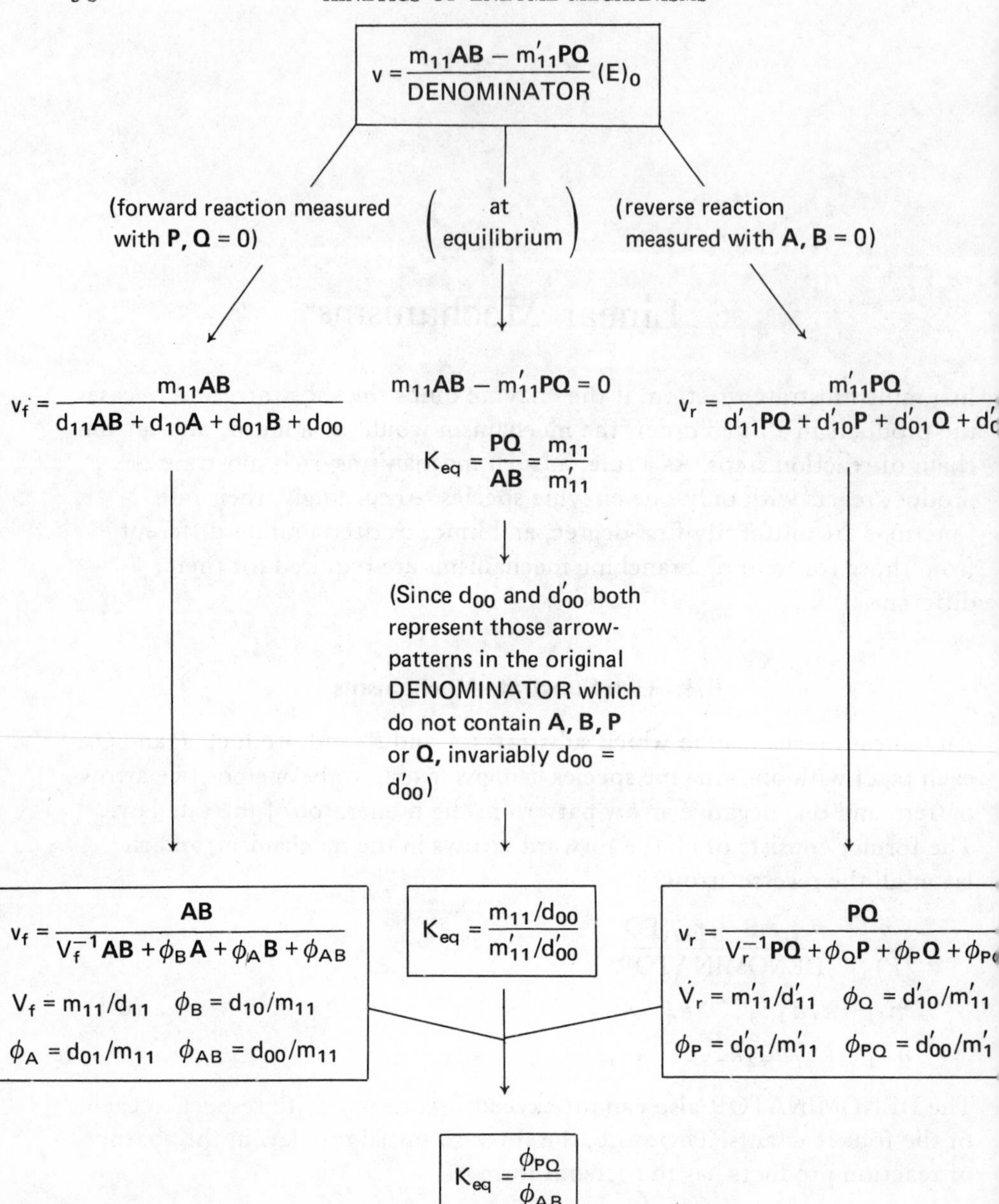


Fig. 6.1. General rate law for hyperbolic two-substrate mechanisms

The derivation of Eqns (6.2) and (6.3) from (6.1) is charted in Fig. 6.1. If substrate $A$ is varied while its cosubstrate $B$ is maintained at some fixed concentration, Eqn (6.2) predicts a hyperbolic relationship between $v_f$ and **A**:

$$
\begin{aligned}
v_f &= \frac{V_{(A)} \cdot \mathrm{A}}{\mathrm{A} + K^A_{(m)}} \\
V_{(A)} &= \frac{\mathrm{B}}{V_f^{-1}\mathrm{B} + \phi_B} \\
K^A_{(m)} &= \frac{\phi_A\ \mathrm{B} + \phi_{AB}}{V_f^{-1}\mathrm{B} + \phi_B} \\
\frac{V_{(A)}}{K^A_{(m)}} &= \frac{\mathrm{B}}{\phi_A \mathrm{B} + \phi_{AB}} \,.
\end{aligned}
\tag{6.4}
$$

The *conditional saturational velocity* $V_{(A)}$, and the *conditional Michaelis constant* $K^A_{(m)}$, vary with **B**. At any fixed concentration of $B$ they may be estimated from a primary plot of the velocity measurements in the form of the Eadie, Lineweaver-Burk, or Hanes plot (Fig. 1.3). In turn, estimates of $V_{(A)}$ and $K^A_{(m)}$ obtained at different fixed concentrations of $B$ may be analysed to yield the four kinetic parameters of Eqn (6.2) on the basis of secondary plots, as shown for example in Fig. 6.2. The parameters of Eqn (6.3) for the reverse reaction may be obtained in comparable fashion.

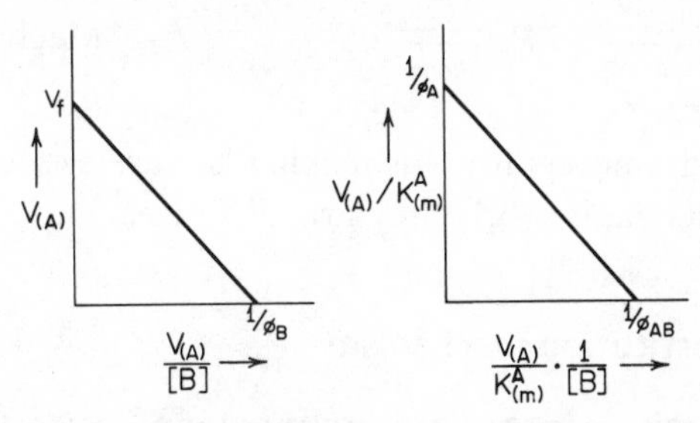


Fig. 6.2. Estimation of kinetic parameters for hyperbolic two-substrate mechanisms.

The graphical estimation of four independent kinetic parameters for either the forward or reverse reaction was first achieved by Florini and Vestling (1957), who employed $V_f$ along with three $K$-type parameters, and by Dalziel (1957), who employed four $\phi$-type parameters. The present form of Eqn (6.2) is a midway combination of these two systems of nomenclature. The $V_f$ of Florini and Vestling is retained as an appropriate symbol for the maximum forward velocity attainable upon saturation by both substrates; the reciprocal maximum velocity $V_f^{-1}$ replaces the $\phi_0$ of Dalziel's terminology. However, the other $\phi$-type designations are adopted in preference to the $K$-type ones in order to avoid confusion with Michaelis constants and dissociation constants.

### 6.1.1. Sequential mechanism with ternary complexes

$$
E \underset{k_{-1}}{\overset{k_1\mathbf{A}}{\rightleftharpoons}} EA \underset{k_{-2}}{\overset{k_2\mathbf{B}}{\rightleftharpoons}} \boxed{EAB \rightleftharpoons EPQ} \underset{k_{-3}\mathbf{Q}}{\overset{k_3}{\rightleftharpoons}} EP \underset{k_{-4}\mathbf{P}}{\overset{k_4}{\rightleftharpoons}} E
$$

Mechanism 6.I

This mechanism describes a sequential addition of *A* before *B* to the enzyme in the forward direction, and *P* before *Q* in the reverse direction. The ternary complexes *EAB* and *EPQ* may be contracted into a single central complex in the topology, or expanded to portray additional intermediate complexes, without affecting the nature of the concentration terms found in the rate law. So long as none of the ternary complexes reacts with any nonenzymic reactant, their contraction or expansion will not alter the topological relationships of the various reactant arrows, and it is the character of these relationships that determines the algebraic form of the rate law.

The forward and reverse kinetic parameters for this mechanism are expressed in terms of the elementary rate constants in Table 6.1.

### 6.1.2. Two binary complexes

If the ternary complexes interconvert and breakdown so rapidly that they have negligible existence, Mechanism 6.I is reduced to 6.II, which is known as the Theorell-Chance mechanism (Theorell and Chance, 1951):

$$E \underset{k_{-1}}{\overset{k_1\mathbf{A}}{\rightleftharpoons}} EA \underset{k_{-2}\mathbf{Q}}{\overset{k_2\mathbf{B}}{\rightleftharpoons}} EP \underset{k_{-3}\mathbf{P}}{\overset{k_3}{\rightleftharpoons}} E \qquad \text{Mechanism 6.II}$$

The eight kinetic parameters for this mechanism are expressed in Table 6.1. As in the case of Mechanism 6.I, they are all finite.

### 6.1.3. Enzyme-substitution mechanism

In some group-transfer reactions, a functional moiety on the enzyme molecule may serve as an intermediate acceptor in a double transfer, e.g. the amination of an enzyme-bound pyridoxal in transaminase reactions. Occurrence of such enzyme-substitution may then give rise to a substituted-enzyme, *E–G*, as a discrete enzyme species. Mechanism 6.III represents the case where *E–G* is formed via an unbranching pathway:

$$E \underset{k_{-1}}{\overset{k_1\mathbf{A}}{\rightleftharpoons}} EA \underset{k_{-2}\mathbf{Q}}{\overset{k_2}{\rightleftharpoons}} E{-}G \underset{k_{-3}}{\overset{k_3\mathbf{B}}{\rightleftharpoons}} EP \underset{k_{-4}\mathbf{P}}{\overset{k_4}{\rightleftharpoons}} E \qquad \text{Mechanism 6.III}$$

In order to form the substituted enzyme, *A* and *P* in this mechanism must represent the group-donors, and *B* and *Q* the group-acceptors. Moreover, $\phi_{AB}$ and $\phi_{PQ}$ are both equal to zero (Table 6.1). The topological reason for this is straightforward. In the initial absence of the products *P* and *Q*, the $k_{-2}$Q and $k_{-4}$**P** arrows are insignificant, leaving only six significant arrows:

$$E \underset{k_{-1}}{\overset{k_1\mathbf{A}}{\rightleftharpoons}} EA \xrightarrow{k_2} E{-}G \underset{k_{-3}}{\overset{k_3\mathbf{B}}{\rightleftharpoons}} EP \xrightarrow{k_4} E.$$

TABLE 6.1

Kinetic parameters for hyperbolic two-substrate mechanisms in the absence of reaction products

| Mechanism | $V_f$ | $\phi_B$ | $\phi_A$ | $\phi_{AB}$ | $V_r$ | $\phi_Q$ | $\phi_P$ | $\phi_{PQ}$ |
|---|---|---|---|---|---|---|---|---|
| 6.I | $\frac{k_3k_4}{k_3+k_4}$ | $\frac{k_{-2}+k_3}{k_2k_3}$ | $\frac{1}{k_1}$ | $\frac{k_{-1}(k_{-2}+k_3)}{k_1k_2k_3}$ | $\frac{k_{-1}k_{-2}}{k_{-1}+k_{-2}}$ | $\frac{k_{-2}+k_3}{k_{-2}k_{-3}}$ | $\frac{1}{k_{-4}}$ | $\frac{k_4(k_{-2}+k_3)}{k_{-2}k_{-3}k_{-4}}$ |
| 6.II | $k_3$ | $\frac{1}{k_2}$ | $\frac{1}{k_1}$ | $\frac{k_{-1}}{k_1k_2}$ | $k_{-1}$ | $\frac{1}{k_{-2}}$ | $\frac{1}{k_{-3}}$ | $\frac{k_3}{k_{-2}k_{-3}}$ |
| 6.III, 6.IV | $\frac{k_2k_4}{k_2+k_4}$ | $\frac{k_{-3}+k_4}{k_3k_4}$ | $\frac{k_{-1}+k_2}{k_1k_2}$ | 0 | $\frac{k_{-1}k_{-3}}{k_{-1}+k_{-3}}$ | $\frac{k_{-1}+k_2}{k_{-1}k_{-2}}$ | $\frac{k_{-3}+k_4}{k_{-3}k_{-4}}$ | 0 |
| 6.Va | $k_0$ | $K'_B/k_0$ | $K'_A/k_0$ | $K_AK'_B/k_0$ | $k_{-0}$ | $K'_Q/k_{-0}$ | $K'_P/k_{-0}$ | $K_PK'_Q/k_{-0}$ |
| 6.Vb | $k_0$ | $K_B/k_0$ | $K_A/k_0$ | $K_AK_B/k_0$ | $k_{-0}$ | $K_Q/k_{-0}$ | $K_P/k_{-0}$ | $K_PK_Q/k_{-0}$ |
| 6.VI | $k_3$ | $\frac{k_{-2}+k_3}{k_2k_3}$ | $\frac{1}{k_1}$ | $\frac{k_{-1}(k_{-2}+k_3)}{k_1k_2k_3}$ | — | — | — | — |
| 6.VII | $\infty$ | $\frac{1}{k_2}$ | $\frac{1}{k_1}$ | $\frac{k_{-1}}{k_1k_2}$ | — | — | — | — |

Free $E$ has only the solitary $k_1$A to contribute to any arrow-pattern, and $E$–$G$ has only $k_3$B to contribute. Since all but one enzyme species have to contribute to each denominator arrow-pattern, these patterns must include $k_1$A, or $k_3$B, or both. The zero-degree term $\phi_{AB}$ is therefore absent. Addition of either $P$ or $Q$ to the reaction system will eliminate this topological constraint and *recall* back to the rate law a finite $\phi_{AB}$-term. The same applies to the reverse reaction; the zero-degree term $\phi_{PQ}$ is equal to zero if both $A$ and $B$ are absent, but takes on a finite value if either $A$ or $B$ is added.

### 6.1.4. Induced release of product

In Mechanism 6.I the first substrate must be bound to the enzyme to permit the binding of the second substrate. Presumably some conformational adjustment at the active center of the enzyme, induced by one substrate, is essential for the binding of the other substrate. Similarly, some substrate-induced adjustment might be essential for the release of a reaction product, e.g. in Mechanism 6.IV, substrate $B$ must be present on the enzyme to induce the release of product $P$, and *vice versa:*

$$EB \underset{k_{-1}}{\overset{k_1\mathrm{A}}{\rightleftharpoons}} \boxed{EAB \rightleftharpoons EPQ} \underset{k_{-2}\mathrm{Q}}{\overset{k_2}{\rightleftharpoons}} EP \underset{k_{-3}}{\overset{k_3\mathrm{B}}{\rightleftharpoons}} EBP \underset{k_{-4}\mathrm{P}}{\overset{k_4}{\rightleftharpoons}} EB$$

Mechanism 6.IV

($E + B \rightarrow EB$ and $E + P \rightarrow EP$ occur only slowly). The additions of the two substrates occur sequentially in Mechanisms 6.I and 6.II, but occur alternately with the release of the two products in Mechanisms 6.III and 6.IV. The latter type of mechanisms are called *ping-pong* mechanisms by Cleland (1963a). Since the topology of 6.IV is similar to that of 6.III, $\phi_{AB}$ is again equal to zero but becomes finite in the presence of either $P$ or $Q$; also $\phi_{PQ}$ is equal to zero but becomes finite in the presence of either $A$ or $B$.

### 6.1.5. Quasi-equilibrium random mechanism

If substrates $A$ and $B$ add to the enzyme in random sequences, the branching topology of the mechanism would yield, at steady-state, velocity-substrate functions that are nonhyperbolic. However, if quasi-equilibrium conditions prevail, the functions will be reduced to hyperbolic, and it becomes necessary to differentiate such quasi-equilibrium cases from linear mechanisms which also predict hyperbolic profiles.

Under quasi-equilibrium conditions, the substrate-binding steps are rapid relative to the catalytic breaking and making of covalent bonds. Mechanism 6.Va represents an adequate general model, and Mechanism

6.Vb is an important limiting case wherein the substrates bind to the enzyme independently of one another.

$$\begin{array}{ccccccccc} & & EB & & & & & EQ & \\ & \mathbf{B}/K_B \nearrow\swarrow & & \searrow\nwarrow \mathbf{A}/K'_A & & & \mathbf{P}/K'_P \nearrow\swarrow & & \searrow\nwarrow \mathbf{Q}/K_Q \\ E & & & & EAB \underset{k_{-0}}{\overset{k_0}{\rightleftharpoons}} EPQ & & & & E \\ & \mathbf{A}/K_A \searrow\nwarrow & & \nearrow\swarrow \mathbf{B}/K'_B & & & \mathbf{Q}/K'_Q \searrow\nwarrow & & \nearrow\swarrow \mathbf{P}/K_P \\ & & EA & & & & & EP & \end{array}$$

Mechanism 6.Va: $K'_A \neq K_A, \quad K'_B \neq K_B$

$K'_P \neq K_P, \quad K'_Q \neq K_Q$

Mechanism 6.Vb: $K'_A = K_A, \quad K'_B = K_B$

$K'_P = K_P, \quad K'_Q = K_Q$

$K_A, K_B, K_P$ and $K_Q$ are the dissociation constants for the binary complexes, and $K'_A, K'_B, K'_P$ and $K'_Q$ are the dissociation constants for the ternary complexes. In Mechanism 6.Va no special restriction is imposed on these constants other than the thermodynamic requirement for *microscopic reversibility*, i.e. the product of equilibrium constants taken around a closed cycle must equal one (Hearon, 1953):

$$(K_B)^{-1}(K'_A)^{-1}K'_BK_A = 1 \qquad \text{or} \qquad \frac{K'_B}{K_B} = \frac{K'_A}{K_A}$$

$$(K_Q)^{-1}(K'_P)^{-1}K'_QK_P = 1 \qquad \text{or} \qquad \frac{K'_Q}{K_Q} = \frac{K'_P}{K_P}$$

In Mechanism 6.Vb, the substrates bind to the enzyme independently of one another in both directions of the reaction. The corresponding dissociation constants for the binary and the ternary complexes are identical:

$$\frac{K'_B}{K_B} = \frac{K'_A}{K_A} = \frac{K'_Q}{K_Q} = \frac{K'_P}{K_P} = 1$$

i.e. from Table 6.1 and Eqn (6.4),

$$K^A_{(m)} = \frac{\phi_A \mathbf{B} + \phi_{AB}}{V_f^{-1}\mathbf{B} + \phi_B} = K_A.$$

The Michaelis constant $K^A_{(m)}$ does not vary with **B**. Therefore the $v_f(\mathrm{A})$ profiles obtained at different levels of **B** give straight lines which intersect on the 1/A-axis in the Lineweaver-Burk plot, or straight lines which run parallel to each other in the Eadie plot. Similarly, $K^B_{(m)}$ is predicted not to vary with **A**, $K^P_{(m)}$ not to vary with **Q**, and $K^Q_{(m)}$ not to vary with **P**.

### 6.1.6. Irreversible mechanisms

When the reaction is largely irreversible, the linear mechanisms 6.VI and 6.VII define two very simple models:

$$E \underset{k_{-1}}{\overset{k_1\mathbf{A}}{\rightleftharpoons}} EA \underset{k_{-2}}{\overset{k_2\mathbf{B}}{\rightleftharpoons}} EAB \xrightarrow{k_3} E + \text{PRODUCTS}$$

Mechanism 6.VI

$$E \underset{k_{-1}}{\overset{k_1\mathbf{A}}{\rightleftharpoons}} EA \xrightarrow{k_2\mathbf{B}} E + \text{PRODUCTS}$$

Mechanism 6.VII

A ternary complex with a significant half-life is formed in 6.VI but not in 6.VII. The latter was used by Chance (1943) to describe the kinetics of horseradish peroxidase. With only two enzyme species, its rate law cannot contain any **AB**-pattern in the denominator. Accordingly, $V_f^{-1}\mathbf{AB}$ equals zero, and $V_f$ must be boundless: it is impossible to saturate the enzyme at once with respect to both of the substrates, because increases in **A** pushes the enzyme over to $EA$, and increases in **B** pushes it back to $E$. Thus, in the example of cytochrome $C$ peroxidase, $V_f$ was found to be finite by Yonetani and Ray (1966), and Mechanism 6.VII was ruled out in favour of 6.VI.

## 6.2. Numerical Relations Between Parameters

Alberty (1953) discovered that the *Haldane relations* between the equilibrium constant and the kinetic parameters (cf. Section 2.5) are not the same for all mechanisms, and therefore may be employed to distinguish between some of the mechanisms. Where the equilibrium constant is not accurately known, numerical relations recognized by Dalziel (1957) to exist amongst the kinetic parameters themselves may also be employed for the same purpose. Both types of numerical relations can be derived from the kinetic parameters presented in Table 6.1, and they are summarized in Table 6.2.

At equilibrium, net reaction velocity is zero, which according to Eqn (6.1) means that

$$k_1 k_2 \ldots k_n \mathbf{AB} - k_{-1} k_{-2} \ldots k_{-n} \mathbf{PQ} = 0$$

or

$$K_{eq} = \frac{\mathbf{P} \cdot \mathbf{Q}}{\mathbf{A} \cdot \mathbf{B}} = \frac{k_1 k_2 \ldots k_n}{k_{-1} k_{-2} \ldots k_{-n}} .$$

This equation is general to all the linear mechanisms and forms the basis of the predicted Haldane relations. For instance, there are four forward steps and four reverse steps in Mechanism 6.I, and

TABLE 6.2

Haldane and Dalziel relations for hyperbolic two-substrate mechanisms

| Mechanism | Haldane relations | Dalziel relations |
|---|---|---|
| 6.I | $K_{eq} = \phi_{PQ}/\phi_{AB}$ | $\phi_{AB} > V_r\phi_A\phi_B$, $\phi_{PQ} > V_f\phi_P\phi_Q$ |
| 6.II | $K_{eq} = \phi_{PQ}/\phi_{AB} = V_f\phi_P\phi_Q/V_r\phi_A\phi_B$ | $\phi_{AB} = V_r\phi_A\phi_B$, $\phi_{PQ} = V_f\phi_P\phi_Q$ |
| 6.III, 6.IV | $K_{eq} = \phi_P\phi_Q/\phi_A\phi_B$ | – |
| 6.Va | $K_{eq} = \phi_{PQ}/\phi_{AB}$ | – |
| 6.Vb | $K_{eq} = \phi_{PQ}/\phi_{AB} = V_r\phi_P\phi_Q/V_f\phi_A\phi_B$ | $\phi_{AB} = V_f\phi_A\phi_B$, $\phi_{PQ} = V_r\phi_P\phi_Q$ |
| 6.VI, 6.VII | – | – |

$$K_{eq} = \frac{k_1k_2k_3k_4}{k_{-1}k_{-2}k_{-3}k_{-4}}.$$

From Table 6.1 we have

$$\frac{\phi_{PQ}}{\phi_{AB}} = \frac{k_4(k_{-2}+k_3)}{k_{-2}k_{-3}k_{-4}} \cdot \frac{k_1k_2k_3}{k_{-1}(k_{-2}+k_3)} = \frac{k_1k_2k_3k_4}{k_{-1}k_{-2}k_{-3}k_{-4}}.$$

Consequently the predicted Haldane relation is that $K_{eq} = \phi_{PQ}/\phi_{AB}$. The Haldane relations predicted for the other mechanisms may be verified in like manner. The *Dalziel relations* may also be derived from the expressions for the various rate parameters. For instance, for Mechanism 6.I.

$$\phi_{AB} = \frac{k_{-1}(k_{-2}+k_3)}{k_1k_2k_3}$$

$$V_r\phi_A\phi_B = \frac{k_{-1}(k_{-2}+k_3)}{k_1k_2k_3} \cdot \frac{k_{-2}}{k_{-1}+k_{-2}}.$$

Since $k_{-2}/(k_{-1}+k_{-2})$ must be less than unity, this mechanism predicts that $\phi_{AB} > V_r\phi_A\phi_B$.

On the other hand, the parameters for Mechanism 6.II are such that

$$\phi_{AB} = \frac{k_{-1}}{k_1k_2}$$

$$V_r\phi_A\phi_B = \frac{k_{-1}}{k_1k_2}$$

and therefore

$$\phi_{AB} = V_r\phi_A\phi_B.$$

For each mechanism then, the predicted Haldane and Dalziel relations constitute specific statements regarding the relative numerical magnitudes which must be fulfilled by the experimental parameters. Otherwise the model is not valid for the experimental system in question. These relations, which are highly sensitive to variations in mechanistic details, thus furnish a means for testing the validity of the proposed mechanisms. In fact Mahler, Baker and Shiner (1962) found that, in Mechanism 6.II for example, simple isomerizations of the binary complexes will transform the predicted equality between $\phi_{AB}$ and $V_r \phi_A \phi_B$ into an inequality. This sensitivity is both a disadvantage and an advantage; a disadvantage in that it would be erroneous to rule out Mechanism 6.II in favour of 6.I solely on the basis of an observed inequality, but an advantage in that an observed inequality would pinpoint an isomerization step if Mechanism 6.II could be established through other kinetic or nonkinetic methods.

## 6.3. Method of Alternative Substrates

### 6.3.1. Induction of nonhyperbolic behaviour

In Mechanisms 6.I, 6.II, 6.VI and 6.VII, there is a sequential addition of $A$ before $B$ to the enzyme. If two alternative substrates $A_1$ and $A_2$ are presented simultaneously to the system, there will be two different enzyme species $EA_1$ and $EA_2$ to react with $B$:

$$E \begin{array}{l} \xrightarrow{A_1} EA_1 \xrightarrow{B} \cdots \\ \xrightarrow{A_2} EA_2 \xrightarrow{B} \cdots \end{array}$$

Accordingly *the* $v_f(\mathrm{B})$ *profile will become second-degree, nonhyperbolic.* In contrast, if two alternative substrates $B_1$ and $B_2$ are presented, there remains only the one enzyme species $E$ to react with $A$, and $v_f(\mathrm{A})$ remains first-degree, hyperbolic:

$$E \rightarrow EA \begin{array}{l} \xrightarrow{B_1} \cdots \\ \xrightarrow{B_2} \cdots \end{array}$$

This mode of behaviour does not extend to Mechanism 6.III, which provides only $E{-}G$ to react with $B$ in the presence of $A_1$ and $A_2$, and only $E$ to react with $A$ in the presence of $B_1$ and $B_2$. Therefore the presence of $B_1$ and $B_2$ does not induce a nonhyperbolic $v_f(\mathrm{A})$, and the presence of $A_1$ and $A_2$ does not induce a nonhyperbolic $v_f(\mathrm{B})$.

In Mechanism 6.IV, the alternative substrates $A_1$ and $A_2$ may produce the same $P$ but different $Q_1$ and $Q_2$, or the same $Q$ but different $P_1$ and $P_2$. In the former instance, $B$ reacts only with $EP$ even in the presence of

$A_1$ and $A_2$, and $v_f$(B) remains hyperbolic. In the latter instance, $B$ reacts with $EP_1$ and $EP_2$ in the presence of $A_1$ and $A_2$, and $v_f$(B) becomes nonhyperbolic. However, the presence of $B_1$ and $B_2$ always gives rise to two distinct species, $EB_1$ and $EB_2$, to react with $A$ and cause $v_f$(A) to become nonhyperbolic.

In Mechanisms 6.Va and 6.Vb, the condition of quasi-equilibrium causes $v_f$(A) and $v_f$(B) to remain strictly hyperbolic as long as only one molecule of each substrate can bind to the enzyme, regardless of the number of enzyme species with which each substrate reacts.

These predictions, presented in Table 6.3, can be employed to establish the sequence of substrate additions in the sequential Mechanisms 6.I, 6.II,

TABLE 6.3

Induction of nonhyperbolic velocity-substrate relations

| Mechanism | $v_f$(A) in presence of $B_1$ and $B_2$ | $v_f$(B) in presence of $A_1$ and $A_2$ |
|---|---|---|
| 6.I, 6.II, 6.VI, 6.VII | Hyperbolic | Nonhyperbolic |
| 6.III, 6.Va, 6.Vb | Hyperbolic | Hyperbolic |
| 6.IV | Nonhyperbolic | Hyperbolic if $A_1$ (or $A_2$) + $B \rightarrow P + Q_1$ (or $Q_2$)<br>Nonhyperbolic if $A_1$ (or $A_2$) + $B \rightarrow P_1$ (or $P_2$) + $Q$ |

6.VI and 6.VII. The two ping-pong mechanisms 6.III and 6.IV, which have identical rate laws and are inseparable on the basis of cosubstrate and product rate effects, may also be separated by this method.

This method of analysis, proposed by Wong and Hanes (1962) and again by Rudolph and Fromm (1970), was applied to the glucose-6-phosphate dehydrogenase from *Leuconostoc mesenteroides* by Olive, Geroch and Levy (1971). As the data in Fig. 6.3 indicate, $v_f$(NAD) was hyperbolic with glucose-6-phosphate or 2-deoxyglucose-6-phosphate as cosubstrate, or with both of these alternative cosubstrates present. On the other hand, $v_f$(G6P) was hyperbolic with either NAD or NADP as cosubstrate, but became nonhyperbolic when both NAD and NADP were present. This mode of behaviour points to the sequential addition of NAD (or NADP) to the enzyme before G6P (or dG6P).

Ricard, Noat, Got and Borel (1972) have pointed out two important aspects of this method. First, if substrate $A$ binds before substrate $B$ in a sequential mechanism, the presence of $A_1$ and $A_2$ induces a nonhyperbolic $v_f$(B). However, the deviation would be negligible if the rate parameters for the mechanism happen to fulfil any of a number of simplifying relationships established by these workers. As always then, a nonhyperbolic observation disproves a hyperbolic prediction, but a seemingly hyperbolic observa-

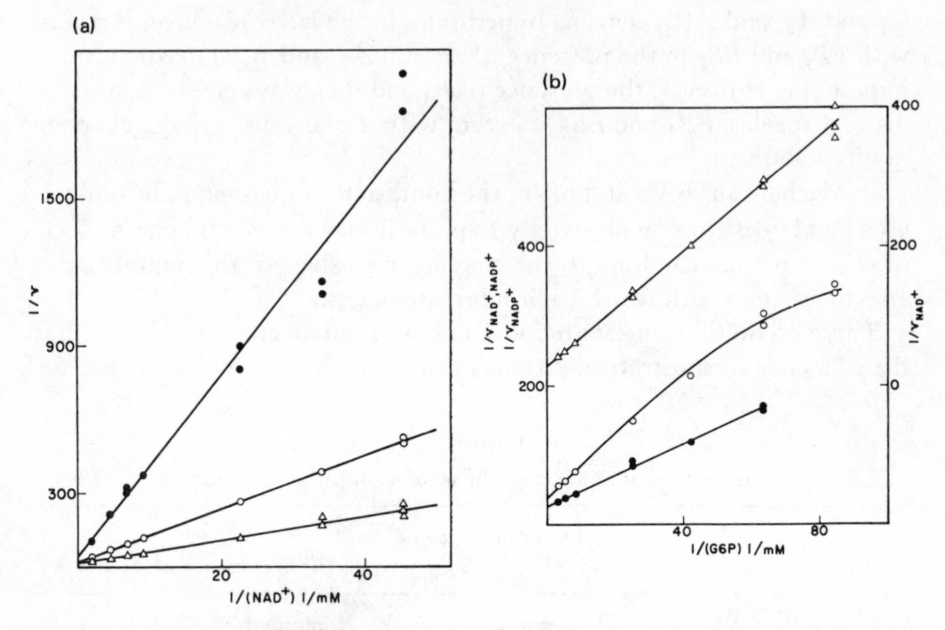


Fig. 6.3. Induction of nonhyperbolic behaviour by two alternative cosubstrates in the glucose-6-phosphate dehydrogenase system from *L. mesenteroides.* In (a), the Lineweaver-Burk plot for $NAD^+$ was linear (i.e., hyperbolic behaviour) in the presence of G6P (△), dG6P (●), or both G6P and dG6P (○). In (b), the Lineweaver-Burk plot for G6P was linear in the presence of $NAD^+$ (△) or $NADP^+$ (●), but became nonlinear (i.e., nonhyperbolic behaviour) in the presence of both $NAD^+$ and $NADP^+$(○). (From Olive, Geroch and Levy, 1971.)

tion does not necessarily disprove a nonhyperbolic prediction. Second, the rate parameters for the two alternative substrates added to the system must be significantly different before they can induce a nonhyperbolic rate dependence on their cosubstrate. Thus the suggestion made by Rudolph and Fromm (1970) that two isotopically distinct forms of the same substrate are serviceable as alternative substrates is mostly invalid. Apart from perhaps the hydrogen atom, the substitution of any one atom in the substrate by its isotope usually does not result in sufficiently large differences in the rate parameters.

### 6.3.2. Effects on kinetic parameters

In the foregoing method, it is assumed that the same mechanism holds for the alternative substrates added simultaneously to the reaction system. If this is so, differentiation between mechanisms also could be achieved by comparing the kinetic parameters obtained separately with the alternative substrates. To do so, alternative forms of $B$ are allowed to react separately with the same form of $A$, or *vice versa.* In either case, the alternative forms of substrate may lead to different forms of $Q$ (plus the same form of $P$), or different forms of $P$ (plus the same form of $Q$). Altogether then, there are four possible sets of reactant relationships:

$$\text{Set 1: } A + B_1 (\text{or } B_2) \to P + Q_1 (\text{or } Q_2)$$
$$\text{Set 2: } A + B_1 (\text{or } B_2) \to Q + P_1 (\text{or } P_2)$$
$$\text{Set 3: } B + A_1 (\text{or } A_2) \to P + Q_1 (\text{or } Q_2)$$
$$\text{Set 4: } B + A_1 (\text{or } A_2) \to Q + P_1 (\text{or } P_2)$$

Consider Mechanism 6.I. In Set 1, a change from $B_1$ to $B_2$ alters the nature of $Q$. Accordingly, $k_2, k_{-2}, k_3$ and $k_{-3}$, but not $k_1, k_{-1}, k_4$ and $k_{-4}$ may be altered. From the forward rate parameters tabulated in Table 6.1, it is evident that $\phi_A$ $(=1/k_1)$ and $\phi_B/\phi_{AB}$ $(=k_1/k_{-1})$ must remain invariant to the change. In Set 2, a change from $B_1$ to $B_2$ alters the nature of $P$. As a result, $k_2, k_{-2}, k_3, k_{-3}, k_4$ and $k_{-4}$, but not $k_1$ and $k_{-1}$, may be altered. Again, $\phi_A$ and $\phi_B/\phi_{AB}$ must remain invariant. The parameters predicted by the various hyperbolic mechanisms to remain invariant in each of the four sets can be worked out in like manner, and the predictions summarized in Table 6.4 provide useful criteria for testing the various mechanisms.

TABLE 6.4

Kinetic parameters predicted to be invariant for two alternative forms of a substrate

| | Alternative substrates and products | | | |
|---|---|---|---|---|
| Mechanism | Set 1: $B_1(B_2) \to Q_1(Q_2)$ | Set 2: $B_1(B_2) \to P_1(P_2)$ | Set 3: $A_1(A_2) \to Q_1(Q_2)$ | Set 4: $A_1(A_2) \to P_1(P_2)$ |
| 6.I | $\phi_A, \phi_B/\phi_{AB}$ | $\phi_A, \phi_B/\phi_{AB}$ | None | None |
| 6.II | $V_f, \phi_A, \phi_B/\phi_{AB}$ | $\phi_A, \phi_B/\phi_{AB}$ | $V_f$ | None |
| 6.III | Implausible† | $\phi_A$ | $\phi_B$ | None |
| 6.IV | None | None | $\phi_B$ | None |
| 6.Va | $\phi_B/\phi_{AB}$ | $\phi_B/\phi_{AB}$ | $\phi_A/\phi_{AB}$ | $\phi_A/\phi_{AB}$ |
| 6.Vb | $\phi_B/\phi_{AB}, \phi_A/V_f$ | $\phi_B/\phi_{AB}, \phi_A/V_f$ | $\phi_A/\phi_{AB}, \phi_B/V_f$ | $\phi_A/\phi_{AB}, \phi_B/V_f$ |
| 6.VI | $\phi_A, \phi_B/\phi_{AB}$ | $\phi_A, \phi_B/\phi_{AB}$ | None | None |
| 6.VII | $\phi_A, \phi_B/\phi_{AB}$ | $\phi_A, \phi_B/\phi_{AB}$ | None | None |

† In Mechanism 6.III, both $B$ and $Q$ must be group-acceptors in the reversible group-transfer. Therefore it is implausible that a chemical change in $B$ would lead to a chemical change in $Q$.

For example, if $B_1$ and $B_2$ in Set 1 yield different experimental values of $\phi_A$, Mechanism 6.I would be excluded. The two ping-pong mechanisms also become separable by this method, since in Set 2 $\phi_A$ remains invariant according to Mechanism 6.III but not according to Mechanism 6.IV. Furthermore, it becomes feasible to establish by this method the sequence of substrate additions in Mechanisms 6.I and 6.II; in both cases a change in the second-reacting $B$ leaves $\phi_A$ and $\phi_B/\phi_{AB}$ invariant, but a change in the first-reacting $A$ causes these parameters to vary.

A related method of exploiting rate differences between alternative substrates is to compare substrates which differ only in the isotopic form of a constituent hydrogen atom. Since the rate effects due to the replacement of hydrogen by deuterium or tritium are well quantitated, it becomes possible to interpret the observed rate differences between such alternative substrates in terms of the rate constants of the reaction mechanism (Melander, 1960; Rose, 1970). In the case of triosephosphate isomerase, analysis of tritium exchange between substrate and solvent had led Rose (1962) to propose the formation of an enzyme-bound enediol intermediate. From a quantitative analysis of such exchanges and the observable hydrogen-isotopic rate effects, Albery and Knowles (1974) succeeded in determining the elementary rate constants of the mechanism (DHAP stands for dihydroxyacetone phosphate, and GAP for glyceraldehyde-3-phosphate):

$$E \underset{k_{-1}}{\overset{k_1(\text{DHAP})}{\rightleftharpoons}} E/\text{DHAP} \underset{k_{-2}}{\overset{k_2}{\rightleftharpoons}} \boxed{E/\text{enediol} \leftrightarrow E/\text{GAP}} \underset{k_{-3}(\text{GAP})}{\overset{k_3}{\rightleftharpoons}} E$$

$$k_1 = 3 \times 10^7\ \text{M}^{-1}\ \text{sec}^{-1} \quad k_2 = 2 \times 10^3\ \text{sec}^{-1} \quad k_3 = 2 \times 10^3\ \text{sec}^{-1}$$

$$k_{-1} = 2 \times 10^4\ \text{sec}^{-1} \quad k_{-2} = 7 \times 10^3\ \text{sec}^{-1} \quad k_{-3} = 4 \times 10^8\ \text{M}^{-1}\ \text{sec}^{-1}$$

## 6.4. Types of Cosubstrate and Product Effects

If the substrate function $v_f(\text{A})$ remains hyperbolic at all levels of a second ligand (e.g. $B$, $P$, $Q$, or an inhibitor), the rate effects of the second ligand on $v_f(\text{A})$ will be classifiable into one of four types (cf. Section 4.3.2):

*Competitive:* Saturation velocity $V_{(A)}$ does not vary with the ligand concentration.

*Uncompetitive:* The ratio $K^A_{(m)}/V_{(A)}$ does not vary with the ligand concentration.

*Classic-noncompetitive:* $K^A_{(m)}$ does not vary with the ligand concentration.

*Mixed-noncompetitive:* $V_{(A)}$, $K^A_{(m)}$ as well as $K^A_{(m)}/V_{(A)}$ all vary with the ligand concentration.

As Fig. 6.4 shows, graphical distinction between them can be achieved on the basis of the Eadie or Lineweaver-Burk plot. The Hanes plot also lends itself toward the same end. All the cosubstrate and product effects predicted by the hyperbolic two-substrate mechanisms fall into these four types:

(a) For the ping-pong Mechanisms 6.III and 6.IV, $\phi_{AB}$ is zero. Because of this, Eqn (6.4) requires that $K^A_{(m)}/V_{(A)} = \phi_A$, which does not vary with **B**. Therefore **B** affects $v_f(\text{A})$ uncompetitively. Analogously, $K^B_{(m)}/V_{(B)} = \phi_B$, which does not vary with **A**. Therefore **A** also affects $v_f(\text{B})$ uncompetitively. This kind of uncompetitive cosubstrate effects, caused by $\phi_{AB}$

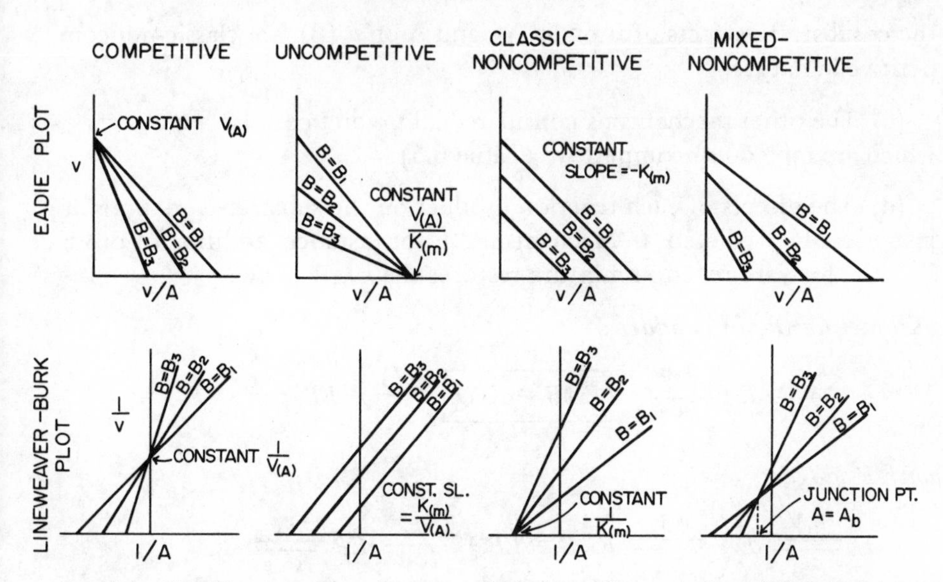


**Fig. 6.4. Classification of cosubstrate and product effects on the $v_f$(A) function.**

being zero, is revealed by parallel Lineweaver-Burk plots, or Eadie plots that converge on the $v$/A axis (Fig. 6.4). It also may be revealed by adding to the enzyme a constant mixture of **A** and **B**, so that $\mathbf{B} = r\mathbf{A}$. Under these conditions, rate law (6.2) becomes

$$v_f = \frac{r\mathbf{A}^2}{rV_f^{-1}\mathbf{A}^2 + (\phi_B + r\phi_A)\mathbf{A} + \phi_{AB}}$$

and $v_f$ is expected to vary nonhyperbolically with **A**, yielding a nonlinear Eadie plot. However, if $\phi_{AB}$ is zero and uncompetitive effects are predicted, the above equation is simplified to

$$v_f = \frac{r\mathbf{A}}{rV_f^{-1}\mathbf{A} + (\phi_B + r\phi_A)}$$

and $V_f$ is expected to vary hyperbolically with **A**, yielding a linear Eadie plot. Thus a linear Eadie plot, observed when **A** and **B** are jointly varied, will be a criterion for the ping-pong mechanisms and their uncompetitive behaviour. In fact, it might prove easier in many instances to test for linearity in the Eadie plot than to decide whether or not $K_{(m)}^A/V_{(A)}$ varies with **B**.

(b) For Mechanism 6.Vb, we have $\phi_{AB} = V_f\phi_A\phi_B$. Because of this, Eqn (6.4) requires that

$$K_{(m)}^A = \frac{\phi_A\mathbf{B} + V_f\phi_A\phi_B}{V_f^{-1}\mathbf{B} + \phi_B} = V_f\phi_A, \text{ which does not vary with } \mathbf{B};$$

and analogously, $K_{(m)}^B = V_f\phi_B$, which does not vary with **A**. Accordingly, the cosubstrate effects of **B** on $v_f(\mathbf{A})$, and **A** on $v_f(\mathbf{B})$, are classic-noncompetitive in nature.

(c) The other mechanisms considered all predict cosubstrate effects which are mixed-noncompetitive (Table 6.5).

(d) The effects of each reaction product on the substrate functions are just as easily predicted. In Mechanism 6.I, for instance, addition of product $P$ to the forward reaction mixture restores the $k_{-4}\mathbf{P}$ arrow:

*with neither P nor Q added:*

$$E \underset{k_{-1}}{\overset{k_1\mathbf{A}}{\rightleftharpoons}} EA \underset{k_{-2}}{\overset{k_2\mathbf{B}}{\rightleftharpoons}} \boxed{EAB \rightleftharpoons EPQ} \xrightarrow{k_3} EP \xrightarrow{k_4} E$$

*with P added:*

$$E \underset{k_{-1}}{\overset{k_1\mathbf{A}}{\rightleftharpoons}} EA \underset{k_{-2}}{\overset{k_2\mathbf{B}}{\rightleftharpoons}} EAB \rightleftharpoons EPQ \xrightarrow{k_3} EP \underset{k_{-4}\mathbf{P}}{\overset{k_4}{\rightleftharpoons}} E.$$

The extra $k_{-4}\mathbf{P}$ arrow in the topology introduces into the denominator of the rate-law new arrow-patterns containing this arrow:

$$v_f = \frac{m_{11}\mathbf{AB}}{d_{11}\mathbf{AB} + d_{10}\mathbf{A} + d_{01}\mathbf{B} + d_{011}\mathbf{BP} + d_{00} + d_{001}\mathbf{P}}.$$

The new **BP**-patterns, since they contain **B** but not **A**, are included into $\phi_A\mathbf{B}$, so that

$$\phi_A\mathbf{B} = \frac{d_{01}\mathbf{B}}{m_{11}},$$

which upon adding $P$ becomes

$$\frac{(d_{01} + d_{011}\mathbf{P})\mathbf{B}}{m_{11}}.$$

The new **P**-patterns, since they contain neither **B** nor **A**, are included into $\phi_{AB}$, so that

$$\phi_{AB} = \frac{d_{00}}{m_{11}}$$

which upon adding $P$ becomes

$$\frac{d_{00} + d_{001}\mathbf{P}}{m_{11}}.$$

On the other hand, since the $k_{-4}\mathbf{P}$ and $k_1\mathbf{A}$ arrows both leave from $E$, they cannot appear in the same arrow patterns. Therefore no **ABP**- or **AP**-patterns are introduced into the rate law; accordingly, $V_f$ and $\phi_B$ are

TABLE 6.5

Cosubstrate and product effects on substrate functions

| Mechanism | Cosubstrate effect of **B** on $v_f$(**A**) | Cosubstrate effect of **A** on $v_f$(**B**) | Product effect of **P** on $v_f$(**A**) | Product effect of **Q** on $v_f$(**A**) | Product effect of **P** on $v_f$(**B**) | Product effect of **Q** on $v_f$(**B**) |
|---|---|---|---|---|---|---|
| 6.I | Mixed-noncompetitive | Mixed-noncompetitive | Competitive | Mixed-noncompetitive | Mixed-noncompetitive | Mixed-noncompetitive |
| 6.II | Mixed-noncompetitive | Mixed-noncompetitive | Competitive | Mixed-noncompetitive | Mixed-noncompetitive | Competitive |
| 6.III | Uncompetitive | Uncompetitive | Competitive | Mixed-noncompetitive | Mixed-noncompetitive | Competitive |
| 6.IV | Uncompetitive | Uncompetitive | Competitive | Mixed-noncompetitive | Mixed-noncompetitive | Competitive |
| 6.Va | Mixed-noncompetitive | Mixed-noncompetitive | Competitive | Competitive | Competitive | Competitive |
| 6.Vb | Classic-noncompetitive | Classic-noncompetitive | Competitive | Competitive | Competitive | Competitive |
| 6.VI | Mixed-noncompetitive | Mixed-noncompetitive | None | None | None | None |
| 6.VII | Mixed-noncompetitive | Mixed-noncompetitive | None | None | None | None |

unchanged in the presence of $P$. It follows from Eqn (6.4) that $V_{(A)}$ is also unchanged:

$$V_{(A)} = \frac{\mathrm{B} \cdot V_f}{\mathrm{B} + V_f \phi_B},$$

which remains unchanged upon adding $P$. The effects of **P** on $v_f(\mathrm{A})$ are accordingly competitive. By inspecting the various mechanisms and applying topological reasoning, their product effects are all readily determined. The results are summarized in Table 6.5. Usefully, the predicted cosubstrate and product effects are different for different mechanisms, and also for the different cosubstrates and products within the same mechanism.

This method for identifying a reaction mechanism, based on the observed patterns of cosubstrate and product effects, was first proposed by Alberty (1953, 1958). Experimentally, *uncompetitive* cosubstrate effects were exemplified elegantly by the observations of Velick and Vavra (1962) on glutamate-aspartate transaminase (Fig. 6.5a) and *classic-noncompetitive* cosubstrate effects by the observations of Craine, Hall and Kaufman (1972) on dihydropteridine reductase (Fig. 6.5b). Carbamate kinase, on the other hand, exhibited *mixed-noncompetitive* cosubstrate effects, which will be considered in conjunction with product effects in the next section.

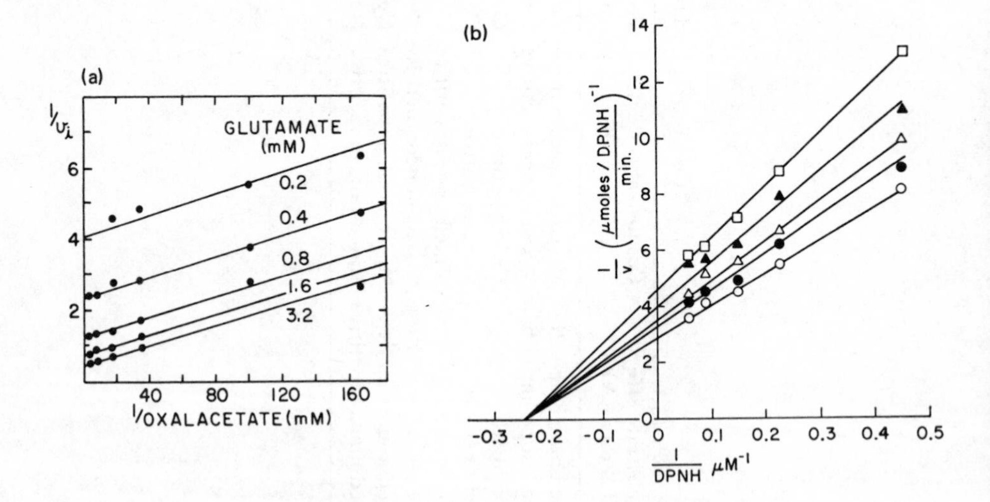


Fig. 6.5. (a) Uncompetitive cosubstrate effects of glutamate on the $v_f$(oxalacetate) function in the glutamate-aspartate transminase reaction. (From Velick and Vavra, 1962.) (b) Classic-noncompetitive cosubstrate effects on the $v_f$(DPNH) function; the different symbols refer to different concentrations of quinonoid-dihydrobiopterin. (From Craine, Hall and Kaufman, 1972.)

## 6.5. Carbamate Kinase

The carbamate kinase from *Streptococcus faecalis* catalyses the reaction

ATP + Carbamate (C) ⇌ ADP + Carbamylphosphate (CP)

Marshall and Cohen (1966) have identified its reaction mechanism at non-excess reactant concentrations as formally identical to 6.I:

$$E \underset{k_{-1}}{\overset{k_1(\text{MgATP})}{\rightleftharpoons}} E/\text{MgATP} \underset{k_{-2}}{\overset{k_2(\text{C})}{\rightleftharpoons}} \begin{matrix}\text{Central}\\ \text{Complex}\end{matrix} \underset{k_{-3}(\text{CP})}{\overset{k_3}{\rightleftharpoons}} E/\text{MgADP} \underset{k_{-4}(\text{MgADP})}{\overset{k_4}{\rightleftharpoons}} E.$$

The forward initial rate law for this mechanism is†:

$$v_f = \frac{(\text{MgATP})(\text{C})}{V_f^{-1}(\text{MgATP})(\text{C}) + \phi_{\text{MgATP}}(\text{C}) + \phi_{\text{C}}(\text{MgATP}) + \phi_{\text{MgATP}\cdot\text{C}}}$$

or,

$$\frac{1}{v_f} = \frac{1}{V_f} + \frac{\phi_{\text{MgATP}}}{(\text{MgATP})} + \frac{\phi_{\text{C}}}{(\text{C})} + \frac{\phi_{\text{MgATP}\cdot\text{C}}}{(\text{MgATP})(\text{C})}. \tag{6.5}$$

At the virtual concentration point of (MgATP) = $-\phi_{\text{MgATP}\cdot\text{C}}/\phi_{\text{C}}$, the last two terms in the above equation will cancel out each other, and $v_f$ becomes independent of the carbamate concentration. Consequently the linear Lineweaver-Burk plots of $1/v_f$ versus 1/(MgATP) obtained at all levels of carbamate should converge through this point; the same is expected of the Eadie or Hanes plot. Analogously, the Lineweaver-Burk plots of $1/V_f$ versus 1/(carbamate) obtained at all levels of MgATP should converge through the virtual concentration point of (carbamate) = $-\phi_{\text{MgATP}\cdot\text{C}}/\phi_{\text{MgATP}}$. The reverse initial rate law for the mechanism is:

$$\frac{1}{v_r} = \frac{1}{V_r} + \frac{\phi_{\text{MgADP}}}{(\text{MgADP})} + \frac{\phi_{\text{CP}}}{(\text{CP})} + \frac{\phi_{\text{MgADP}\cdot\text{CP}}}{(\text{MgADP})(\text{CP})}. \tag{6.6}$$

† In the original paper of Marshall and Cohen, the parameter $\phi_{\text{MgATP}\cdot\text{C}}$ was replaced with $\phi_{\text{C}}$ multiplied by a new $\phi$-parameter, and $\phi_{\text{MgADP}\cdot\text{CP}}$ was replaced with $\phi_{\text{CP}}$ multiplied by another new $\phi$-parameter. The number of independent parameters remained four in each direction.

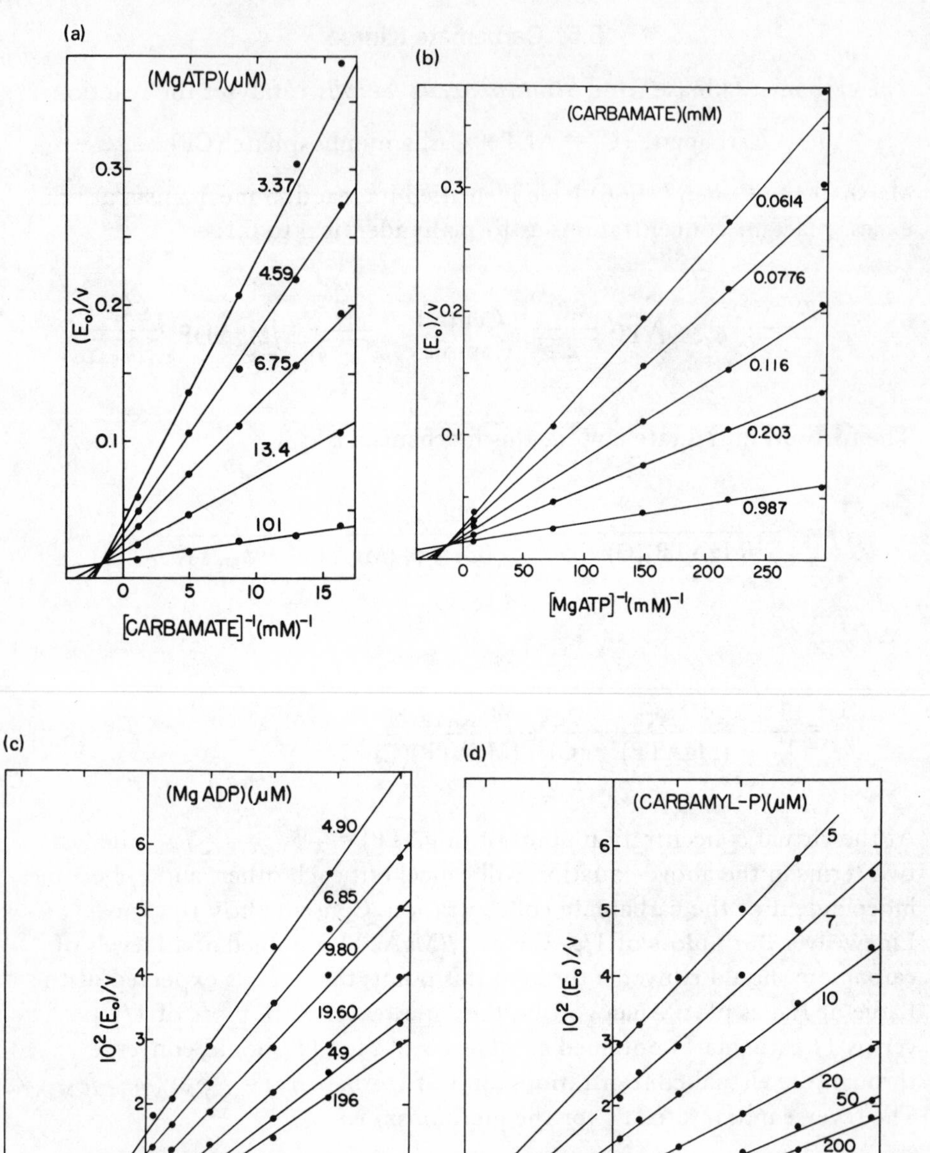


Fig. 6.6. Mixed-noncompetitive cosubstrate effects in the carbamate kinase reaction. (From Marshall and Cohen, 1966.)

Again, the Lineweaver-Burk plots for MgADP obtained at different levels of CP should be convergent at some negative (MgADP), and those for CP obtained at different levels of MgADP convergent at some negative (CP). These predictions are borne out by the four sets of *convergent Lineweaver-Burk plots* presented in Fig. 6.6. When the slopes and intercepts of these primary plots were replotted, the resultant secondary plots all exhibited extensive linearity, providing important confirmation for the proposed linear mechanism and yielding estimates for all eight parameters from Eqns (6.5) and (6.6). The calculation of elementary rate constants from these parameters will be continued in Section 10.3.2. It might be noted at this point that for the proposed mechanism the ratio $\phi_{MgADP \cdot CP}/\phi_{CP}$ is equal to $k_4/k_{-4}$, namely the dissociation constant for the $E$/MgADP complex. The ratio determined kinetically was 5.2 $\mu$M, and the dissociation constant determined from binding measurements was 6.6 $\mu$M. The satisfactory agreement between these independent estimates lent further support to the mechanism.

Instability of carbamyl phosphate prevented a rigorous test of the predicted Haldane relation, but the pattern of the product inhibitions, shown in Fig. 6.7, was instrumental in establishing the sequence of substrate and product reactions. As was expected from the topology of the mechanism, MgATP and MgADP, both of which reacted with free E, gave mutually competitive product inhibitions. In contrast, carbamate and carbamyl-phosphate gave mutually mixed-noncompetitive inhibitions.

## 6.6. Ligand Effects on Slopes and Intercepts

Alberty's (1953, 1958) analysis left no doubt that the classification of cosubstrate and product rate effects is an invaluable method for distinguishing between hyperbolic mechanisms. It would add to the usefulness if the type of rate effects exerted by any ligand could be established directly from an inspection of the mechanism. Toward this end, the system of rules developed by Hearon (1952, 1958) is more adapted for the analysis of inhibition mechanisms (Section 4.7.1). The system of rules developed by Cleland (1963b), on the other hand, is entirely suited for the analysis of multisubstrate mechanisms; the two basic rules are:

*Rule 1:* $V_{(A)}$, the reciprocal intercept of the Lineweaver-Burk plot, is affected by ligand $X$ if $X$ combines reversibly with an enzyme species (form) other than the one with which $A$ combines, thereby changing the reaction velocity in a manner which cannot be eliminated by saturation with $A$.

*Rule 2:* $K^A_{(m)}/V_{(A)}$, the slope of the Lineweaver-Burk plot, is affected by ligand $X$ if $X$ and $A$ either combine with the same enzyme species, or are separated in the reaction sequence by a series of reversible steps, along which they can interact such that the rate effect due to a change in $X$ can be eliminated by an appropriate change in $A$.

Although formulated in terms of ligand effects on the slope and intercept of the Lineweaver-Burk plot, Rule 1 in effect determines whether any ligand acts on $v$(A) *competitively*, and Rule 2 whether any ligand acts on $v$(A) *uncompetitively*. There is also a third rule for compound affects arising from two or more sites of ligand actions. These rules provide an alternative basis for deriving the rate predictions summarized in Table 6.5. For example, in Mechanism 6.III, both $A$ and $P$ react with the free enzyme, and both $B$ and $Q$ react with the substituted enzyme. Accordingly, Rule 1 predicts that $V_{(A)}$ is affected by **Q** but not **P**, whereas $V_{(B)}$ is affected by **P** but not **Q**. The application of Rule 2 is equally straightforward for each of the three different sets of experimental conditions:

*With neither P nor Q added,*

$$E \underset{k_{-1}}{\overset{k_1\mathbf{A}}{\rightleftharpoons}} EA \xrightarrow{k_2} E\text{–}G \underset{k_{-3}}{\overset{k_3\mathbf{B}}{\rightleftharpoons}} EP \xrightarrow{k_4} E$$

*With P added,*

$$E \underset{k_{-1}}{\overset{k_1\mathbf{A}}{\rightleftharpoons}} EA \xrightarrow{k_2} E\text{–}G \underset{k_{-3}}{\overset{k_3\mathbf{B}}{\rightleftharpoons}} EP \underset{k_{-4}\mathbf{P}}{\overset{k_4}{\rightleftharpoons}} E$$

*With Q added,*

$$E \underset{k_{-1}}{\overset{k_1\mathbf{A}}{\rightleftharpoons}} EA \underset{k_{-2}\mathbf{Q}}{\overset{k_2}{\rightleftharpoons}} E\text{–}G \underset{k_{-3}}{\overset{k_3\mathbf{B}}{\rightleftharpoons}} EP \xrightarrow{k_4} E$$

With neither $P$ nor $Q$ added, $E$ and $E$–$G$ are not connected by reversible steps, because both the $k_2$ and $k_4$ steps are irreversible. If either $P$ or $Q$ is added, one of these irreversible gaps becomes reversible. Accordingly, Rule 2 predicts that $K^A_{(m)}/V_{(A)}$ is not affected by **B** in the absence of $P$ and $Q$ (i.e., uncompetitive cosubstrate effects), but it is affected by **B** if either $P$ or $Q$ is added (i.e., noncompetitive cosubstrate effects). Moreover, since the reaction of $Q$ is reversibly connected to that of $A$, the action of $Q$ on $v_f$(A) is predictedly noncompetitive. Since $P$ and $A$ react with the same enzyme species, the action of $P$ on $v_f$(A) is predictedly competitive.

Unlike the structural rule (Section 2.3) or Hearon's rules (Section 4.7.1),

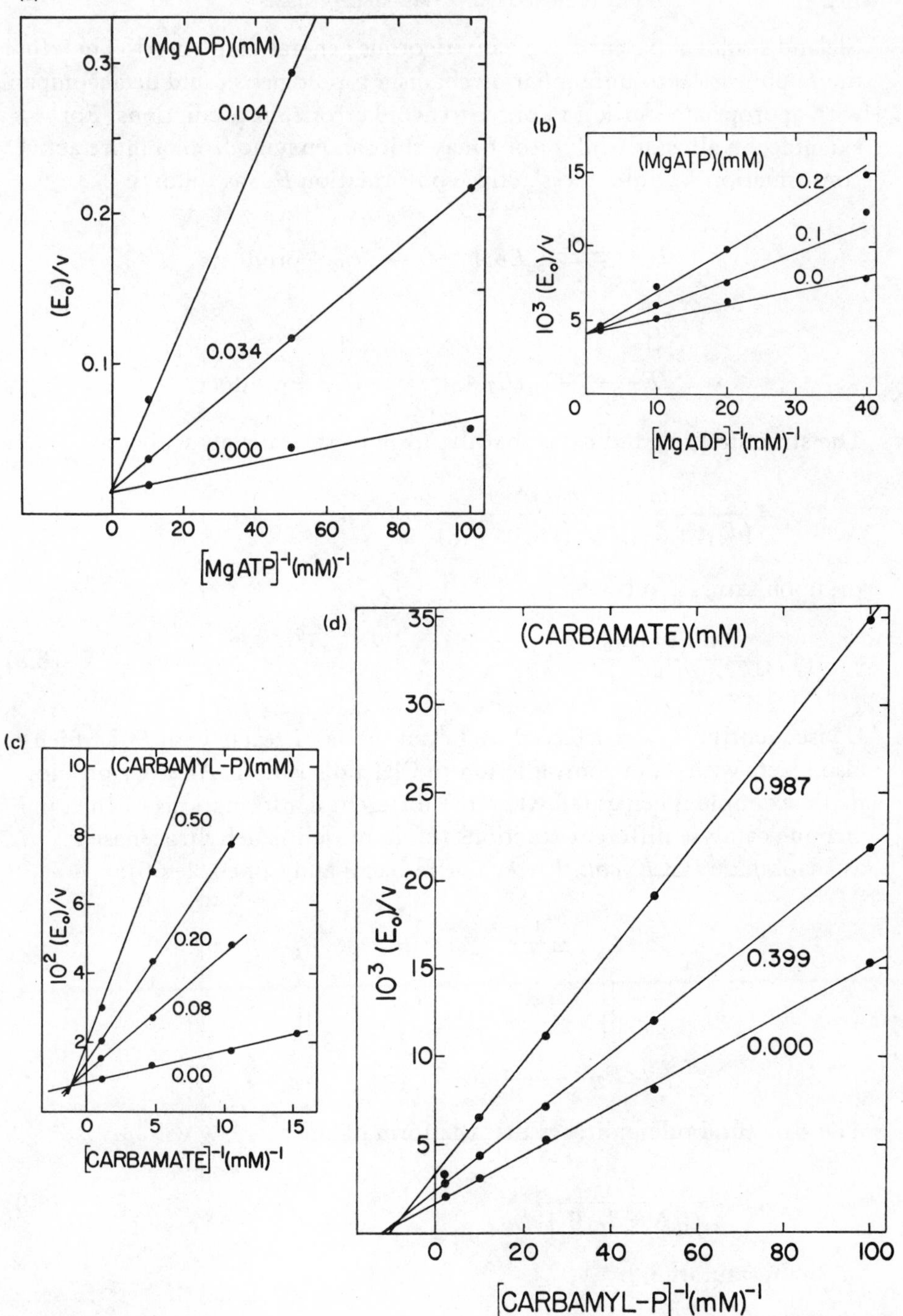


Fig. 6.7. Product effects in the carbamate kinase system. (a) Competitive effects of MgADP on $v_f$(MgATP); (b) Competitive effects of MgATP on $v_r$(MgADP); (c) Mixed-noncompetitive effects of carbamyl-phosphate on $v_f$(carbamate); and (b) Mixed-noncompetitive effects of carbamate on $v_r$(carbamyl-phosphate). (From Marshall and Cohen, 1966.)

Cleland's rules apparently lack any rigorous general derivation†. Therefore their application to unfamiliar mechanism topologies should be accompanied with appropriate caution in order to avoid erroneous predictions. For example, an allosteric inhibitor $I$ may shift an enzyme from a more active conformation $E_R$ into a less active conformation $E_T$, as follows:

$$\begin{array}{ccccc} E_R & \underset{}{\overset{A}{\rightleftharpoons}} & E_RA & \longrightarrow & E_R + \text{products} \\ {\scriptstyle I}\downarrow\uparrow & & & & {\scriptstyle I}\downarrow\uparrow \\ IE_T & \underset{}{\overset{A}{\rightleftharpoons}} & IE_TA & \longrightarrow & IE_T + \text{products} \end{array}$$

The structural rule indicates that the form of the rate law will be:

$$v_f = \frac{(m_{11}\mathbf{I} + m_{10})\mathbf{A}}{(d_{11}\mathbf{I} + d_{10})\mathbf{A} + (d_{01}\mathbf{I} + d_{00})} \tag{6.7}$$

or, upon saturation by **A**,

$$V_{(A)} = \frac{m_{11}\mathbf{I} + m_{10}}{d_{11}\mathbf{I} + d_{10}}. \tag{6.8}$$

Consequently, $V_{(A)}$ is affected by **I** even though $I$ reacts with $E_R$, which also reacts with $A$, in contradiction to Cleland's Rule 1. Another problematic example is generated when the different conformations of the same enzyme catalyse different reactions (cf. homoserine dehydrogenase *I*-aspartokinase *I* of *E. coli*, Patte, Truffa-Bachi and Cohen, 1966):

$$\begin{array}{ccccc} E_R & \underset{}{\overset{A}{\rightleftharpoons}} & E_RA & \longrightarrow & E_R + P \\ \downarrow\uparrow & & & & \downarrow\uparrow \\ E_T & \underset{}{\overset{B}{\rightleftharpoons}} & E_TB & \longrightarrow & E_T + Q \end{array}$$

The structural rule indicates that the form of the rate law will be:

$$v_{A \to P} = \frac{m_{10}\mathbf{A}}{d_{10}\mathbf{A} + d_{01}\mathbf{B} + d_{00}} \tag{6.9}$$

or, upon saturation by **A**,

$$V_{(A)} = \frac{m_{10}}{d_{10}}. \tag{6.10}$$

† However, as Bjorksten (1968) has demonstrated, where Cleland's rules are valid, the validity can be proven on the basis of the structural rule.

Consequently, **B** cannot affect $V_{(A)}$ even though $A$ and $B$ react with different enzyme species, both within a viable reaction pathway. Again Rule 1 has to be interpreted liberally to obtain the correct prediction.

In general then, Cleland's rules should be applied with caution when unfamiliar topologies are considered. In such instances, it would be safe to check their predictions by means of either the *structural rule* or a detailed analysis of the *rate law.* For example, the apparently uncompetitive cosubstrate kinetics of muscle phosphoglycerate mutase, with partially parallel Lineweaver-Burk plots but replete with substrate inhibitions, were interpreted by Grisolia and Cleland (1968) in favour of a ping-pong mechanism over a sequential mechanism. However, Mantle and Garfinkel (1969) demonstrated, by a thorough investigation of the pertinent rate laws, that there was in fact insufficient basis for choosing between these alternatives.

## 6.7. Three-Substrate Reactions

The three main classes of three-substrate reactions all involve the scission of one set of covalent bonds coupled to the creation of a new set:

*Class 1:* e.g. amino acid-transfer RNA ligase:

$$\text{AMP—PP} + \text{amino acid} + \text{tRNA} \rightleftharpoons \text{AMP} + \text{PP} + \text{aminoacyl-tRNA}$$

*Class 2:* e.g. glutamate dehydrogenase (OKG stands for α-ketoglutarate, and $NH_2$—KGH stands for glutamate):

$$\text{NADH} + \text{OKG} + \text{NH}_3 \rightleftharpoons \text{NAD}^+ + \text{NH}_2\text{—KGH}$$

*Class 3:* e.g. glyceraldehydephosphate dehydrogenase ($GPH_2$ stands for glyceraldehydephosphate, and GP—P for diphosphoglycerate):

$$\text{GPH}_2 + \text{NAD}^+ + \text{P}_\text{i} \rightleftharpoons \text{GP—P} + \text{NADH}$$

There are three products in a Class 1 reaction, but only two products in a Class 2 or Class 3 reaction. Since the nature of the reaction varies for these three classes of reactions, the range of model mechanisms appropriate to each class also varies, and there is never need to distinguish between mechanisms that belong to different classes.

### 6.7.1. Model mechanisms

For a mechanism to yield only hyperbolic velocity functions, either each substrate and product must react with just one enzyme species in a linear sequence, or else their randomized reactions with the enzyme must be so rapid relative to other reaction steps that quasi-equilibrium prevails.

Representing the substrates by $A$, $B$ and $C$, and the products by $P$, $Q$ and $R$, a linear mechanism appropriate to a Class 1 reaction would be:

$$E \underset{}{\overset{A}{\rightleftharpoons}} EA \overset{B}{\rightleftharpoons} EAB \underset{P}{\rightleftharpoons} E' \overset{C}{\rightleftharpoons} E'C \underset{Q}{\rightleftharpoons} ER \underset{R}{\rightleftharpoons} E$$

Mechanism 6.VIIIa

There are three random, quasi-equilibrium models closely related to this mechanism:

$$E \xrightleftharpoons[B/K_B]{A/K_A} \{EA, EB\} \xrightleftharpoons[A/K'_A]{B/K'_B} EAB \underset{k_{-3}P}{\overset{k_3}{\rightleftharpoons}} E' \underset{k_{-4}}{\overset{k_4C}{\rightleftharpoons}} E'C \underset{k_{-5}Q}{\overset{k_5}{\rightleftharpoons}} ER \underset{k_{-6}R}{\overset{k_6}{\rightleftharpoons}} E$$

Mechanism 6.VIIIb

$$E \overset{A}{\rightleftharpoons} EA \overset{B}{\rightleftharpoons} EAB \underset{P}{\rightleftharpoons} E' \overset{C}{\rightleftharpoons} E'C \xrightleftharpoons[R]{Q} \{ER, EQ\} \xrightleftharpoons[Q]{R} E$$

Mechanism 6.VIIIc

$$E \xrightleftharpoons[B]{A} \{EA, EB\} \xrightleftharpoons[A]{B} EAB \underset{P}{\rightleftharpoons} E' \overset{C}{\rightleftharpoons} E'C \xrightleftharpoons[R]{Q} \{ER, EQ\} \xrightleftharpoons[Q]{R} E$$

Mechanism 6.VIIId

There has been no nomenclature that can distinguish completely between four such mechanisms. One possible solution is to order the substrate and product symbols according to their reaction sequence, but cover with a bar those symbols which belong to a random, quasi-equilibrium segment in the mechanism. On this basis, Mechanism 6.VIIIa becomes $ABPCQR$, 6.VIIIb becomes $\overline{AB}PCQR$, 6.VIIIc becomes $ABPC\overline{QR}$, and 6.VIIId becomes $\overline{AB}PC\overline{QR}$. All the hyperbolic three-substrate mechanisms to be considered can be represented in this manner:†

† This list of model mechanisms is meant to be illustrative rather than exhaustive. For instance, an $APQBCR$ mechanism is omitted, because it is not easily compatible with a ligase-type reaction. An $\overline{AB}CPQR$ case is also omitted, because the random additions of $A$ and $B$ to form an $EAB$ species will not conform easily to quasi-equilibrium in the face of high concentrations of $C$, which will render the conversion of $EAB$ to $EABC$ no longer rate-limiting relative to the enzyme binding of $A$ and $B$.

*Class-1 Mechanisms*

6.VIIIa $ABPCQR$
6.VIIIb $\overline{AB}PCQR$
6.VIIIc $ABPC\overline{QR}$
6.VIIId $\overline{AB}PC\overline{QR}$

6.IXa $APBCQR$
6.IXb $AP\overline{BC}QR$
6.IXc $APBC\overline{QR}$
6.IXd $AP\overline{BC}\,\overline{QR}$

6.X $APBQCR$

6.XIa ABCPQR
6.XIb $\overline{ABC}PQR$
6.XIc $\overline{ABC}\,\overline{PQ}R$
6.XId $\overline{ABC}\,\overline{PQR}$
6.XIe $A\overline{BC}PQR$
6.XIf $A\overline{BC}\,\overline{PQ}R$
6.XIg $A\overline{BC}\,\overline{PQR}$
6.XIh $ABC\overline{PQ}R$
6.XIi $ABC\overline{PQR}$

*Class-2 and Class-3 Mechanisms*

6.XIIa $ABPCQ$
6.XIIb $\overline{AB}PCQ$

6.XIIIa $APBCQ$
6.XIIIb $AP\overline{BC}Q$

6.XIVa $ABCPQ$
6.XIVb $\overline{AB}CPQ$
6.XIVc $\overline{ABC}\,\overline{PQ}$
6.XIVd $A\overline{BC}PQ$
6.XIVe $A\overline{BC}\,\overline{PQ}$
6.XIVf $ABC\overline{PQ}$

### 6.7.2. General rate law

If reaction velocity cannot exceed first degree with respect to each substrate, the rate law for a three-substrate system cannot be more complex than

$$v_f = \frac{m_{111}\mathrm{ABC}}{d_{111}\mathrm{ABC} + d_{110}\mathrm{AB} + d_{101}\mathrm{AC} + d_{011}\mathrm{BC} + d_{100}\mathrm{A} + d_{010}\mathrm{B} + d_{001}\mathrm{C} + d_{000}}.$$

The coefficients $d_{111}, d_{110}, \ldots$, are not determinable experimentally, but their ratios over $m_{111}$, which can be expressed in the form of $V_f$ and seven $\phi$-parameters, are:

$$v_f = \frac{\mathrm{ABC}}{V_f^{-1}\mathrm{ABC} + \phi_C\mathrm{AB} + \phi_B\mathrm{AC} + \phi_A\mathrm{BC} + \phi_{BC}\mathrm{A} + \phi_{AC}\mathrm{B} + \phi_{AB}\mathrm{C} + \phi_{ABC}} \quad (6.11)$$

Graphical methods for estimating the eight parameters from Eqn (6.11) have been described by Dalziel (1969) and Wong and Hanes (1969). One way to proceed is as follows:

(1) Hold **B** and **C** constant and vary **A**. Under these conditions rate-law (6.11) becomes a hyperbolic function of **A**:

$$v_f = \frac{V_{(A)} \cdot \mathbf{A}}{\mathbf{A} + K^A_{(m)}}$$

$$V_{(A)} = \frac{\mathbf{BC}}{V_f^{-1}\mathbf{BC} + \phi_C\mathbf{B} + \phi_B\mathbf{C} + \phi_{BC}} \tag{6.12}$$

$$K^A_{(m)} = \frac{\phi_A\mathbf{BC} + \phi_{AC}\mathbf{B} + \phi_{AB}\mathbf{C} + \phi_{ABC}}{V_f^{-1}\mathbf{BC} + \phi_C\mathbf{B} + \phi_B\mathbf{C} + \phi_{BC}}$$

As usual, the *conditional saturation velocity* $V_{(A)}$ and the *conditional Michaelis constant* $K^A_{(m)}$ may be determined by the Eadie plot of $v_f$ versus $v_f/\mathbf{A}$. Either the Lineweaver-Burk or the Hanes plot also may be employed for treating the rate data at this primary stage.

(2) Still hold **C** at the same fixed level, but repeat the estimation of $V_{(A)}$ and $K^A_{(m)}$ at different levels of **B**. Plot $1/V_{(A)}$ versus $1/\mathbf{B}$, and also $K^A_{(m)}/V_{(A)}$ versus $1/\mathbf{B}$. Both of these secondary plots should be linear, yielding the four intercepts $\text{Int}_1$, $\text{Int}_2$, $\text{Int}_3$ and $\text{Int}_4$ (Fig. 6.8):

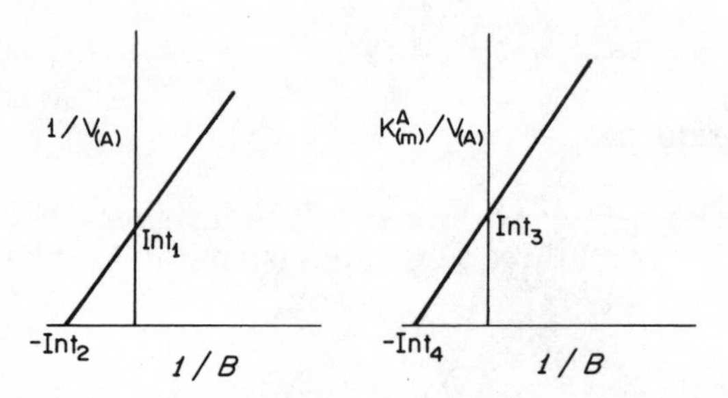


Fig. 6.8. Secondary plots for hyperbolic three-substrate mechanisms.

$$\text{Int}_1 = V_f^{-1} + \phi_C/\mathbf{C}$$

$$\text{Int}_2 = \frac{V_f^{-1} + \phi_C/\mathbf{C}}{\phi_B + \phi_{BC}/\mathbf{C}}$$

$$\text{Int}_3 = \phi_A + \phi_{AC}/\mathbf{C}$$

$$\text{Int}_4 = \frac{\phi_A + \phi_{AC}/\mathbf{C}}{\phi_{AB} + \phi_{ABC}/\mathbf{C}}.$$

(3) The preceding steps are repeated at different fixed levels of **C**, and the $\text{Int}_1$ to $\text{Int}_4$ so obtained are analysed on the basis of the four tertiary plots indicated in Fig. 6.9. All eight parameters from Eqn (6.11) are given

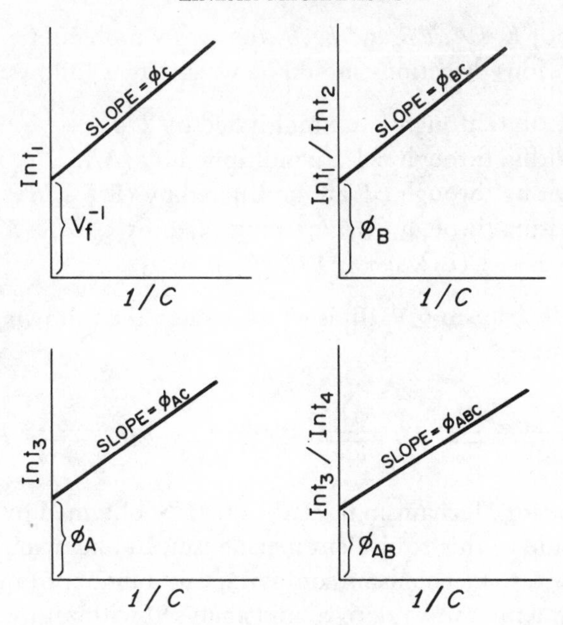


Fig. 6.9. Tertiary plots for hyperbolic three-substrate mechanisms.

by the slopes and intercepts of these tertiary plots, all of which again should be linear. If any of the secondary or tertiary plots were nonlinear, Eqn (6.11) itself would be invalid.

### 6.7.3. Quasi-equilibrium segments

The indivudal rate laws for the range of hyperbolic three substrate mechanisms are best derived by three different methods. The steady-state mechanisms, like 6.VIIIa and 6.XIa, can be analysed solely by means of the structural rule. The quasi-equilibrium mechanisms, like 6.XId and 6.XIVc, can be analysed solely on the basis of equilibrium considerations. Those mechanisms, like 6.VIIIb–d, that contain both steady-state and quasi-equilibrium segments can be analysed by means of the method of Cha (1968). In using this method, equilibrium considerations are first applied to each quasi-equilibrium segment, which may then be represented as a single enzyme species for the purpose of steady-state derivation. For example, in Mechanism 6.VIIIb, the four enzyme species *E, EA, EB* and *EAB* belong to a quasi-equilibrium segment. Their relative concentrations are fixed by the equilibrium dissociation constants for *EA, EB* and *EAB,* viz. $K_A$, $K_B$ and $K'_B$:

$$E:EA:EB:EAB = 1:\mathbf{A}/K_A:\mathbf{B}/K_B:\mathbf{AB}/K_AK'_B.$$

With their relative concentrations so prescribed, the four can be *lumped* together and represented as a single enzyme species, $E_{\text{sum}}$, in deriving the steady-state rate law, e.g. by using the structural rule. In view of the relative concentrations of $E$, $EA$, $EB$ and $EAB$, the arrow-probabilities leaving from $E_{sum}$ in various directions should be weighted as follows:

Arrows exiting through $E$ are multiplied by $1/\sigma$;
Arrows exiting through $EA$ are multiplied by $(\mathbf{A}/K_A)/\sigma$;
Arrows exiting through $EB$ are multiplied by $(\mathbf{B}/K_B)/\sigma$;
Arrows exiting through $EAB$ are multiplied by $(\mathbf{AB}/K_A K'_B)/\sigma$;
$\sigma = 1 + (\mathbf{A}/K_A) + (\mathbf{B}/K_B) + (\mathbf{AB}/K_A K'_B)$.

On this basis, Mechanism 6.VIIIb is equivalent to the following steady-state mechanism:

$$E_{sum} \underset{k_{-3}\mathbf{P}}{\overset{k_3(\mathbf{AB}/K_A K'_B)/\sigma}{\rightleftharpoons}} E' \underset{k_{-4}}{\overset{k_4\mathbf{C}}{\rightleftharpoons}} E'C \underset{k_{-5}\mathbf{Q}}{\overset{k_5}{\rightleftharpoons}} ER \underset{k_{-6}\mathbf{R}/\sigma}{\overset{k_6}{\rightleftharpoons}} E_{sum}$$

and the rate law for Mechanism 6.VIIIb is readily obtained by applying the structural rule to this equivalent mechanism. In like manner, the rate laws for steady-state mechanisms containing any number of quasi-equilibrium segments can be derived and analysed without hindrance.

### 6.7.4. Recall of missing parameters

The derived rate laws for the various hyperbolic three-substrate mechanisms can be extensively differentiated from one another by the criteria already described for the two-substrate cases in the earlier sections of this chapter. There are in addition a number of criteria more specifically useful for the three-substrate cases. One of these is the recall of missing parameters.

Equation (6.11), with its eight parameters, is the most complex three-substrate rate law to remain hyperbolic with respect to each of the substrates. Frieden (1959) discovered that, in some mechanisms, one or more of the parameters may be missing from the actual rate law. The *missing parameter* in some cases may be *recalled* back into the rate law by the addition of a reaction product and, usefully, different mechanisms predict different patterns of recall (Wong and Hanes, 1969).

Consider the example of Mechanism 6.VIIIa, when the reaction products $P$, $Q$ and $R$ are initially absent from the reaction system. The topological constraints due to the insignificance of the $P$, $Q$ and $R$ reaction steps render it impossible to construct any arrow-pattern belonging to the $d_{100}\mathbf{A}$, $d_{010}\mathbf{B}$, or $d_{000}$ term in Eqn (6.11). Accordingly, the three parameters $\phi_{BC}$, $\phi_{AC}$ and $\phi_{ABC}$ are equal to zero, i.e. missing from the rate law. However, if $P$ is added to the system, the topological constraints will be relieved sufficiently to allow the construction of the $d_{100}\mathbf{A}$- and $d_{000}$-types of arrow-patterns, e.g.

$$E \xrightarrow{A} EA \longleftarrow EAB \underset{P}{\longleftarrow} E' \qquad E'C \longrightarrow ER \longrightarrow E$$

TABLE 6.6

Recall of missing parameters for three-substrate mechanisms by reaction products.

+: Parameter is present in the rate law even in the absence of all the reaction products.

None: It is impossible to recall the missing parameter by adding any *one* of the reaction products.

| Mechanism | | Product capable of recalling a missing parameter | | | | | | | |
|---|---|---|---|---|---|---|---|---|---|
| | | $V_f^{-1}$ | $\phi_C$ | $\phi_B$ | $\phi_A$ | $\phi_{BC}$ | $\phi_{AC}$ | $\phi_{AB}$ | $\phi_{ABC}$ |
| 6.VIIIa | $ABPCQR$ | + | + | + | + | $P$ | $R$ | + | $P$ |
| 6.VIIIb | $\overline{AB}PCQR$ | + | + | None | None | None | None | + | $P$ |
| 6.VIIIc | $ABPC\overline{QR}$ | + | + | + | + | $P$ | $R$ | + | $P$ |
| 6.VIIId | $\overline{AB}PC\overline{QR}$ | + | + | None | None | None | None | + | $P$ |
| 6.IXa | $APBCQR$ | + | + | + | + | + | $R$ | $P$ | $P$ |
| 6.IXb | $AP\overline{BC}QR$ | + | None | None | + | + | None | None | $P$ |
| 6.IXc | $APBC\overline{QR}$ | + | + | + | + | + | $R$ | $P$ | $P$ |
| 6.IXd | $AP\overline{BC}\,\overline{QR}$ | + | None | None | + | + | None | None | $P$ |
| 6.X | $APBQCR$ | + | + | + | + | $Q$ | $R$ | $P$ | $Q$ |
| 6.XIa | $ABCPQR$ | + | + | + | + | + | $R$ | + | + |
| 6.XIb | $\overline{ABC}PQR$ | + | None | None | None | None | None | None | + |
| 6.XIc | $\overline{ABC}\,\overline{PQ}R$ | + | None | None | None | None | None | None | + |
| 6.XId | $\overline{ABC}\,\overline{PQR}$ | + | None | None | None | None | None | None | + |
| 6.XIe | $A\overline{BC}PQR$ | + | None | None | + | + | None | None | + |
| 6.XIf | $A\overline{BC}\,\overline{PQ}R$ | + | None | None | + | + | None | None | + |
| 6.XIg | $A\overline{BC}\,\overline{PQR}$ | + | None | None | + | + | None | None | + |
| 6.XIh | $ABC\overline{PQ}R$ | + | + | + | + | + | $R$ | + | + |
| 6.XIi | $ABC\overline{PQR}$ | + | + | + | + | + | $R$ | + | + |
| 6.XIIa | $ABPCQ$ | + | + | + | + | $P$ | $Q$ | + | $P$ |
| 6.XIIb | $\overline{AB}PCQ$ | + | + | None | None | None | None | + | $P$ |
| 6.XIIIa | $APBCQ$ | + | + | + | + | + | $Q$ | $P$ | $P$ |
| 6.XIIIb | $AP\overline{BC}Q$ | + | None | None | + | + | None | None | $P$ |
| 6.XIVa | $ABCPQ$ | + | + | + | + | + | $Q$ | + | + |
| 6.XIVb | $\overline{AB}CPQ$ | + | None | None | None | None | None | None | + |
| 6.XIVc | $\overline{ABC}\,\overline{PQ}$ | + | None | None | None | None | None | None | + |
| 6.XIVd | $A\overline{BC}PQ$ | + | None | None | + | + | None | None | + |
| 6.XIVe | $A\overline{BC}\,\overline{PQ}$ | + | None | None | + | + | None | None | + |
| 6.XIVf | $ABC\overline{PQ}$ | + | + | + | + | + | $Q$ | + | + |

(This arrow-pattern falls under $d_{100}$A in Eqn (6.11) because it contains **A**, but neither **B** nor **C**.)

$$E \longleftarrow EA \longleftarrow EAB \underset{P}{\longleftarrow} E' \qquad E'C \longrightarrow ER \longrightarrow E$$

(This arrow-pattern falls under $d_{000}$ in Eqn (6.11) because it contains neither **A** nor **B** nor **C**.)

As a result, the missing parameters $\phi_{BC}$ and $\phi_{ABC}$ become finite in the presence of $P$, i.e., recalled by $P$ back into the rate law. In contrast, the addition of $P$ fails to allow the construction of any $d_{010}$B-type arrow-pattern. Therefore $\phi_{AC}$ is not recalled by $P$. The addition of either $Q$ or $R$ also fails to recall any of the three missing parameters. This recall pattern as well as those for other model mechanisms are summarized in Table 6.6. They provide yet another basis for differentiating between the different mechanisms.

### 6.7.5. Variation of two cosubstrates

In a three-substrate reaction, the $v_f(\mathbf{A})$ function is affected by each of the two cosubstrates $B$ and $C$. Purich and Fromm (1972) suggested that a particularly simple method for extracting useful information is to vary at once both of the cosubstrates and observe the effects on $v_f(\mathbf{A})$. From the expressions for $V_{(A)}$ and $K^A_{(m)}$ given in Eqn (6.12), we have

$$\frac{K^A_{(m)}}{V_{(A)}} = \phi_A + \frac{\phi_{AC}}{\mathbf{C}} + \frac{\phi_{AB}}{\mathbf{B}} + \frac{\phi_{ABC}}{\mathbf{BC}} .$$

If **B** and **C** are both varied, $K^A_{(m)}/V_{(A)}$ will remain constant (giving an uncompetitive pattern of cosubstrate effects) only if $\phi_{AC}$, $\phi_{AB}$ and $\phi_{ABC}$ are all zero. This provides a rapid means to identify mechanisms for which all three of these parameters are zero in the absence of products, e.g. Mechanisms 6.IX, 6.X and 6.XIII from Table 6.6.

Analogously, if **A** and **C** are both varied, $K^B_{(m)}/V_{(B)}$ will remain constant only if $\phi_{AB}$, $\phi_{BC}$ and $\phi_{ABC}$ are all zero, e.g. for Mechanism 6.X. If **A** and **B** are both varied, $K^C_{(m)}/V_{(C)}$ will remain constant only if $\phi_{AC}$, $\phi_{BC}$ and $\phi_{ABC}$ are all zero, e.g., for Mechanisms 6.VIII, 6.X, and 6.XII.

### 6.7.6. Loci of junction points

If the cosubstrate effects of **B** on $v_f(\mathbf{A})$ are mixed-noncompetitive, the linear plots (Eadie, Lineweaver-Burk or Hanes) obtained at different fixed levels of **B** will meet at the *junction point* $\mathbf{A} = \mathbf{A}_b$, at which the reaction velocity becomes invariant with cosubstrate **B** (see Fig. 6.4). To derive the expression for $\mathbf{A}_b$, Eqn (6.11) may be divided through by **ABC** to give

$$v_f = \frac{1}{V_f^{-1} + \frac{\phi_C}{\mathrm{C}} + \frac{\phi_A}{\mathrm{A}} + \frac{\phi_{AC}}{\mathrm{AC}} + \frac{1}{\mathrm{B}}\left(\phi_B + \frac{\phi_{BC}}{\mathrm{C}} + \frac{\phi_{AB}}{\mathrm{A}} + \frac{\phi_{ABC}}{\mathrm{AC}}\right)}.$$

Clearly, the velocity will be invariant with **B** at $\mathbf{A} = \mathbf{A}_b$ if and only if $\mathbf{A}_b$ satisfies the relation

$$\phi_B + \frac{\phi_{BC}}{\mathrm{C}} + \frac{\phi_{AB}}{\mathrm{A}_b} + \frac{\phi_{ABC}}{\mathrm{A}_b\mathrm{C}} = 0$$

from which the expression for $\mathbf{A}_b$ can be derived:

$$-\mathrm{A}_b = \frac{\phi_{AB}\mathrm{C} + \phi_{ABC}}{\phi_B\mathrm{C} + \phi_{BC}}. \tag{6.13}$$

By determining $\mathbf{A}_b$ at different levels of **C**, the variation of $\mathbf{A}_b$ with **C** can be defined experimentally. The locus of the variation reflects on the four parameters $\phi_{AB}$, $\phi_{ABC}$, $\phi_B$ and $\phi_{BC}$, and Fig. 6.10 shows the *predicted loci of* $A_b$ when some of these parameters are equal to zero. This method for revealing zero parameters was suggested by Keleti (1972).

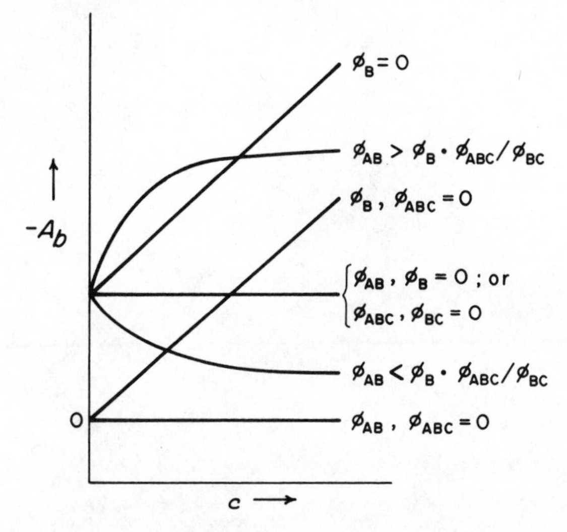


Fig. 6.10. Loci of junction point from linear plots for $v_f(\mathbf{A})$ obtained at different fixed levels of **B**. The value of $-\mathbf{A}_b$ varies with **C** in different ways depending on the values of various kinetic parameters. (After Keleti, 1972.)

## Problem 6.1

Mechanism 6.II predicts the Dalziel relation of $\phi_{AB} = V_r\phi_A\phi_B$ when $\phi_{AB}$, $\phi_A$ and $\phi_B$ have been evaluated in the absence of the two products *P* and *Q*. Is this prediction altered if these parameters have been evaluated in the presence of some fixed concentration of *P*, or of *Q*?

# 7 Cooperative Interactions

## 7.1. The Allosteric Concept

Like man in a complex society, no ligand is an island within the dynamic matrix of a protein molecule. Many types of interactions are possible between ligands, and Fig. 7.1 shows a variety of ligands that may interact with substrate molecule $S$ bound to one of the *protomers*, or identical

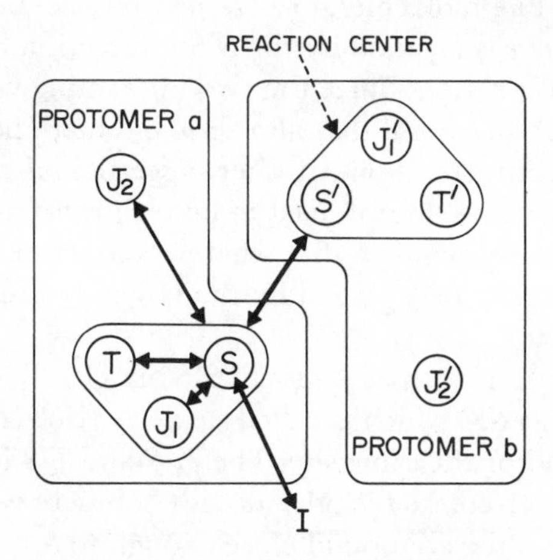


Fig. 7.1. Interactions between substrate $S$ and various ligands. The competition between $I$ and $S$ for the same binding site is *isosteric*; the effects of $T$ and $J_1$, situated alongside $S$, are *parasteric;* the effects of $S'$ and $J_2$, situated at some distance from $S$, are *allosteric*.

subunits, of an oligomeric protein. Of these ligands, $S'$ represents a molecule of the same substrate bound to a neighbouring protomer. $T$ is a cosubstrate, and $J_1$ and $J_2$ are modifiers (activators or inhibitors); $T$ and $J_1$ bind close to $S$, but $J_2$ binds some distance away. $I$ is a competitive inhibitor that binds to the same site as $S$.

The first enzyme in a biosynthetic pathway is very often sensitive to feedback regulation by the eventual end product of the pathway. The end product, as well as other regulatory ligands of the enzyme, are referred to as *effectors*. With aspartate transcarbamylase, Gerhardt and Pardee (1962) suggested that the effector CTP, bearing so little chemical resemblance to the substrates of the enzyme, probably binds to some site on the enzyme distinct from the substrate sites. This and kindred observations on other enzyme systems underlined the regulatory significance of the $J_2$-$S$ type of interactions (Fig. 7.1), i.e. effector action at some distance from the substrate. With threonine deaminase, Umbarger (1956) observed that the kinetic response of the enzyme to substrate and effector exhibited sigmoidal characteristics, later recognized to be shared by many regulatory enzymes. The sigmoidal response also reminded one of the sigmoidal binding of oxygen to hemoglobin, a process in which the $S'$-$S$ type of interaction, i.e. between substrate sites on different protomers, was known to play a prominent role. The similarity between hemoglobin, a tetrameric protein, and regulatory enzymes could only be strengthened by the oligomeric nature of many of these enzymes. To achieve unity of focus, Monod, Changeux and Jacob (1963) introduced the concept of *allosteric* (allo = other) interactions to describe those interactions between ligands bound to the protein at some distance from one another. Allosteric interactions can be either *homotropic,* i.e. between like molecules, or *heterotropic,* i.e. between unlike molecules. The former is typified by the $S'$-$S$ interaction in Fig. 7.1, and the latter by the $J_2$-$S$ interaction. Since the two interacting ligands were not close together, it was suggested that allosteric interactions are mediated by changes in protein conformation. Another suggestion also made, namely that allosteric interactions do not need to be reciprocal, is unsupportable; microscopic reversibility requires that whatever effect exerted by an effector on the binding of $S$ must be equalled by the effect exerted by $S$ on the binding of the effector.

Allosteric interactions such as $J_2$-$S$ and $S'$-$S$ are clearly distinguished from an interaction such as $I$-$S$, which is called *isosteric* (iso = equal) because $I$ and $S$ are competing for the same site. The $J_1$-$S$ and $T$-$S$ interactions fall between these two extremes: $J_1$ and $T$ do not compete with $S$ for the same binding site, and yet they are bound close enough to $S$ to permit some direct intermolecular interactions not mediated by protein conformation. "*Parasteric*" (para = alongside) might be one possible designation for such interactions. It may be noted that, since $T$ and $S$ are cosubstrates, eventually they will enter into chemical reaction. Before they do so, however, they can affect on another's binding to the protein. They may bind independently (e.g. with dihydropteridine reductase, Section 6.4), one may alter the binding constant of the other (e.g. with liver alcohol dehydrogenase, Section 10.3.3), or the binding of the one may be prerequisite to the binding of the other (e.g. with carbamate kinase, Section 6.5).

Effector action from a distance is not easy to adhere to as a criterion for allostery, because the exact locations of the substrate and effector sites are not known prior to a complete elucidation of the three-dimensional structure of the protein. Possible substitute criteria, such as lack of chemical resemblance between substrate and effector, sigmoidal responses, changes in protein conformation, or the oligomeric nature of the protein, are just as problematic. In the $J_1$-$S$ type of parasteric interactions, there also can be a total lack of chemical resemblance between substrate and effector, as well as changes in protein conformation; allosteric interactions need not involve sigmoidal responses; and the protomers of an oligomeric protein may act independently. Furthermore, when the "concerted-transition" model was proposed by Monod, Wyman and Changeux (1965) as a plausible model to explain allosteric interactions (Section 8.1.3), the postulated concerted transition between two protein conformations was designated an *allosteric transition*. Consequently, the allosteric term becomes associated with not only the general phenomenon of effector action at a distance, but also one (of several) particular model mechanism for explaining the phenomenon. However, in spite of these difficulties, allostery has been a singularly heuristic concept that brings into focus the major role of ligand-ligand interactions in a broad range of biological processes. Since no single definition of the allosteric concept is easily practised, we may best regard it, not unlike other elusive but magnificent human notions, as a many-splendoured thing!

## 7.2. Uses of Ligand-Ligand Interactions

The question of why protein molecules have come to be so big has no simple answer. However, one significant factor might be that bigness enhances the versatility of ligand-ligand interactions within the framework of the protein molecule. Hemoglobin is about four times the size of myoglobin. It also permits homotropic (oxygen-oxygen) and heterotropic (oxygen-diphosphoglycerate) interactions which are beyond the capability of myoglobin. In general, the usefulness of ligand-ligand interactions in metabolic control, especially in the form of feedback inhibitions, is well established. There are other potential uses as well, in connection with processes of energy transformation and memory recording.

### 7.2.1. Metabolic control

Enzymic proteins catalyse chemical reactions, transport-carrier proteins distribute solutes, repressor proteins modulate gene expression, and structural proteins determine morphology. With each of these varied protein functions, allosteric and parasteric effectors can control the binding and catalytic activities of the proteins. Since such effectors need not bear any chemical resemblance to the substrates, they also need not originate from the same domain of cellular metabolism as their target proteins. This freedom has paved the way toward an almost unlimited flexibility in the evolved designs of metabolic circuitry, as the elegant expositions of Atkinson (1970), Sanwal (1970) and Stadtman (1970) have made very plain to us. The chemical resemblance between CTP and aspartate in the aspartate transcarbamylase system (Yates and Pardee, 1956), between 2,3-diphosphoglycerate and oxygen in the hemoglobin system (Benesch and Benesch, 1969), or between β-galactosides and the *lac*-operator DNA in the lactose operon (Gilbert and Muller-Hill, 1966) is quite negligible, and yet the ligand-ligand interaction between the pair is crucial to the metabolic control of pyrimidine synthesis, oxygen transport, or β-galactosidase induction, respectively.

### 7.2.2. Energy transformation

In metabolic control mechanisms, the effector affects the binding or the rate of catalysed reaction of the substrate, and the task of the effector-substrate interactions is to modulate protein activity. However, such interactions may also be useful in other ways, e.g. the *free energy* of the interaction itself may be an intermediary in some energy-transformation process, or the *duration* of the interaction itself may be a component in some memory-storage mechanism.

Figure 7.2 presents a model mechanism suggested by Weber (1972) for the generation of a high-energy chemical bond. Protein $E$ is confined to a

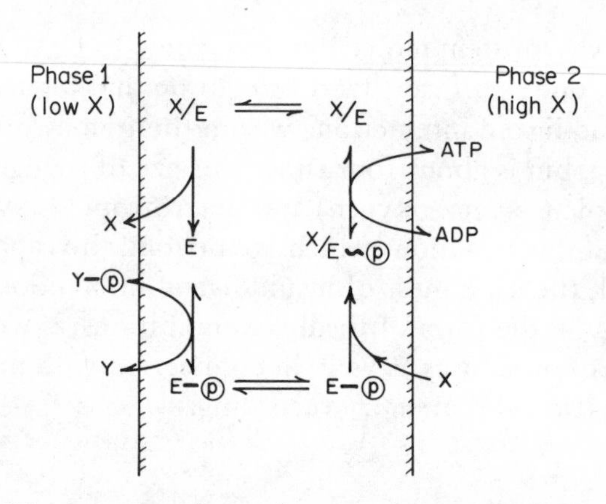


Fig. 7.2. A model mechanism that makes use of a negative interaction between two ligands to generate a "high-energy" bond. (After Weber, 1972.)

membrane separating two liquid phases, 1 and 2, which contain respectively low and high concentrations of ligand $X$. $E$ is phosphorylated to yield a low-energy bond $E$-Ⓟ on the phase-1 side of the membrane. When $E$-Ⓟ migrates to the phase-2 side, the high concentrations of $X$ in the liquid medium forces the formation of the $X/E\sim$ Ⓟ complex, within which a strong negative interaction between $X$ and Ⓟ moieties propels the $E\sim$ Ⓟ linkage to such a high potential for phosphate-transfer that it can react with ADP to generate ATP. Afterwards, the $X/E$ complex diffuses back to the phase-1 side and releases $X$ to the liquid medium, which is low in $X$, to complete the reaction cycle. In the process, free-energy contributions from the low energy *E*-Ⓟ bond, *and* from the concentration gradient of *X* across the membrane, are combined to produce the high-energy phosphorylation of ADP. The negative interaction between ligands *X* and Ⓟ constitutes the immediate driving force for scaling the high-energy barrier. Alternative models for translating the energy of some form of ligand-ligand interaction into chemical bond energy also have been formulated by Chance and Williams (1955), Hill (1969), and Boyer, Cross and Momsen (1973).

### 7.2.3. Memory recording

The cellular mechanism for learning and memory remains largely a miraculous black box in the living world as yet little illuminated by biochemical insight. Broadly, the engramming in the brain of a pattern of nerve impulses may be a multistage recording process that progresses from labile electrical reverberations on to permanent storage in the form of macromolecular structures (e.g. Lorente de No 1938; Hebb, 1949; Sachs, 1961; Griffith, 1967; Katchalsky and Neumann, 1972). Aside from any regulation of membrane-transport functions by ligand-ligand interactions, it might be noted that the duration of such interactions may add to the temporal dimension of this recording process. Figure 7.3 shows a model situation in which the passage of a nerve impulse is accompanied by a transient change in the concentration of an effector $X$ from $X_1$ to $X_2$ over the time interval from $t_1$ to $t_2$. The conformation of a responsive protein may undergo one of three different types of behaviour. In the type-1 behaviour, the conformation changes from $E_1$ to $E_2$, and back to $E_1$,

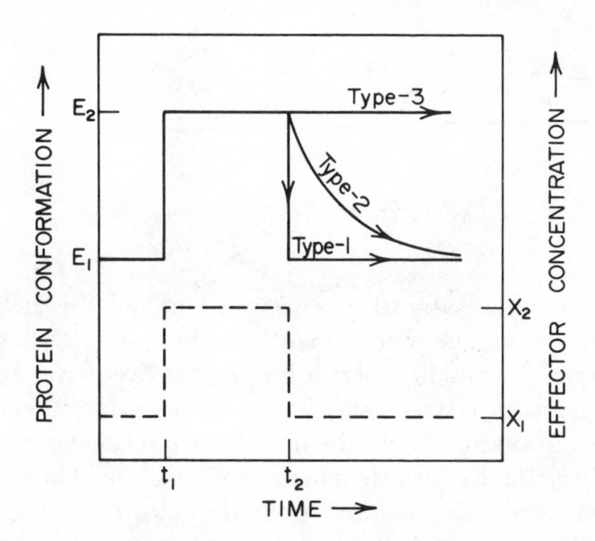


Fig. 7.3. Responses of protein conformation to a transient pulse of ligand *X*. The type-2 and type-3 responses can contribute to the memory recording of the passage of the pulse.

perfectly in phase with the effector change. In the type-2 behaviour, response lags behind stimulus, and the system exhibits hysteresis (Everett, 1967). Finally, in the type-3 behaviour, the new conformation $E_2$, once it is switched on, does not revert back to the initial conformation $E_1$ even after the passage of the stimulus. The types 2 and 3 behaviour prolong the after-effects of the stimulus, and so introduce an element of memory that could be an important link in the orderly transformation of a transient electrical pulse into short-term, and finally long-term, neural memory. It is noteworthy that protein-ligand and ligand-ligand interactions may form the basis of a type-2 or type-3 response.

The type-2, or *hysteretic*, response calls for a conformation change in the protein that is slow relative to the duration of the stimulus (in the form of a concentration change in $X$):

$$E_1 \underset{k_{-1}}{\overset{k_1X}{\rightleftharpoons}} E_2X$$

For this purpose, a fast $k_1$ and a slow $k_{-1}$ would allow the protein conformation to respond rapidly to a transient increase in $X$, but revert slowly. Conversely, a slow $k_1$ and a fast $k_{-1}$ would allow the protein conformation to respond rapidly to a transient decrease in $X$, but revert slowly. Frieden (1970) has listed many hysteretic enzymes with a half-time of seconds or even minutes.

The type-3, or *switching*, response could be satisfied through the interaction of two ligands $S$ and $X$:

$$\begin{array}{ccccc} E_1 & \overset{S}{\rightleftharpoons} & E_1S & & \\ X \Big\updownarrow & & & & \\ E_2X & \overset{S}{\rightleftharpoons} & E_2XS & \underset{X}{\rightleftharpoons} & E_2S \end{array}$$

If ligand $S$ is present at saturating concentrations and $X$ is lacking, initially at $t_1$ the protein would be predominantly in the species $E_1S$; but a strong transient pulse of $X$ would convert it into $E_2XS$. Even after $X$ has vanished, the protein nevertheless would stay in the new conformation as $E_2S$. In effect, the saturation by $S$ locks the protein into its new conformation and prevents its reversion back to the initial conformation. This switch-lock type of behaviour was encountered in the liver alcohol dehydrogenase mechanism, where an excess of ethanol locks up a new enzyme conformation induced by the binding of NADH (Wong, Gurr, Bronskill and Hanes, 1972):

$$
\begin{array}{ccccc}
E_1 & \xrightleftharpoons{\text{ethanol}} & E_1/\text{ethanol} & & \\
\text{NADH}\updownarrow & & & & \\
E_2/\text{NADH} & \xrightleftharpoons{\text{ethanol}} & E_2/\text{NADH/ethanol} & \xrightleftharpoons[\text{NADH}]{} & E_2/\text{ethanol}
\end{array}
$$

If a dehydrogenase can switch and lock, might not proteins involved in the memory machinery be capable of similar behaviour?

## 7.3. Concept of Cooperativity

### 7.3.1. Hemoglobin-oxygen reaction

Although the concept of cooperativity between protomeric subunits is fundamental to many regulatory protein systems, for long it had evolved upon the foundation of oxygen binding to hemoglobin. Heme-heme interaction, in its modern allosteric guise, poses as vibrant a theoretico-experimental problem to-day as it had confronted Hill (1910) and Adair (1925) decades ago.

Figure 7.4 compares the oxygen binding profiles of myoglobin and hemoglobin. The fractional saturation of myoglobin varies hyperbolically

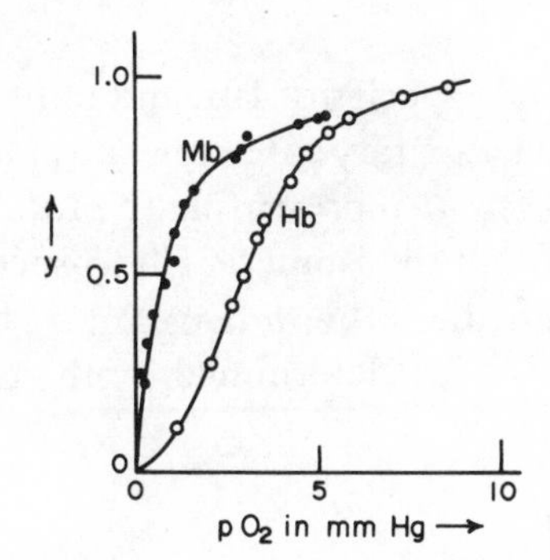


Fig. 7.4. The oxygen-binding curves for myoglobin (Mb: from Rossi Fanelli and Antonini, 1958) and hemoglobin (Hb: from Roughton, 1964).

with oxygen tension, as predicted by Eqn (7.1) for the equilibration of a monomeric protein with oxygen:

$$\text{Mb} + \text{O}_2 \rightleftharpoons \text{Mb}\cdot\text{O}_2$$

$$\bar{K} = \frac{(\text{MbO}_2)}{(\text{Mb})\cdot\text{O}_2}$$

$$y,\ \text{fractional saturation} = \frac{(\text{MbO}_2)}{(\text{MbO}_2) + (\text{Mb})} = \frac{\bar{K}\cdot\text{O}_2}{\bar{K}\cdot\text{O}_2 + 1} \tag{7.1}$$

The symbol $\bar{K}$ is used to designate the stability (binding) constant, i.e. reciprocal of the dissociation constant, for ligand binding to protein. The form of Eqn (7.1) can be generalized by replacing $O_2$ with the general substrate symbol of $S$:

$$y = \frac{\bar{K} \cdot \mathrm{S}}{\bar{K}\mathrm{S} + 1}. \tag{7.2}$$

The tetrameric hemoglobin does not conform to this hyperbolic, first-degree form. Instead, it shows a sigmoid profile. The appropriate equation for its behaviour emerges from a consideration of the equilibria between the different species of hemoglobin carrying different numbers of bound oxygen molecules:

$$\begin{bmatrix} \Sigma(\mathrm{Hb}) \\ \text{i.e., all species} \\ \text{without } O_2 \end{bmatrix} \underset{u_{-1}}{\overset{u_1\mathbf{S}}{\rightleftharpoons}} \begin{bmatrix} \Sigma(\mathrm{Hb} \cdot S) \\ \text{i.e., all species} \\ \text{with one } O_2 \end{bmatrix} \underset{u_{-2}}{\overset{u_2\mathbf{S}}{\rightleftharpoons}} \begin{bmatrix} \Sigma(\mathrm{Hb} \cdot S_2) \\ \text{i.e., all species} \\ \text{with two } O_2 \end{bmatrix}$$

$$\underset{u_{-3}}{\overset{u_3\mathbf{S}}{\rightleftharpoons}}$$

$$\begin{bmatrix} \Sigma(\mathrm{Hb} \cdot S_3) \\ \text{i.e., all species} \\ \text{with three } O_2 \end{bmatrix} \underset{u_{-4}}{\overset{u_4\mathbf{S}}{\rightleftharpoons}} \begin{bmatrix} \Sigma(\mathrm{Hb} \cdot S_4) \\ \text{i.e., all species} \\ \text{with four } O_2 \end{bmatrix}$$

If there occurred only a single species of Hb, and only a single species of $\mathrm{Hb} \cdot S$, $u_1$ would be a true elementary rate constant. More generally, it is merely an *apparent rate constant* representing the total conversion of all the isomeric Hb-species into all the isomeric $\mathrm{Hb} \cdot S$-species. A similar reservation applies to each of the other $u$-constants. The *apparent stability constants*, $U_1$, $U_2$, $U_3$ and $U_4$, are determined by the corresponding pairs of apparent rate constants:

$$U_1 = \frac{u_1}{u_{-1}} = \frac{\Sigma(\mathrm{Hb} \cdot S)}{\mathrm{S} \cdot \Sigma(\mathrm{Hb})}$$

$$U_2 = \frac{u_2}{u_{-2}} = \frac{\Sigma(\mathrm{Hb} \cdot S_2)}{\mathrm{S} \cdot \Sigma(\mathrm{Hb} \cdot S)}$$

$$U_3 = \frac{u_3}{u_{-3}} = \frac{\Sigma(\mathrm{Hb} \cdot S_3)}{\mathrm{S} \cdot \Sigma(\mathrm{Hb} \cdot S_2)}$$

$$U_4 = \frac{u_4}{u_{-4}} = \frac{\Sigma(\mathrm{Hb} \cdot S_4)}{\mathrm{S} \cdot \Sigma(\mathrm{Hb} \cdot S_3)}$$

In turn, these expressions enable the relative concentrations of the various hemoglobin species to be defined in terms of the apparent stability constants:

$$\Sigma(\mathrm{Hb}) = 1$$

$$\Sigma(\mathrm{Hb}\cdot S) = U_1\mathrm{S}$$

$$\Sigma(\mathrm{Hb}\cdot S_2) = U_1U_2\mathrm{S}^2$$

$$\Sigma(\mathrm{Hb}\cdot S_3) = U_1U_2U_3\mathrm{S}^3$$

$$\Sigma(\mathrm{Hb}\cdot S_4) = U_1U_2U_3U_4\mathrm{S}^4$$

Therefore, the fractional saturation of all the heme sites, four per hemoglobin molecule, is

$$y = \frac{\Sigma(\mathrm{Hb}\cdot S) + 2\Sigma(\mathrm{Hb}\cdot S_2) + 3\Sigma(\mathrm{Hb}\cdot S_3) + 4\Sigma(\mathrm{Hb}\cdot S_4)}{4\Sigma(\mathrm{Hb}) + 4\Sigma(\mathrm{Hb}\cdot S) + 4\Sigma(\mathrm{Hb}\cdot S_2) + 4\Sigma(\mathrm{Hb}\cdot S_3) + 4\Sigma(\mathrm{Hb}\cdot S_4)}$$

$$= \frac{U_1\mathrm{S} + 2U_1U_2\mathrm{S}^2 + 3U_1U_2U_3\mathrm{S}^3 + 4U_1U_2U_3U_4\mathrm{S}^4}{4(1 + U_1\mathrm{S} + U_1U_2\mathrm{S}^2 + U_1U_2U_3\mathrm{S}^3 + U_1U_2U_3U_4\mathrm{S}^4)}\,. \qquad (7.3)$$

This fourth degree equation was introduced by Adair (1925). It is entirely compatible with the sigmoid saturation curve recorded in Fig. 7.4, for cooperativity between the four subunits in the protein is already presupposed in its derivation. If the four hemes had been *independent as well as equivalent* in oxygen binding, $u_1$, $u_2$, $u_3$ and $u_4$ would be multiples of a single *intrinsic combination rate constant* $k_a$, the multiplicity being determined by the number of empty hemes on the different protein species:

$$u_1 = 4k_a, \quad u_2 = 3k_a, \quad u_3 = 2k_a, \quad u_4 = k_a. \qquad (7.4)$$

Similarly, $u_{-1}, u_{-2}, u_{-3}$ and $u_{-4}$ would be multiples of a single *intrinsic dissociation rate constant* $k_d$, the multiplicity being determined by the number of liganded hemes on the protein species:

$$u_{-1} = k_d, \quad u_{-2} = 2k_d, \quad u_{-3} = 3k_d, \quad u_{-4} = 4k_d \qquad (7.5)$$

$\bar{K}$, the single *intrinsic stability constant* applicable to all four subunits, would be equal to $k_a/k_d$; and the $U$-constants would become

$$U_1 = \frac{u_1}{u_{-1}} = \frac{4k_a}{k_d} = 4\bar{K}$$

$$U_2 = \frac{u_2}{u_{-2}} = \frac{3k_a}{2k_d} = \frac{3\bar{K}}{2} \qquad (7.6)$$

$$U_3 = \frac{u_3}{u_{-3}} = \frac{2k_a}{3k_d} = \frac{2\bar{K}}{3}$$

$$U_4 = \frac{u_4}{u_{-4}} = \frac{k_a}{4k_d} = \frac{\bar{K}}{4}$$

The fourth-degree, nonhyperbolic Eqn (7.3) would then be reduced to a hyperbola:

$$y = \frac{4\bar{K}S + 12\bar{K}^2S^2 + 12\bar{K}^3S^3 + 4\bar{K}^4S^4}{4(1 + 4\bar{K}S + 6\bar{K}^2S^2 + 4\bar{K}^3S^3 + \bar{K}^4S^4)}$$

$$= \frac{4\bar{K}S(1 + \bar{K}S)^3}{4(1 + \bar{K}S)^4}$$

$$= \frac{\bar{K}S}{1 + \bar{K}S} \cdot \qquad (7.7)$$

Accordingly, if the four subunits of hemoglobin were independent and equivalent, the $y(S)$ function would be hyperbolic like that of myoglobin. Therefore, the sigmoidal $y(S)$ function exhibited by hemoglobin rules out independence and equivalence between the subunits and establishes the occurrence of some sort of homotropic cooperation between them.†

### 7.3.2. Discrete cooperativity

The fourth-degree equation (7.3), known as the *Adair equation*, can be generalized to $n$th degree for a protein with $n$ protomers:

$$y = \frac{U_1S + 2U_1U_2S^2 + \cdots + nU_1U_2 \cdots U_nS^n}{n(1 + U_1S + U_1U_2S^2 + \cdots + U_1U_2 \cdots U_nS^n)} \cdot \qquad (7.8)$$

The apparent stability constants, $U_1, U_2, \ldots, U_n$, describe the successive additions of ligand $S$ to the protein. When there is no homotropic interaction between the protomers, such that the successive additions are independent as well as equivalent, the ratio between $U_{j+1}$ and $U_j$ depends only on the values of $n$ and $j$, and not on the intrinsic stability constant $\bar{K}$ (Bjerrum, 1941; Edsall and Wyman, 1958):

$$U_j = \frac{n - j + 1}{j} \times \bar{K}$$

$$U_{j+1} = \frac{n - j}{j + 1} \times \bar{K}$$

and therefore

$$\frac{U_{j+1}}{U_j} = \frac{n - j}{n - j + 1} \cdot \frac{j}{j + 1} \cdot \qquad (7.9)$$

† If the four subunits were independent but nonequivalent, the $y(S)$ function could be nonhyperbolic but could not be sigmoidal (Weber and Anderson, 1965).

For a tetramer with independent and equivalent subunits, Eqn (7.9) requires the noncooperative ratios between $U_1$, $U_2$, $U_3$ and $U_4$ to be:

$$\frac{U_2}{U_1} = \frac{4-1}{4-1+1} \cdot \frac{1}{1+1} = \frac{3}{8}$$

$$\frac{U_3}{U_2} = \frac{4-2}{4-2+1} \cdot \frac{2}{2+1} = \frac{4}{9}$$

$$\frac{U_4}{U_3} = \frac{4-3}{4-3+1} \cdot \frac{3}{3+1} = \frac{3}{8}.$$

These ratios, so derived, are easily confirmed by Eqn (7.6). Experimentally, if any of the three $U_{j+1}/U_j$ ratios for an experimental system deviates from Eqn (7.9), the cooperativity indicated by the deviation becomes calculable in terms of a *cooperativity coefficient*, $\gamma_j$, defined by Eqn (7.10) (Endrenyi, Chan and Wong, 1971):

$$\gamma_j = \log\left(\frac{U_{j+1}}{U_j} \cdot \frac{n-j+1}{n-j} \cdot \frac{j+1}{j}\right). \tag{7.10}$$

This coefficient is zero if $U_{j+1}/U_j$ conforms to the noncooperative ratio defined by Eqn (7.9), and a complete lack of cooperativity between the $j$th and $(j+1)$th ligand additions is indicated. If this coefficient is positive, *positive cooperativity* is indicated. Finally, if this coefficient is negative, *negative cooperativity* is indicated. For example, the cooperativity between the first and second, second and third, and third and fourth, oxygen additions to hemoglobin may be calculated, using Eqn (7.10), from the apparent stability constants estimated by Roughton (1963):

*Apparent stability constants for experimental sheep No. A38, in (mm Hg)$^{-1}$*

$U_1 = 0.326$, $U_2 = 0.114$, $U_3 = 0.268$, $U_4 = 2.0$

*Calculation of three cooperativity coefficients from $U_1 - U_4$:*

$$\gamma_1 = \log\left(\frac{0.114}{0.326} \cdot \frac{4}{3} \cdot \frac{2}{1}\right) = -0.03$$

$$\gamma_2 = \log\left(\frac{0.268}{0.114} \cdot \frac{3}{2} \cdot \frac{3}{2}\right) = +0.72$$

$$\gamma_3 = \log\left(\frac{2.0}{0.268} \cdot \frac{2}{1} \cdot \frac{4}{3}\right) = +1.30$$

Accordingly, there was close to zero cooperativity between the first and second additions of oxygen, but strongly positive cooperativity between the second and third, and the third and fourth, additions.

For any homotropic system then, the generalized Adair equation (7.8) provides a basis for the evaluation of the *U*-constants for the successive bindings of ligand. Curve-fitting procedures useful for such evaluation have been developed by Sillen (1956), Cornish-Bowden and Koshland (1970a), and Atkins (1971). Once the *U*-constants are obtained, the cooperativity between any two discrete stages of ligand addition may be calculated using Eqn (7.10).

### 7.3.3. Continuous cooperativity

The concept of *discrete cooperativity,* discussed in the preceding section, is mostly useful for the purpose of mechanistic analysis. However, it would be equally useful to be able to consider cooperativity without first having to resolve the ligand-addition process into discrete stages. Such a concept of *continuous cooperativity* has been formulated by Wyman (1967) on the basis of the continuous *Hill plot* relating fractional saturation to ligand concentration; it is particularly suited to the purpose of thermodynamic analysis. These two concepts of cooperativity are different but complementary, one stressing the mechanistic viewpoint, and the other the thermodynamic viewpoint.

Fig. 7.5 shows the Hill plot for protein saturation by substrate. According to Wyman (1963), the Hill curve for an oligomeric protein with $n$ equivalent protomers will approach asymptotes with unit slope at both ends of the substrate concentration scale. The perpendicular separation between the initial and final asymptotes is proportional to the average free energy of interaction realized per binding site.† The interaction energy is positive if the final asymptote lies above the initial one but negative if the reverse is true. At any point along the curve, $H$, the value of the slope, indicates the direction of the free energy gradient: $H > 1$ means a positive gradient, and $H < 1$ a negative gradient. The continuous cooperativity exhibited at any point along the Hill curve can thus be characterized solely on the basis of $H$ (Whitehead, 1970, 1973):

$H = 1$ *zero cooperativity*

$H > 1$ *positive cooperativity*

$H < 1$ *negative cooperativity*

The concepts of discrete and continuous cooperativity are closely related. $H \geqslant 1$ is expected throughout the Hill curve if all the $\gamma_j s$ in the system

† Saroff and Minton (1972) suggested that the separation might be more precisely interpreted as the difference between the free energies of interaction on binding the first and the last ligands.

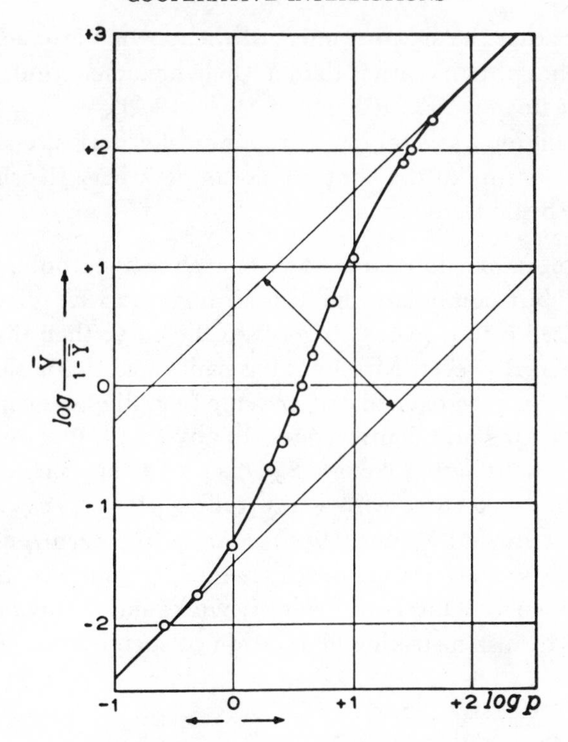


Fig. 7.5. The Hill plot for the binding of oxygen to hemoglobin ($p$ stands for concentration of oxygen measured in terms of partial pressure). (From Wyman, 1963.) In many enzyme systems, saturations at very low and very high substrate concentrations are often not readily measured: consequently only the middle portion of the Hill plot, and not the two asymptotes with unit slope, would be observed.

are positive, and $H \leqslant 1$ is expected throughout if all the $\gamma_j s$ are negative. Observation of both $H > 1$ and $H < 1$ regions would suggest that there are both positive and negative $\gamma_j s$. However, the absence of a $H < 1$ region does not rule out the occurrence of a weakly negative $\gamma_j$ being masked by other more strongly positive $\gamma_j s$. Similarly, the absence of a $H > 1$ region does not rule out the occurrence of a weakly positive $\gamma_j$ being masked by other more strongly negative $\gamma_j s$. In this connection it is especially worth remembering that $H < 1$ may also reflect binding of ligand to independent but non-equivalent sites.

The Hill slope serves the additional purpose of defining the minimum number of interacting protein protomers, which determine the degree of the Adair equation (see Section 3.2). For example, aspartate transcarbamylase has six protomeric units, but its maximal Hill slope, or $H_m$, is only 1.55 (Changeux, Gerhart and Schachman, 1968). The experimental data in fact appear compatible with a second-degree function on the strength of the standardized Scatchard plot, the $H_m$ versus log $(S_{0.7}/S_{0.3})$ plot, or the $H_m$ versus log $(S_{0.9}/S_{0.1})$ plot, all three of which attest to second-degree rather than sixth-degree behaviour (Endrenyi *et al.*, 1971). Accordingly, the possibility arises that the six protomers of the enzyme interact in the form of three independent pairs rather than a single hexameric unit (Markus, McClintock and Bussel, 1971; Chan and Mort, 1973).

Besides the Hill plot, several familiar graphical criteria are also useful for revealing the nature of the continuous cooperativity (Koshland, 1970; Kirschner, 1971b; Endrenyi *et al.*, 1971):

(a) The simple saturation curve of $y$ versus S is hyperbolic when cooperativity is lacking, but sigmoidal when it is strongly positive. Negative effects may be recognized by a slower tailing-off in the curve than that expected from a hyperbola. However, Kirschner has cautioned that a slow tailing-off might be caused by heterogeneity of enzyme (e.g. due to ageing of enzyme preparations) creating nonidentical sites. In any event, it is useful to remember that in a hyperbolic curve $S_{0.9}/S_{0.1} = 81$; in a sigmoid curve, $S_{0.9}/S_{0.1} < 81$; and in a curve with a slow tailing-off, $S_{0.9}/S_{0.1} > 81$ (Koshland, Nemethy and Filmer, 1966). Finally, the occurrence of negative together with positive effects may cause *bumps,* or multiple inflections, in the saturation curve. Observations by Levitzki and Koshland (1969) on CTP synthetase furnish a striking illustration of such bumps (Fig. 7.6).

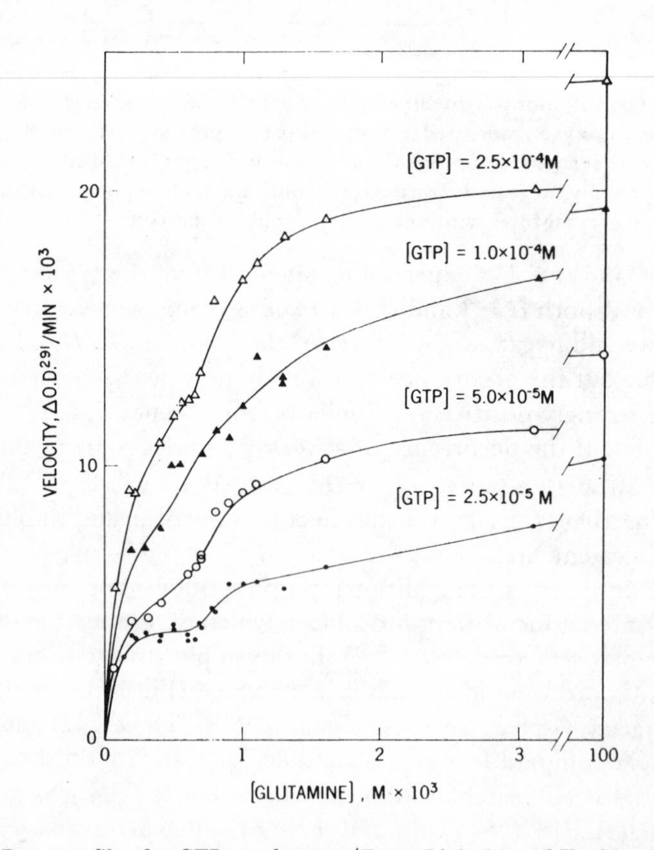


Fig. 7.6. Rate profiles for CTP synthetase. (From Livitzki and Koshland, 1969.)

(b) The Eadie plot of $y$ versus $y/S$ plot is curved upwards when cooperativity is negative, but downwards when positive. (In practice, the Eadie plot is applied mostly to kinetic data; its inverse, the Scatchard plot of $y/S$ versus $y$, is applied mainly to binding data.)

(c) The Lineweaver-Burk plot of $1/y$ versus $1/S$ is curved upwards when cooperativity is positive, but downwards when negative.

(d) The Hanes plot of $S/y$ versus S is curved upwards when cooperativity is positive, but downwards when negative.

### 7.3.4. Equilibrium vs. kinetic cooperativity

The Hill plot is applicable to the initial velocity $v_0$ as well as to the fractional saturation $y$. The same is true of the other graphical indicators of cooperativity. However, the cooperativities established from kinetic and from ligand-binding measurements need not be equivalent.

When the rate function $v_0(S)$ for an enzyme system displays the nonhyperbolic traits (e.g., sigmoidicity, or $H > 1$) so characteristic of cooperative processes, it is necessary to decide whether or not the binding function $y(S)$ is likewise nonhyperbolic. If binding measurements indicate a hyperbolic $y(S)$, the nonhyperbolic $v_0(S)$ is more likely due to a steady-state branching mechanism in which more than one enzyme species reacts with $S$ (Section 5.1.1), rather than to homotropic interactions between subunit protomers of the enzyme. Although homotropic interactions that affect only catalysis but not substrate binding can also account for the combination of a nonhyperbolic $v_0(S)$ with a hyperbolic $y(S)$, this type of homotropic interactions (i.e. a *V-system* as opposed to the commonly encountered $K$- or $KV$-system: see below) appears to be rare. It is moreover useful to note that the nonhyperbolic $v_0(S)$ due to homotropic interactions are often dependent on, or abolished by, some allosteric modifier (Sanwal, 1970). However, the nonhyperbolic $v_0(S)$ due to a branching mechanism likewise may be dependent on some modifier if the modifier is responsible for creating the branching pathways in the mechanism. It also may be abolished by some modifier if the modifier is capable of converting a branching multisubstrate mechanism into a linear one: Sanwal and Cook (1966) have suggested that AMP acts on isocitrate dehydrogenase in this manner.

If $y(S)$ shows nonhyperbolic symptoms of positive cooperativity, homotropic interactions between subunits would be indicated. Strictly negative cooperativity, on the other hand, may arise from either homotropic interactions or heterogeneity amongst enzyme sites. The latter possibility may be tested in part through further purification of the enzyme. In general, a nonhyperbolic $y(S)$ may be established by means of binding measurements, or by means of a method suggested by Engers, Bridger and Madsen (1970), which uses *isotopic exchanges* to test for quasi-equilibrium in the mechanism.

If quasi-equilibrium conditions are indicated by equality between the rates of different isotopic fluxes, steady-state mechanisms would be eliminated as a plausible explanation for the nonhyperbolic $v_0(\mathrm{S})$, and a nonhyperbolic $y(\mathrm{S})$ would have to be implicated (Section 9.2.3).

Even if a nonhyperbolic $y(\mathrm{S})$ and hence the occurrence of homotropic interactions could be established, it remains to be analysed whether such interactions affect only affinity for substrate (i.e. a *K-system*), or both affinity for substrate and catalytic efficiency (i.e. a *KV-system*). Since the kinetic system may be working under either quasi-equilibrium or steady-state conditions, there are altogether four possible combinations to consider:

1. *K*-system + quasi-equilibrium
2. *KV*-system + quasi-equilibrium
3. *K*-system + steady-state
4. *KV*-system + steady-state

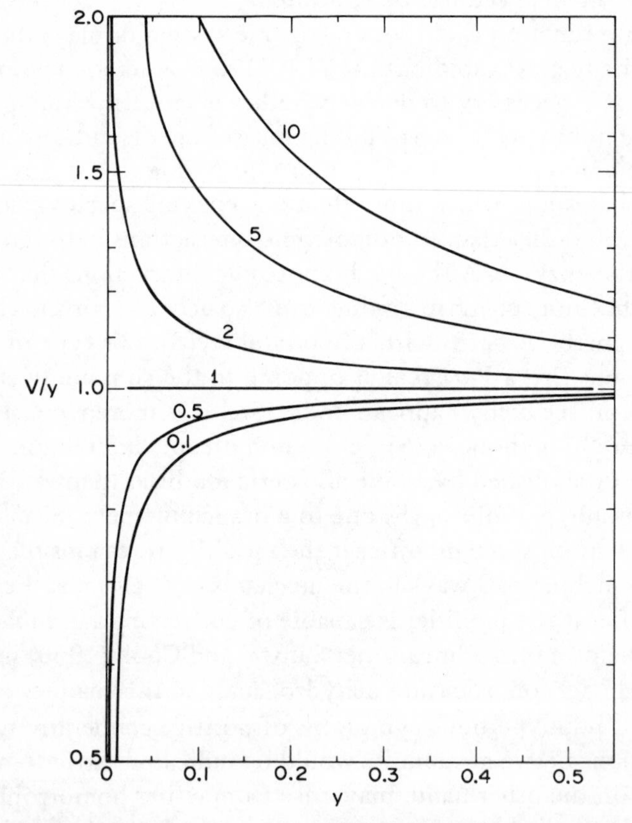


Fig. 7.7. Detection of *KV*-effects. The phosphofructokinase mechanism of Blangy, Buc and Monod (1968) was employed as test system. Numbers on the curves indicate the ratio between the catalytic rate constants for the *T* and *R* states of the protein, and hence the magnitude of the *V*-effect. A ratio of 1 implies that the two catalytic rate constants are equal, so that the system exhibits only *K*-effects and no *V*-effects. Even though the *K*-effects are about $2 \times 10^3$ for this system, a *V*-effect of 2 or 0.5 already results in large variations of $v/y$ with $y$. (From Wong and Endrenyi, 1971.)

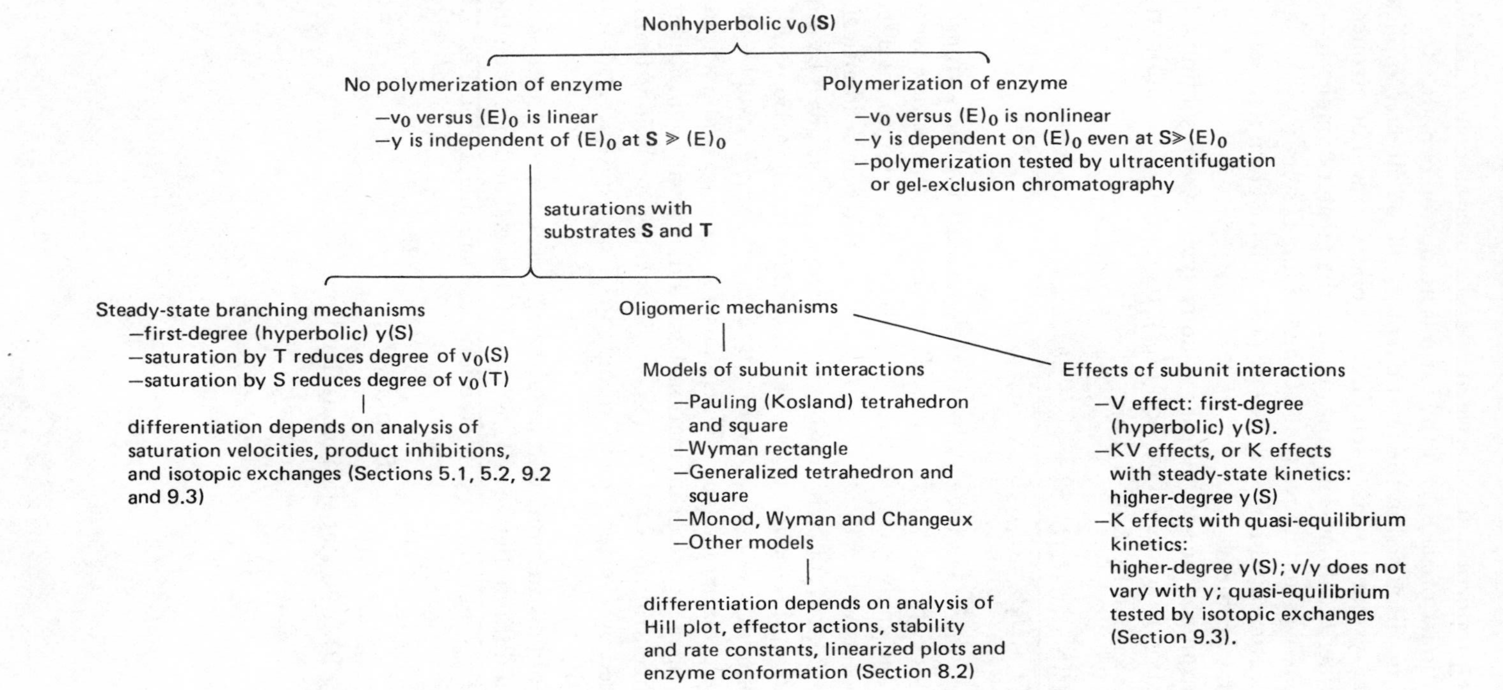


Fig. 7.8. Differentiation between nonhyperbolic mechanisms.

There will be proportionality between $v_0$ and $y$ over all concentrations of $S$ if combination-1 prevails, and the various homotropic models for the binding system, discussed in the next chapter, will be directly applicable to the kinetic system as well. Such direct application will be erroneous if any of the other combinations prevails, and the greater complexity of the kinetic system over and above that of the binding system must be taken into account. Ricard, Mouttet and Nari (1974) have given a penetrating analysis of this problem.

Take for example the general case of an enzyme with $n$ subunits. Its binding function is described by Eqn (7.8). Its rate law, if combination-2 prevails, is described by Eqn (7.11):

$$v_0 = \frac{k_1 U_1 \mathrm{S} + 2k_2 U_1 U_2 \mathrm{S}^2 + \cdots + nk_n U_1 U_2 \ldots U_n \mathrm{S}^n}{n(1 + U_1 \mathrm{S} + U_1 U_2 \mathrm{S}^2 + \cdots + U_1 U_2 \ldots U_n \mathrm{S}^n)} \tag{7.11}$$

where $k_1, k_2, \ldots, k_n$ are the catalytic rate constants for the variously liganded enzyme species. Combination-2 rate laws of this nature were first derived and analysed by Dalziel (1968). Because subunit interactions affect catalysis as well as binding, $k_1, k_2, \ldots, k_n$ will not be identical, and $v_0$ as given by Eqn (7.11) will not be proportional to $y$ as given by Eqn (7.8). In contrast, if combination-1 prevails, subunit interactions will affect only binding and not catalysis. All of $k_1, k_2, \ldots, k_n$ will be identical, and $v_0$ will be proportional to $y$. Thus, a useful test for combination-1 is to plot $v/y$ against $y$. If combination-1 is valid, $v/y$ will be invariant with $y$ over all concentrations of $S$. As Fig. 7.7 illustrates, deviations from invariance caused by $V$-effects equal to only 0.1% of the $K$-effects could be detected on this basis.

The manifold distinctions that require analysis, between steady-state and quasi-equilibrium, between $V$, $K$ and $KV$ effects, and between the various models of homotropic interactions, are summarized in Fig. 7.8.

## Problem 7.1

Derive Eqn (7.9) from statistical considerations.

# 8 Homotropic Mechanisms

## 8.1. Some Simple Models

Cooperative substrate-binding to an oligomeric protein requires some form of mutual influence, or coupling, between the various protomers. One means of coupling would be via neighbourly contact, such that binding of substrate to one protomer modulates *directly* the conformation and reactivity of its adjoining neighbour. Another means would be via a concerted transition in conformation; if the different protomers have to undergo conformational changes concertedly in order to preserve symmetry within the quaternary protein structure, binding of substrate to one protomer would *indirectly* alter the conformation equilibrium of all the protomers. Finally, since the protein oligomer is formed from a polymerization of its protomers, binding of substrate to a protomer may shift the balance between the various states of *polymerization*. These three means of coupling are not mutually exclusive, and may be combined into various hybrid models. They will be illustrated by considering a tetrameric protein with four equivalent protomers.

### 8.1.1. Direct-modulation models

The number of adjoining neighbours a protomer has within an oligomeric protein depends on the geometric arrangement of the protomers. In a tetramer, arrangements which treat the four protomers equally include the tetrahedron, the square and the rectangle:

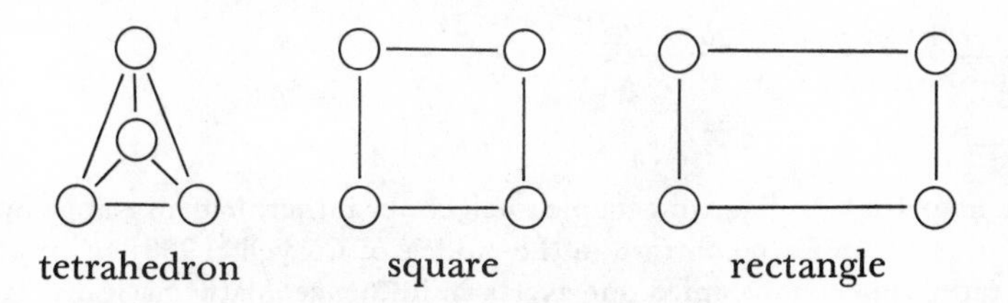


Each protomer has three adjoining neighbours in the tetrahedron, but only two in the square or rectangle. The two neighbours are equally influential in the square, but unequally so in the rectangle. In every case the affinity of a protomer for the substrate depends on the state of occupation of its adjoining neighbours. This influence exerted by neighbours may be formulated in several ways:

(a) In the tetrahedron or square model of Pauling (1935)†, the stability (binding) constant of a protomer for substrate $S$ is modified by the occupation of its neighbours, in accordance with the following rules:

| | *stability constant* |
|---|---|
| when all neighbours are empty | $\bar{K}$ |
| when one neighbour is occupied by $S$ | $\theta\bar{K}$ |
| when two neighbours are occupied by $S$ | $\theta^2\bar{K}$ |
| when three neighbours are occupied by $S$ | $\theta^3\bar{K}$ |

That is, substrate occupation of a neighbouring site always modifies the stability constant by the constant factor of $\theta$. The saturation functions for both the tetrahedron and square models can be derived on the basis of these rules for modifying $\bar{K}$ (Figs. 8.1 and 8.2).

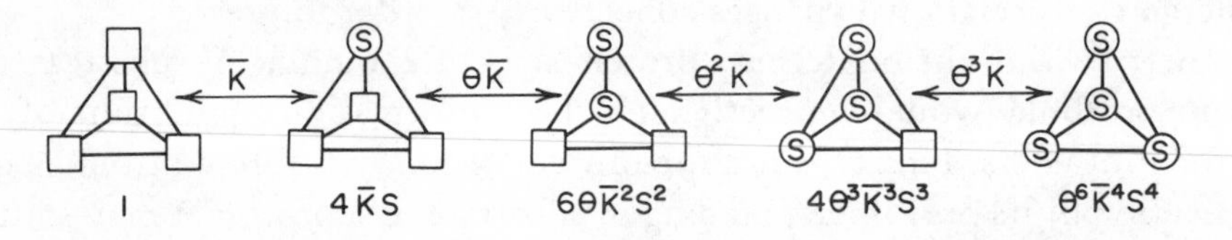


Fig. 8.1. The Pauling (Koshland) tetrahedron. There are only one species each of $E_4(=1)$ and $E_4S_4$ $(=\theta^6\bar{K}^4\mathrm{S}^4)$. However, there are four species of $E_4S(=4\bar{K}\mathrm{S})$, differing from one another with respect to which of the four sites is liganded. Similarly, there are six species of $E_4S_2$ $(=6\theta\bar{K}^2\mathrm{S}^2)$ and four species of $E_4S_3$ $(=4\theta^3\bar{K}^3\mathrm{S}^3)$. Since there are four sites per protein species,

$$\begin{aligned}\text{No. of substrate sites} &= 4(E_4+E_4S+E_4S_2+E_4S_3+E_4S_4)\\ &= 4(1+4\bar{K}\mathrm{S}+6\theta\bar{K}^2\mathrm{S}^2+4\theta^3\bar{K}^3\mathrm{S}^3+\theta^6\bar{K}^4\mathrm{S}^4)\end{aligned}$$

$$\begin{aligned}\text{No. of bound substrates} &= 0\times E_4+1\times E_4S+2\times E_4S_2+3\times E_4S_3+4\times E_4S_4\\ &= 4\bar{K}\mathrm{S}+12\theta\bar{K}^2\mathrm{S}^2+12\theta^3\bar{K}^3\mathrm{S}^3+4\theta^6\bar{K}^4\mathrm{S}^4.\end{aligned}$$

The fractional saturation, $y$, is determined by the number of bound substrates divided by the number of substrate sites:

$$y=\frac{\bar{K}\mathrm{S}+3\theta\bar{K}^2\mathrm{S}^2+3\theta^3\bar{K}^3\mathrm{S}^3+\theta^6\bar{K}^4\mathrm{S}^4}{1+4\bar{K}\mathrm{S}+6\theta\bar{K}^2\mathrm{S}^2+4\theta^3\bar{K}^3\mathrm{S}^3+\theta^6\bar{K}^4\mathrm{S}^4}.$$

† In the models of Pauling, an occupied neighbour rather than an empty one exerts an influence on $\bar{K}$. In contrast, in the models of Coryell (1939) an empty neighbour rather than an occupied one exerts an influence. Mathematically, the two views give indistinguishable predictions, and do not need to be considered separately. Physically there is a distinction. According to the former view the subunit protomers in the completely empty oligomer are expected to be unconstrained and resemble

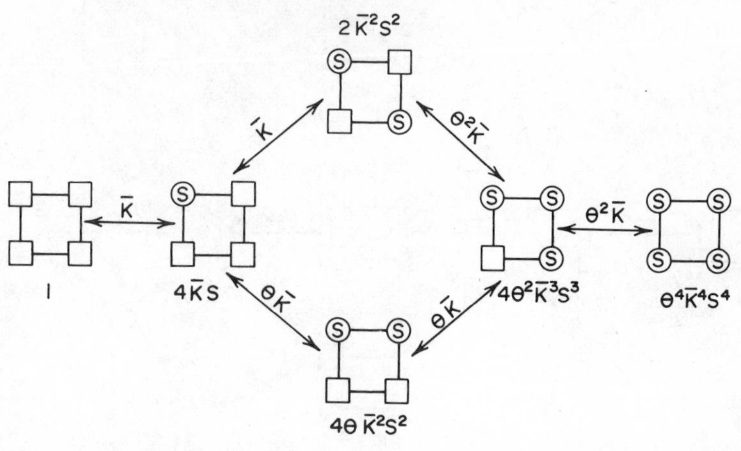


Fig. 8.2. The Pauling (Koshland) square. Its fractional saturation is again obtainable by counting the number of substrates bound to the different protein species and dividing it by the number of substrate sites:

$$y = \frac{\bar{K}S + (2\theta + 1)\bar{K}^2 S^2 + 3\theta^2 \bar{K}^3 S^3 + \theta^4 \bar{K}^4 S^4}{1 + 4\bar{K}S + 2(2\theta + 1)\bar{K}^2 S^2 + 4\theta^2 \bar{K}^3 S^3 + \theta^4 \bar{K}^4 S^4}.$$

(b) In the rectangle model of Wyman (1948), each protomer is joined to two neighbours, which do not affect the protomer equally, one being more intimate than the other. Therefore $\bar{K}$ is modified by two interaction coefficients $\theta_1$ and $\theta_2$ of unequal strength:

| | *stability constant* |
|---|---|
| when all neighbours are empty | $\bar{K}$ |
| when intimate neighbour is occupied by $S$ | $\theta_1 \bar{K}$ |
| when remote neighbour is occupied by $S$ | $\theta_2 \bar{K}$ |
| when both neighbours are occupied by $S$ | $\theta_1 \theta_2 \bar{K}$ |

The coefficients $\theta_1$ and $\theta_2$ may be both greater than one (bringing positive cooperativity), both less than one (bringing negative cooperativity), or one greater and the other less than one (making possible mixed cooperativity). In any event, the saturation function for the model again can be derived from the given rules for modifying $\bar{K}$ (Fig. 8.3).

(c) The assumption made in the Pauling models that the same interaction coefficient $\theta$ pertains to each successive substrate-occupation of neighbouring protomers is a striking restriction. This restriction may be

---

most the physical state of free protomers whereas according to the latter view the protomers in the completely occupied oligomer would be expected to do so (Saroff and Minton, 1972). The models of Thompson (1968) are also close to those of Pauling.

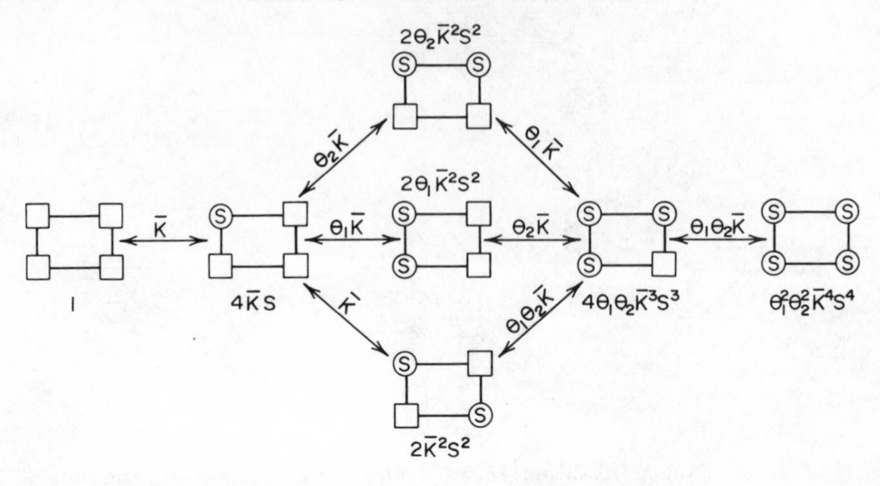


Fig. 8.3. The Wyman rectangle. Its fractional saturation is:

$$y = \frac{\bar{K}S + (\theta_2 + \theta_1 + 1)\bar{K}^2 S^2 + 3\theta_1\theta_2\bar{K}^3 S^3 + \theta_1^2\theta_2^2\bar{K}^4 S^4}{1 + 4\bar{K}S + 2(\theta_2 + \theta_1 + 1)\bar{K}^2 S^2 + 4\theta_1\theta_2\bar{K}^3 S^3 + \theta_1^2\theta_1^2\bar{K}^4 S^4}$$

relaxed by assigning a different interaction coefficient to each of the successive stages of substrate-occupation (Wong and Endrenyi, 1971):

| | *stability constant* |
|---|---|
| when all neighbours are empty | $\bar{K}$ |
| when one neighbour is occupied by *S* | $\theta_1\bar{K}$ |
| when two neighbours are occupied by *S* | $\theta_2\bar{K}$ |
| when three neighbours are occupied by *S* | $\theta_3\bar{K}$ |

These more relaxed rules for enumerating direct modulations form the basis of the generalized tetrahedron and square models, the saturation functions for which are derived in Fig. 8.4 and 8.5.

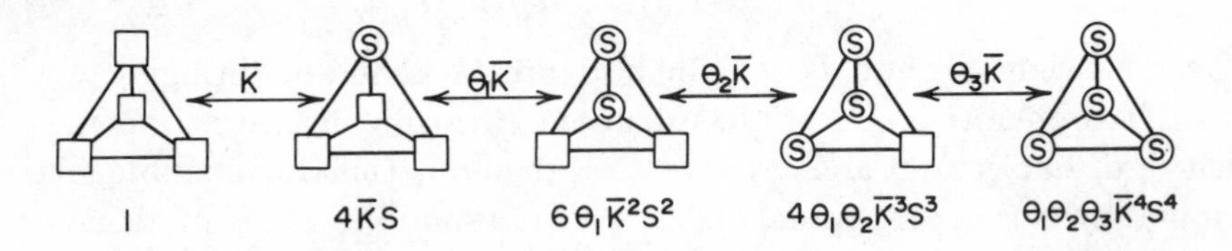


Fig. 8.4. The generalized tetrahedron. Its fractional saturation is:

$$y = \frac{\bar{K}S + 3\theta_1\bar{K}^2 S^2 + 3\theta_1\theta_2\bar{K}^3 S^3 + \theta_1\theta_2\theta_3\bar{K}^4 S^4}{1 + 4\bar{K}S + 6\theta_1\bar{K}^2 S^2 + 4\theta_1\theta_2\bar{K}^3 S^3 + \theta_1\theta_2\theta_3\bar{K}^4 S^4}.$$

### 8.1.2. Two-state vs. many-state models

What happens physically when the activity of a protomer is modulated by a neighbour? It is a truism to say that changes in activity reflect changes in conformation, if by conformation we mean no less than the

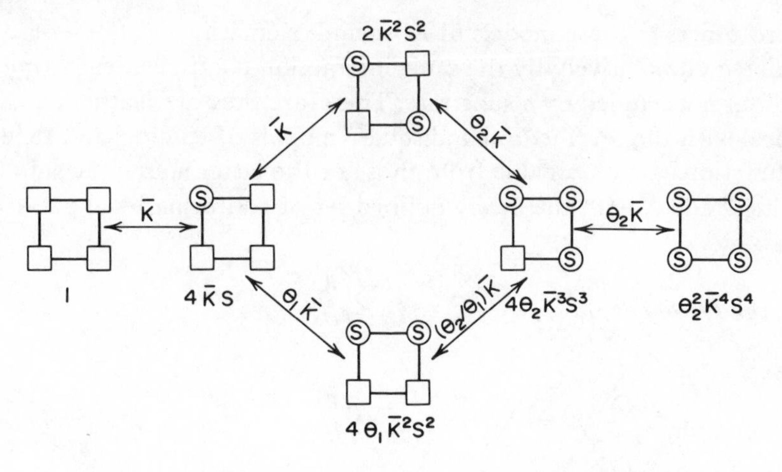


Fig. 8.5. The generalized square. Its fractional saturation is:

$$y = \frac{\bar{K}\mathbf{S} + (2\theta_1 + 1)\bar{K}^2\mathbf{S}^2 + 3\theta_2\bar{K}^3\mathbf{S}^3 + \theta_2^2\bar{K}^4\mathbf{S}^4}{1 + 4\bar{K}\mathbf{S} + 2(2\theta_1 + 1)\bar{K}^2\mathbf{S}^2 + 4\theta_2\bar{K}^3\mathbf{S}^3 + \theta_2^2\bar{K}^4\mathbf{S}^4}$$

total quantum-mechanical configuration of the protomer within the oligomeric framework. Nevertheless, conformation as the basis of activity remains a useful concept that promotes the experimental measurement of protein conformation. In recent years the need to understand allosteric mechanisms has added no small impetus to the ingenious development of many conformational probes which serve this purpose.

Koshland, Némethy and Filmer (1966) have developed a two-state theory of allosteric interaction to interpret the phenomenon of direct modulation in conformational terms. In this theory, each protomer equilibrates between two conformational states, which can be represented as the T- and R-states. Transition between the two states is regarded essentially as being ligand-induced:

$$\square \underset{}{\overset{S}{\rightleftharpoons}} (S)$$

*T*-state *R*-state

The *R*-state is induced by substrate and allosteric activators, and the *T*-state by allosteric inhibitors. The basic parameters for the model include:

| | |
|---|---|
| stability constant between *T*-state and substrate | $= \bar{K}_T$ |
| stabilization factor for a *T–T* neighbour pair | $= 1$ |
| stabilization factor for a *T–R* neighbour pair | $= f_{TR}$ |
| stabilization factor for an *R–R* neighbour pair | $= f_{RR}$ |
| constant for the *T* to *R* transition | $= f_t$ |

The protomers in these models of Koshland, Némethy and Filmer are modulated cumulatively by the same interaction coefficient each time a neighbour is occupied by a substrate. Therefore they are mathematically identical with the tetrahedron and square models of Pauling, and their $y(S)$ functions are obtainable from those of the latter merely by substituting $\theta$ and $\bar{K}$ with the newly defined set of conformational parameters:

*for the tetrahedron:*
$$\theta = f_{RR}/f_{TR}^2, \quad \bar{K} = f_{TR}^3 \bar{K}_T f_t,$$

*for the square:*
$$\theta = f_{RR}/f_{TR}^2, \quad \bar{K} = f_{TR}^2 \bar{K}_T f_t.$$

The mathematical kinship between the Pauling and Koshland models is to be expected. In the Pauling models, the physical significance of the interaction coefficient $\theta$ is left indeterminate. The Koshland models are in effect a subset of the Pauling models, in which $\theta$ is defined explicitly in terms of conformational change. It is of course possible to construct further subsets of the Pauling models. In particular, if a two-state formulation can be shown to be inadequate, it would be necessary to consider the possible existence of three or more conformational states. For example, if an empty protomer enters into a new conformational state every time a neighbour comes to be liganded, the tetrahedron arrangement may generate as many as five recognizably distinct states:

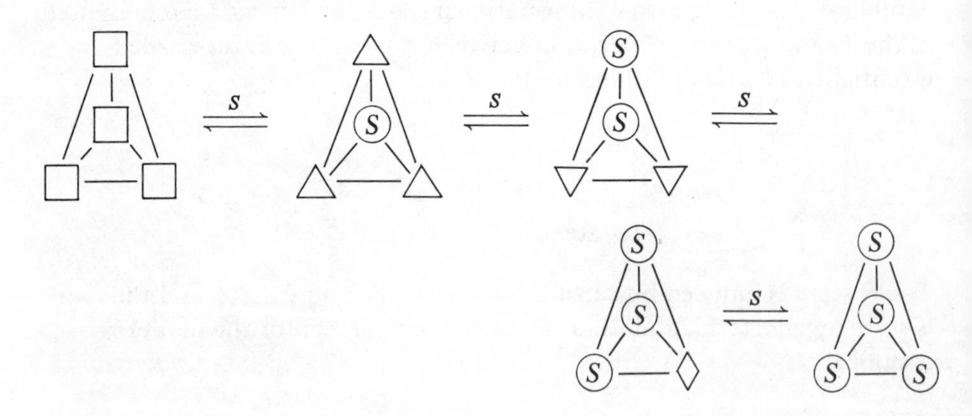


Both the two-state and the many-state formulations are applicable to the Wyman rectangle and the generalized tetrahedron and square no less than to the Pauling tetrahedron and square. The simpler of the direct-modulation models for a tetrameric protein therefore include the following combinations:

$$\begin{bmatrix} \text{Pauling (or Koshland)} \\ \text{Wyman} \\ \text{Generalized} \end{bmatrix} \times \begin{bmatrix} \text{tetrahedron} \\ \text{square} \\ \text{rectangle} \end{bmatrix} \times \begin{bmatrix} \text{2-states} \\ \text{3-states} \\ >\text{3-states} \end{bmatrix}$$

### 8.1.3. Concerted-transition models

If different protomers are constrained to change their conformation in a concerted fashion by reason of preservation of structural symmetry within the oligomer, cooperative behaviour may be achieved even in the absence of direct neighbourly modulations of any kind. This is the basis of the model proposed by Monod, Wyman and Changeux (1965) for cooperative interactions. Consider a tetramer for example. If the four protomers are constrained to be all in the $T$-state or all in the $R$-state, there will be two protein species in the absence of ligands:

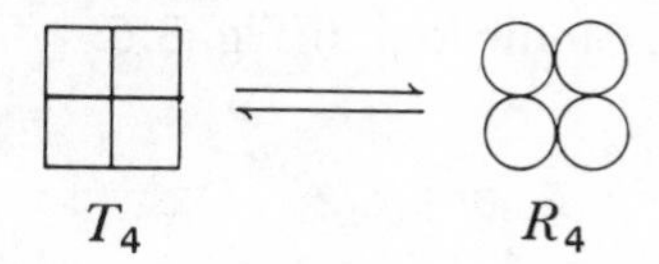


The equilibrium constant for the *allosteric transition* between $T_4$ and $R_4$ is $L$, the *allosteric constant*:

$$L = \frac{T_4}{R_4}.$$

The protomers are completely independent and equivalent within either $R_4$ or $T_4$. The stability constants for substrate binding by the $T$ and $R$ states are respectively $\bar{K}_T$ and $\bar{K}_R$, and the ratio between them is $c$:

$$c = \frac{\bar{K}_T}{\bar{K}_R}.$$

The relaxed $R$-state is considered the more active state than the tense $T$-state, by which convention *c is always smaller than one*. The smaller $c$ is, the less active is the $T$-state relative to the $R$-state. The limit is reached when $c$ equals zero, in which case the substrate binds only to the $R$-state and not at all to the $T$-state. The relative concentrations of the various protein-substrate species are shown in Fig. 8.6, from which the $y(\mathrm{S})$ function may be derived:

$$y = \frac{\bar{K}_R \mathrm{S}(1 + \bar{K}_R \mathrm{S})^3 + Lc\bar{K}_R \mathrm{S}(1 + c\bar{K}_R \mathrm{S})^3}{(1 + \bar{K}_R \mathrm{S})^4 + L(1 + c\bar{K}_R \mathrm{S})^4} \tag{8.1}$$

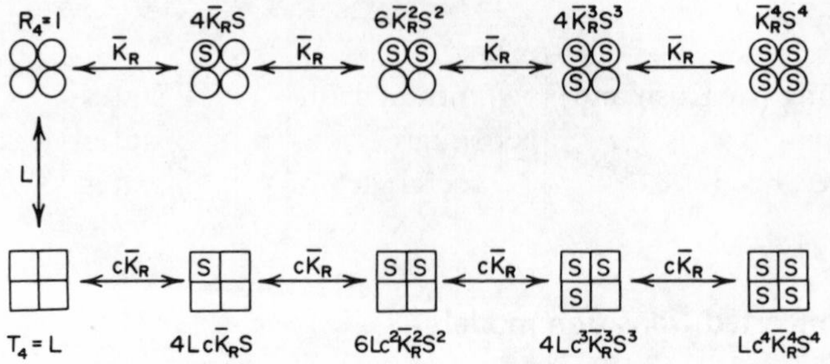

Fig. 8.6. The Monod–Wyman–Changeux concerted transition. Its fractional saturation is:

$$y = \frac{\bar{K}_R \mathrm{S}(1 + \bar{K}_R \mathrm{S})^3 + Lc\bar{K}_R \mathrm{S}(1 + c\bar{K}_R \mathrm{S})^3}{(1 + \bar{K}_R \mathrm{S})^4 + L(1 + c\bar{K}_R \mathrm{S})^4}.$$

Equation (8.1) is a fourth-degree equation, capable of predicting positive cooperativity. The physical basis for the cooperativity becomes evident if we compare the ratios between the corresponding $T$ and $R$ saturation states, whose concentrations are indicated in Fig. 8.6:

$$\frac{T_4}{R_4} = L$$

$$\frac{T_4 S}{R_4 S} = \frac{4Lc\bar{K}_R \mathrm{S}}{4\bar{K}_R \mathrm{S}} = Lc$$

$$\frac{T_4 S_2}{R_4 S_2} = \frac{6Lc^2\bar{K}_R^2 \mathrm{S}^2}{6\bar{K}_R^2 \mathrm{S}^2} = Lc^2$$

$$\frac{T_4 S_3}{R_4 S_3} = \frac{4Lc^3\bar{K}_R^3 \mathrm{S}^3}{4\bar{K}_R^3 \mathrm{S}^3} = Lc^3$$

$$\frac{T_4 S_4}{R_4 S_4} = \frac{Lc^4\bar{K}_R^4 \mathrm{S}^4}{\bar{K}_R^4 \mathrm{S}^4} = Lc^4.$$

Since $c$ is always smaller than one, $Lc^4 < Lc^3 < Lc^2 < Lc < L$. Therefore, as the protein is progressively saturated by substrate, there occurs a continuous shift of the enzyme toward the more active $R$-conformation. The larger the value of $L$ and the smaller the value of $c$, the more dramatic the shift from $T$ to $R$ will be. When $L \gg 1$, and $c \ll 1$, the empty protein that binds the first substrate molecule occurs mainly in the relatively non-reactive $T$-state ($T_4/R_4 = L$), but the tri-liganded protein that binds the last substrate molecule occurs mostly in the highly reactive $R$-state ($T_4S_3/R_4S_3 = Lc^3$). As a result the last substrate molecule will be bound with much greater affinity than the first. This increase in affinity as the protein becomes more saturated represents *positive cooperativity*.

Concerted transition creates positive cooperativity, but it is not necessary that all protomers in the protein form a single interacting group. For example, a tetramer may consist of two independent groups of concerted dimers, which is in fact the mechanism proposed for yeast pyruvate kinase by Johannes and Hess (1973):

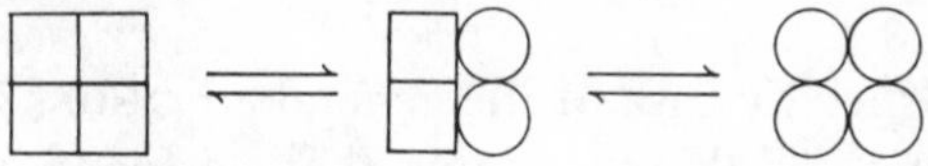

At the same time, the concerted transition also need not be restricted to just two conformation states. A three-state concerted transition model, which is in fact the mechanism proposed for hemoglobin by Minton and Imai (1974), would be as follows:

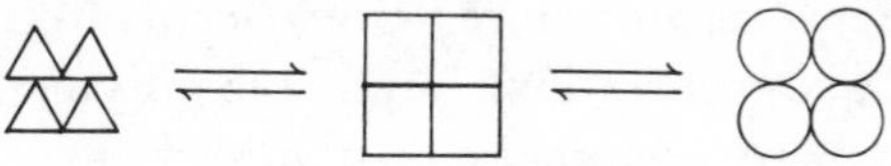

Clearly, the basic idea of a concerted transition is capable of extension in many directions.

### 8.1.4. Polymerization models

Since oligomeric proteins are composed of subunit protomers which are not covalently crosslinked to one another, under appropriate conditions they may undergo a reversible depolymerization-polymerization. Guidotti (1967), Frieden (1967), and Nichol, Jackson and Winzor (1967) proposed that such reversible polymerizations can furnish plausible models to explain allosteric homotropism. Such models are related to the concerted-transition models, insofar as polymerization states are also conformational states, and any polymerization necessarily involves the concerted participation of more than one protomer. Accordingly, Volkenstein and Goldstein (1966) refer to the allosterism in both instances as an *indirect interaction*, as opposed to a *direct interaction* as embodied in a neighbourly modulation.

A simple polymerization model for a tetramer may be constructed by allowing the enzyme to equilibrate between the tetrameric and dimeric states, the protomers within each state being independent as well as equivalent:

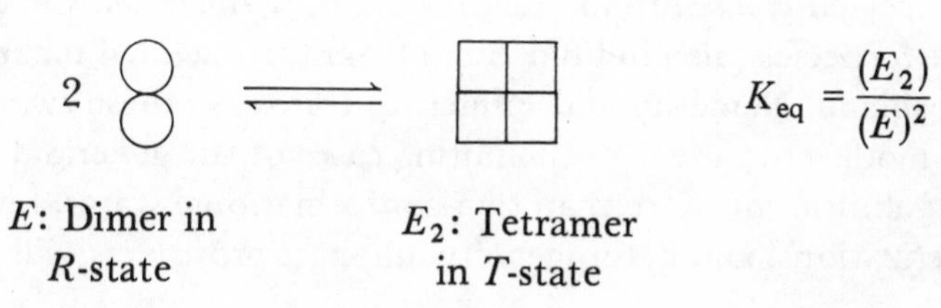

The saturation, or $y(\mathrm{S})$, function for this mechanism is

$$y = \frac{\bar{K}_R\mathrm{S}(1 + \bar{K}_R\mathrm{S}) + 2K_{\mathrm{eq}}(E)\bar{K}_T\mathrm{S}(1 + \bar{K}_T\mathrm{S})^3}{(1 + \bar{K}_R\mathrm{S})^2 + K_{\mathrm{eq}}(E)(1 + \bar{K}_T\mathrm{S})^4}$$

$$E = \frac{-(1 + \bar{K}_R\mathrm{S})^2 + \sqrt{(1 + \bar{K}_R\mathrm{S})^4 + 4(E)_0K_{\mathrm{eq}}(1 + \bar{K}_T\mathrm{S})^4}}{2K_{\mathrm{eq}}(1 + \bar{K}_T\mathrm{S})^4} \tag{8.2}$$

Equation (8.2), a specific case of the general equations given by Frieden, and Nichol, Jackson and Winzor, shows that $y$ varies with the total protein, $(E)_0$. In both direct-modulation or concerted-transition models, the homotropism takes place strictly within the oligomeric protein. Consequently, $y$ is unaffected by protein concentration as long as protein-bound substrate is deducted from total substrate to give the true concentration of free substrate for use in the $y(\mathrm{S})$ function. Only when polymerization (or other interprotein reaction) occurs in the mechanism does $y$ vary with protein concentration. Reversible polymerizations are known to be important in many oligomeric protein systems, e.g. hemoglobin (Guidotti, 1967) and glutamate dehydrogenase (Frieden and Coleman, 1967). Even where other interaction mechanisms may predominate at moderate protein concentrations, polymerization may become a significant factor at more extreme protein concentrations.

Interprotein reactions other than reversible polymerizations may also lead to nonhyperbolic behaviour. An example is the possible transfer of substrate between two protein molecules. Such interprotein transfers may become significant in concentrated protein solutions, and even more so with membrane-bound proteins, e.g. interprotein transfers of hydrogen and electron are likely to be important in the mitochondrial respiratory chain (Chance and Williams, 1955).

### 8.1.5. Hybrid models

Volkenstein and Goldstein (1966) pointed out that the different model mechanisms for homotropic interactions need not be mutually exclusive. Instead they may operate conjointly in hybrid combinations. For a tetrameric protein, the general two-state diagram of Kirschner (1971b) presented in Fig. 8.7 recognizes as many as 25 distinct protein species. The 10 species (in bold type) from the two end columns make up the concerted-transition model of Monod, Wyman and Changeux, whereas the 5 species (also in bold type) from the diagonal make up the model of Koshland, Némethy and Filmer. In this sense these two particular kinds of models represent only limiting cases of the general diagram. The possibilities of more than two conformational states, reversible polymerizations, and heterogeneity amongst protomers will broaden further

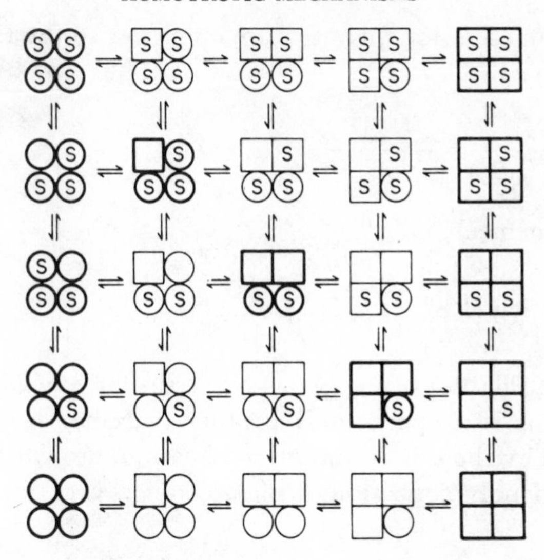

Fig. 8.7. General two-state model (From Kirschner, 1971b).

the possible range of hybrid models. In the light of this potential complexity, the depth of insight that could be achieved in such diverse experimental systems as phosphorylase *b* (Madsen and Shechosky, 1967); Buc and Buc, 1967), phosphofructokinase (Blangy, Buc and Monod, 1968), aspartokinase-homoserine dehydrogenase (Janin and Cohen, 1969), glyceraldehyde-phosphate dehydrogenase (Kirschner *et al.*, 1966), and pyruvate kinase (Johannes and Hess, 1973) is especially encouraging. The mechanistic analysis of allosteric homotropism is undoubtedly a challenging but not impossible task. Nonetheless, the potential complexity due to hybrid models must mean that, as Whitehead (1973) has emphasized, the boundary between different models may become rather indistinct in unfavourable experimental systems. Mechanistic analysis of such systems would have to proceed with considerable circumspection in order to avoid unsupported interpretations.

## 8.2. Differentiation Between Models

### 8.2.1. Maximal Hill slope

For a system of $n$ interacting protomers, cooperation reaches an extreme when all $n$ substrate ligands are bound to the enzyme in virtually a single step:

$$E_n \xrightleftharpoons{+nS} E_nS_n \qquad \text{Mechanism 8.I}$$

$$(E_nS_n) = U_n S^n (E_n)$$

The saturation function for this situation conforms to the Hill equation (1910):

$$y = \frac{(E_n S_n)}{(E_n) + (E_n S_n)} = \frac{U_n \mathbf{S}^n}{1 + U_n \mathbf{S}^n} \tag{8.3}$$

or, upon rearrangement,

$$\log \frac{y}{1 - y} = \log U_n \mathbf{S}^n = \log U_n + n \log \mathbf{S}. \tag{8.4}$$

Accordingly, the Hill plot of $\log (y/1 - y)$ versus $\log \mathbf{S}$ will be a straight line with slope = $n$. If cooperation is less than extreme, the saturation function is given by the $n$th-degree Adair equation, provided that no polymerization-depolymerization of the protomers is taking place (Section 7.3.2):

$$y = \frac{U_1 \mathbf{S} + 2U_1 U_2 \mathbf{S}^2 + \ldots + nU_1 U_2 \ldots U_n \mathbf{S}^n}{1 + U_1 \mathbf{S} + U_1 U_2 \mathbf{S}^2 + \ldots + U_1 U_2 \ldots U_n \mathbf{S}^n} \tag{8.5}$$

The Hill equation (8.3) is in effect a limiting form of the Adair equation (8.5). The latter when plotted in the form of the Hill plot will generate a maximal Hill slope $H_m$, which is also called the *Hill coefficient*, of smaller than $n$. Computational procedures for estimating $H_m$ have been described, either when the saturation binding (or velocity) exhibited by the protein could be defined (Wieker, Johannes and Hess, 1970; Glende, Reich and Wangermann, 1972; Atkins, 1973), or even when it could not be defined (Silanova, Livanova and Kurganov, 1969; Endrenyi, Fajszi and Kwong, 1975).

In the concerted two-state model of Monod, Wyman and Changeux, substrate binding always shifts the protein from the less active toward the more active conformational state. Consequently the *cooperativity is always positive*† over all substrate concentrations. The maximal Hill slope achieved depends on the values of $L$ and $c$. When $c$ is small, substrate binds non-exclusively to both the $T$- and $R$-states, and the protein is shifted from the less active $T$-state toward the more active $R$-state as the protein is progressively saturated by substrate. This cooperative shift cannot be important if $L$ is either very small or very large, since the protein will remain mostly in the $R$ or the $T$ state throughout the saturation process. Accordingly, $H_m$ approaches unity at extreme values of $L$,

† The cooperativity with respect to substrate binding is always positive, but the cooperativity with respect to rate of catalysed reaction might be negative, if the $R$-state, which is more active than the $T$-state in binding substrate, proves to be much less active in catalysis.

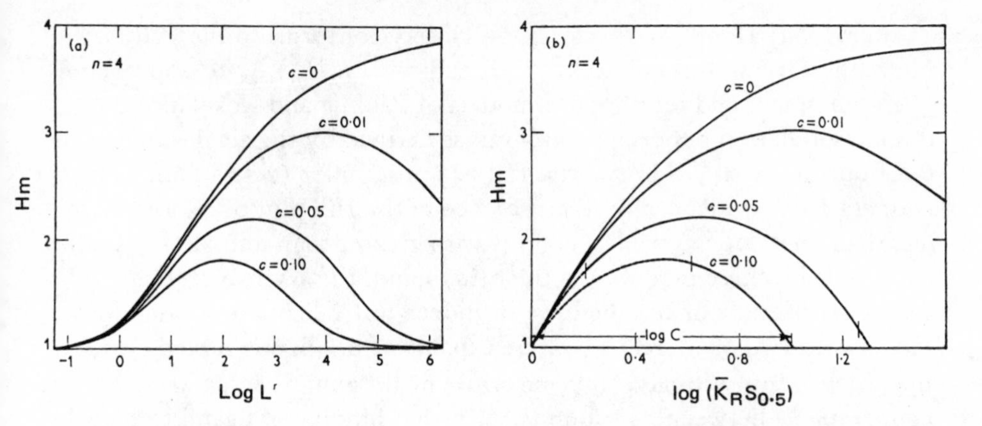


Fig. 8.8. Variation of $H_m$ with $L'$ and $c$ on the basis of the Monod–Wyman–Changeux model. In part (a), the *effective* allosteric constant, $L'$, is equal to $T_n/R_n$ in the absence of allosteric effectors, but varies as follows when allosteric effectors are added:

$$L' = \frac{T_n}{R_n} \cdot \frac{(1 + \bar{K}_I\mathrm{I})^i}{(1 + d\bar{K}_I\mathrm{I})^i} \cdot \frac{(1 + e\bar{K}_M\mathrm{M})^j}{(1 + \bar{K}_M\mathrm{M})^j}$$

where $i$ is the number of binding sites on the protein for $I$, and $j$ the number of sites for $M$. $\bar{K}_I$ is the affinity of $T_n$ for $I$, $d\bar{K}_I$ the affinity of $R_n$ for $I$, $\bar{K}_M$ the affinity of $R_n$ for $M$, and $e\bar{K}_M$ the affinity of $T_n$ for $M$. Both $d$ and $e$ are less than one. Thus the value of $L'$ can be varied continuously, which in turn causes $H_m$ and $\mathrm{S}_{0.5}$, the substrate concentration required to achieve half-saturation, to vary. Part (b) shows the correlated changes in $H_m$ and $\mathrm{S}_{0.5}$. All the curves in part (b) converge at the left-hand end: at this point $L'$ is so negligibly small that all the protein remains in the $R$-state throughout the saturation process, $y(\mathrm{S})$ becomes a hyperbola, $H_m = 1$, $\mathrm{S}_{0.5} = K_R = 1/\bar{K}_R$, and therefore $\log (\bar{K}_R \mathrm{S}_{0.5}) = 0$. (From Rubin and Changeux, 1966.)

but reaches a maximum, $(H_m)_{\max}$, at some intermediate value of $L$ (Fig. 8.8). The height of $(H_m)_{\max}$ depends only on $c$ (J. Wyman, cited in Blangy, Buc and Monod, 1968):

$$(H_m)_{\max} - 1 = (n - 1) \left[\frac{1 - \sqrt{c}}{1 + \sqrt{c}}\right]^2 \qquad (8.6)$$

$(H_m)_{\max}$ increases as $c$ decreases. However, when $c = 1$, the $R$ and $T$ states are equivalent in activity, and $(H_m)_{\max} = 1$. At the other extreme, when $c = 0$, the $T$ state is totally inactive and $(H_m)_{\max}$ is equal to $n$. In this case, $(H_m)_{\max}$ is achieved only as $L$ becomes infinitely large, i.e. as the inactive $T$-state is infinitely favoured over the active $R$-state in the allosteric equilibrium. As Whitehead (1970) has pointed out, the concerted model does not allow hybrid states of the type $T_iR_{n-i}$. The conversion of $T_n \rightarrow R_n$ occurs in a single step; when $L$ becomes infinitely large, this conversion is so rate-limiting that only two protein species are formed in significant amounts throughout the saturation process, namely

$T_n$ and $R_nS_n$. Therefore the situation closely conforms to the Hill Mechanism 8.I and its equation (8.4), and causes $(H_m)_{max}$ to approach $n$.

In the square and tetrahedron models of Pauling and Koshland, the direct modulation between neighbours is defined by the single coefficient $\theta$. Cooperativity is either *positive* ($\theta > 1$) or *negative* ($\theta < 1$). *Mixed* cooperativity, evidenced by the presence in the Hill plot of regions with less than unit slope as well as regions with greater than unit slope, is not explicable by these models. On the other hand, the Wyman rectangle, the generalized square or tetrahedron, or indeed any neighbour-modulation model with two or more $\theta$'s, will be capable of predicting mixed co-operativity. Interestingly, some enzymes might exhibit an *extreme negative cooperativity* between its subunits, such that binding of ligand to half the sites brings about complete inhibition of the remaining sites. This phenomenon, called *half-of-the-sites reactivity* has been investigated by Bernhard, Dunn, Luisi and Schack (1970), Harada and Wolfe (1968) and Stallcup and Koshland (1973). Hijazi and Laidler (1973a) have outlined some easy pitfalls in the identification of such a phenomenon from transient phase studies.

In general, Whitehead (1970) has reasoned that, unlike the concerted-transition models which are compatible with extreme cooperativity ($H_m = n$) under the conditions of $c = 0$ and $L' \to \infty$, direct-modulation models are incapable of predicting such extreme cooperativity. Even the addition of allosteric inhibitors to the system cannot overcome this limitation of the direct-modulation models. In this light, the observation that dimeric phosphorylase $b$ gave an $H_m$ close to 2 for phosphate, or for glucose-1-phosphate in the presence of ATP (Madsen, 1964; Buc, 1967) would favour models with a strong element of concerted transition rather than direct modulation. Conversely, if $c$ is apparently zero, and $H_m$ still falls short of $n$ when an allosteric inhibitor is added to force $L'$ to become infinitely large, then the concerted-transition model can be rejected categorically. In so doing, it should be remembered that $n$ is determined by the number of interacting protomers rather than by the total number of protomers in the oligomeric protein *per se.* For instance, if $H_m$ for a tetrameric protein fails to exceed 2, it is possible that extreme cooperation is not attained. However, it is also possible that extreme cooperation in fact has been attained, only that the four protomers exist as two independent dimeric units, and allosteric interactions operate only within each unit.

### 8.2.2. Action of effectors

Interactions between subunit protomers are often affected by effectors. Negative effectors are inhibitors, and positive effectors are activators. Still others can either inhibit or activate depending on experimental conditions.

Although the line between effectors and simple inhibitors, activators or modifiers is not sharply drawn, generally the effector designation is more suggestive of a regulatory role. Much more interesting than fine points of terminology is the fact that analysis of effector action by a number of different methods can provide valuable mechanistic information.

(a) *Method of Whitehead*

In any two-state mechanism, if an inhibitor $I$ binds *exclusively* to the $T$-state, and an activator $M$ binds *exclusively* to the $R$-state, the addition of $I$ will pull the protein toward the $T$-state, and the addition of $M$ will pull toward the $R$-state. Analysis by Whitehead (1970, 1973) indicates that the simultaneous additions of $I$ and $M$ will give rise to linear and convergent isoactivity plots of **M** versus **I**. Consider a protein-substrate system that exhibits a fractional saturation $y_1$ at effector concentrations of $\mathbf{I}_1$ and $\mathbf{M}_1$. When the activator is changed to $\mathbf{M}_2$, the inhibitor would have to be changed to $\mathbf{I}_2$ in order to maintain the saturation unchanged at $y_1$; and when activator is changed to $\mathbf{M}_3$, inhibitor would have to be changed to $\mathbf{I}_3$, etc. The points $(\mathbf{M}_1, \mathbf{I}_1)$, $(\mathbf{M}_2, \mathbf{I}_2)$, $(\mathbf{M}_3, \mathbf{I}_3)$ etc. define an isoactivity curve for $y_1$. Similar curves can be obtained for other values of $y$, all at the same fixed substrate level. Provided that $M$ and $I$ bind exclusively to the two opposite protein conformations in equal numbers, all these isoactivity curves will be linear and converge to a single point (Fig. 8.9). It is unimportant whether the transition of the different protomers between the two conformations proceeds in a concerted or sequential fashion. Also it applies to any number of binding sites, and to rate as well as binding measurements. However, if there were more than two conformational states, this predicted pattern of linearity and convergence would not hold.

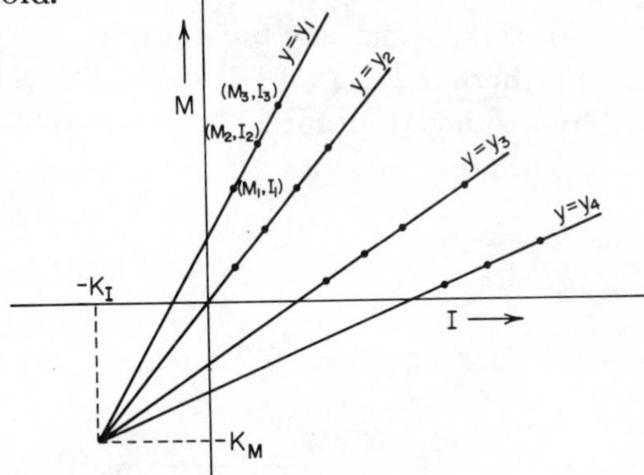


Fig. 8.9. Isoactivity curves for exclusive binding of $I$ and $M$ to two different conformations of the protein. (From Whitehead, 1970.)

(b) *Method of Buc and Buc*

In the concerted-transition model, if substrate $S$ and an activator $M$ both bind *exclusively* to the $R$-state, and an inhibitor $I$ binds *exclusively* to the

$T$-state, the experimental system will conform to linear plots derived by Buc and Buc (1967). Derivation can begin by noting the binomial distributions of the various $R$-state and $T$-state species, which are, provided that $S$, $M$ and $I$ bind independently each to the same number of $n$ sites,

$$\Sigma\ R\text{-state species} = R_n(1 + \bar{K}_R\mathbf{S})^n(1 + \bar{K}_M\mathbf{M})^n$$

$$\Sigma\ T\text{-state species} = LR_n(1 + \bar{K}_I\mathbf{I})^n$$

so that

$$\frac{\Sigma\ R\text{-state}}{\Sigma\ T\text{-state}} = \frac{R_n(1 + \bar{K}_R\mathbf{S})^n(1 + \bar{K}_M\mathbf{M})^n}{LR_n(1 + \bar{K}_I\mathbf{I})^n}.$$

Since only the $R$-state binds substrate, the fractional substrate saturation will be:

$$y = \frac{R\text{-state}}{T\text{-state} + R\text{-state}} \cdot \frac{\bar{K}_R\mathbf{S}}{(1 + \bar{K}_R\mathbf{S})}$$

$$= \frac{(1 + \bar{K}_R\mathbf{S})^n(1 + \bar{K}_M\mathbf{M})^n}{L(1 + \bar{K}_I\mathbf{I})^n + (1 + \bar{K}_R\mathbf{S})^n(1 + \bar{K}_M\mathbf{M})^n} \cdot \frac{\bar{K}_R\mathbf{S}}{(1 + \bar{K}_R\mathbf{S})} \tag{8.7}$$

The expressions containing $y$ and $\mathbf{S}$ can be lumped into a constant term $W$ if $y$ and $\mathbf{S}$ are held constant, and Eqn. (8.7) rearranged into

$$(1 + \bar{K}_M\mathbf{M})^n = W(1 + \bar{K}_I\mathbf{I})^n \tag{8.8}$$

or,

$$\bar{K}_M\mathbf{M} = (\sqrt[n]{W} - 1) + \sqrt[n]{W}\bar{K}_I\mathbf{I}. \tag{8.9}$$

Accordingly, a plot of $\mathbf{M}$ versus $\mathbf{I}$ at constant $y$ and $\mathbf{S}$ will be linear, thus confirming at least for this special case the linear isoactivity curves portrayed in Fig. 8.9. Furthermore, Eqn. (8.7) shows that a limiting value of $y$ may be attained by making $\mathbf{M}$ infinitely large, pushing all the protein into the $R$-state. This limiting value is

$$y_{\text{limit}} = \frac{\bar{K}_R\mathbf{S}}{1 + \bar{K}_R\mathbf{S}} \tag{8.10}$$

Placing this expression back into Eqn (8.7) yields

$$y = \frac{(1 + \bar{K}_R\mathbf{S})^n(1 + \bar{K}_M\mathbf{M})^n}{L(1 + \bar{K}_I\mathbf{I})^n + (1 + \bar{K}_R\mathbf{S})^n(1 + \bar{K}_M\mathbf{M})^n} \cdot y_{\text{limit}}$$

which in turn yields

$$y_{\text{limit}} - y = \frac{L(1 + \bar{K}_I\mathbf{I})^n \cdot y}{(1 + \bar{K}_R\mathbf{S})^n(1 + \bar{K}_M\mathbf{M})^n}$$

and finally

$$\sqrt[n]{\frac{y}{y_{\text{limit}} - y}} = \frac{(1 + \bar{K}_R\mathbf{S})(1 + \bar{K}_M\mathbf{M})}{\sqrt[n]{L} \cdot (1 + \bar{K}_I\mathbf{I})} \tag{8.11}$$

As Buc and Buc suggested, plotting the $n$th-root on the left-hand side against **M** will yield a diagnostic linear plot at any fixed levels of **S** and **I**. Moreover, a family of such plots obtained at different levels of **S** will converge on the **M**-axis at $\mathbf{M} = -1/\bar{K}_M$. The rate observations recorded in Fig. 8.10 for the dimeric phosphorylase $b$ closely conform to this prediction, and constitute evidence for the two-state concerted model, with phosphate (as substrate) and AMP (as allosteric activator) both *exclusively* binding to one of the two enzyme conformations. It might be noted from Eqn. (8.11) that plotting the left-hand side expression against **S** will also yield linear plots at different fixed levels of $M$ and $I$, converging on the **S**-axis at $\mathbf{S} = -1/\bar{K}_R$; and plotting the reciprocal of the expression against **I** will yield linear plots at different fixed levels of $S$ and $M$, converging on the **I**-axis at $\mathbf{I} = -1/\bar{K}_I$.

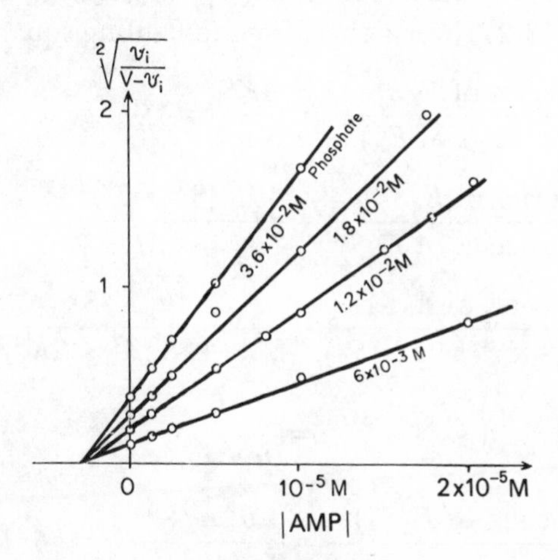


Fig. 8.10. Buc and Buc plot for phosphorylase $b$. For this system, $n = 2$. Since rate rather than binding measurements were used, $v_i$ replaced $y$ in the plot. (From Buc and Buc, 1967.)

(c) *Method of Kurganov*

This method avoids the need to estimate $y_{\text{limit}}$. Instead, it calls for the repeated determinations of $y$ for a *geometric* series of concentrations of the effector $I$, such that all successive concentrations are separated by the constant multiplier $h$:

$$\frac{\mathbf{I}_i}{\mathbf{I}_{i-1}} = \frac{\mathbf{I}_{i+1}}{\mathbf{I}_i} = h$$

This series of determinations provides a basis for calculating a series of $q_i$ by means of the following formula:

$$q_i = \frac{\log\left(\dfrac{1}{y_0 - y_{i-1}} - \dfrac{1}{y_0 - y_i}\right) - \log\left(\dfrac{1}{y_0 - y_i} - \dfrac{1}{y_0 - y_{i+1}}\right)}{\log h}$$

If the concerted model of Monod, Wyman and Changeux is correct, the curve of $q_i$ versus $\log \mathrm{I}_i$ is predicted to be independent of the concentrations of substrate and other effectors. Equations given by Kurganov (1973) may be employed to calculate the predicted curves for any chosen value of $h$.

### 8.2.3. Analysis of stability constants

The general Adair equation (8.5) is applicable to different homotropic models, but the expressions for the apparent stability (or Adair) constants, $U_1, U_2, \ldots, U_n$, vary from model to model. For instance, inspection of Fig. (8.2) reveals that, for the Pauling square model,

$$\begin{aligned}
U_1 &= \frac{(\text{all species of } E_4S)}{\mathrm{S} \times (\text{all species of } E_4)} = \frac{4\bar{K}\mathrm{S}}{\mathrm{S} \times 1} = 4\bar{K} \\
U_2 &= \frac{(\text{all species of } E_4S_2)}{\mathrm{S} \times (\text{all species of } E_4S)} = \frac{4\theta\bar{K}^2\mathrm{S}^2 + 2\bar{K}^2\mathrm{S}^2}{\mathrm{S} \times 4\bar{K}\mathrm{S}} = (\theta + \tfrac{1}{2})\bar{K} \\
U_3 &= \frac{(\text{all species of } E_4S_3)}{\mathrm{S} \times (\text{all species of } E_4S_2)} = \frac{4\theta^2\bar{K}^3\mathrm{S}^3}{\mathrm{S} \times (4\theta\bar{K}^2\mathrm{S}^2 + 2\bar{K}^2\mathrm{S}^2)} \\
&= \frac{\theta^2\bar{K}}{(\theta + \frac{1}{2})} \\
U_4 &= \frac{(\text{all species of } E_4S_4)}{\mathrm{S} \times (\text{all species of } E_4S_3)} = \frac{\theta^4\bar{K}^4\mathrm{S}^4}{\mathrm{S} \times 4\theta^2\bar{K}^3\mathrm{S}^3} = \tfrac{1}{4}\theta^2\bar{K}.
\end{aligned} \tag{8.12}$$

By relating the four $U_i$ to only the two parameters $\bar{K}$ and $\theta$, the Pauling square model in effect *allows only two of the four $U_i$ to be truly independent.* Once any two of the four are known, the other two can be calculated on the basis of Eqn (8.12). There must be satisfactory agreement between these calculated constants and their experimental values if the model is to be considered valid; otherwise the model can be rejected (Wong and Endrenyi, 1970). For example, the experimental $U$-constants for oxygen binding to hemoglobin have been determined by Roughton (1963) to be:

for sheep No. A38: $U_1 = 0.326$, $U_2 = 0.114$, $U_3 = 0.268$, $U_4 = 2.0$.

On the basis of Eqn (8.12) we can use $U_1$ to calculate $\bar{K}$:

$$0.326 = U_1 = 4\bar{K}$$
$$\bar{K} = 0.0815$$

Having derived $\bar{K}$ from $U_1$, $\theta$ can be derived from $U_2$:

$$0.114 = U_2 = (\theta + \tfrac{1}{2})\bar{K} = 0.0815\,(\theta + \tfrac{1}{2})$$
$$\theta = 0.899.$$

Knowing $\bar{K}$ and $\theta$, $U_3$ and $U_4$ can be calculated:

$$U_3 = \frac{\theta^2 \bar{K}}{(\theta + \frac{1}{2})} = 0.0471$$

$$U_4 = \tfrac{1}{4}\theta^2 \bar{K} = 0.0165.$$

These calculated values for $U_3$ and $U_4$ disagree with their experimental values of 0.268 and 2.0. Therefore the Pauling square model can be rejected.

The Pauling tetrahedron also could be rejected by this method. Since the Koshland square and tetrahedron are mathematically the same as those of Pauling, they likewise could be rejected. The Monod–Wyman–Changeux model gave a poor fit, but could not be rejected by this method because the errors in the experimental $U$-constants were compatible with large fluctuations in the calculated parameters for the model. The generalized tetrahedron, which has four model parameters ($\bar{K}$, $\theta_1$, $\theta_2$, and $\theta_3$), could not be tested by this means, because the number of experimental $U$-constants must exceed the number of model parameters before the test can proceed (Wong and Endrenyi, 1971; Roughton, Deland, Kernohan and Severinghaus, 1972).

Magar and Steiner (1971), and Szabo and Karplus (1972) have also proposed test methods related to this method. Cornish-Bowden and Koshland (1970a), on the other hand, proposed using predicted inequalities among the $U$-constants to test models, rather than using their predicted relationships for exact calculations. Their finding that the $y(O_2)$ curve for hemoglobin is compatible with the Pauling (Koshland) square is at variance with the rejection of this model by the above calculations. However, whereas Cornish-Bowden and Koshland relied on a fitting of the entire $y(O_2)$ curve to the model, the above calculations were based on Roughton's $U$-constants derived from an analysis of local segments of the $y(O_2)$ profile. As G. Weber (cited in Cornish-Bowden and Koshland, 1970b) has pointed out, a shortcoming of whole-curve fitting is that good fit between model and experiment over one region of the curve might obscure systematic deviation over another region in terms of total residual errors. One way to overcome this disadvantage of whole-curve fitting would be to examine not only the overall sum of squares, but also the spatial distribution, of the residual errors (see Section 11.6).

### 8.2.4. Analysis of rate constants

The interconversion between protein species carrying different numbers of substrate molecules is characterizable by the two apparent rate constants $u_i$ and $u_{-i}$, whose ratio defines the apparent stability constant $U_i$:

$$\cdots \rightleftharpoons \begin{bmatrix} \text{all protein species} \\ \text{with } i-1 \text{ molecules} \\ \text{of substrate} \end{bmatrix} \underset{u_{-i}}{\overset{u_i}{\rightleftharpoons}} \begin{bmatrix} \text{all protein species with} \\ i \text{ molecules of substrate} \end{bmatrix} \rightleftharpoons \cdots$$

$$U_i = u_i/u_{-i}$$

Just as the experimental $U$-constants provide a basis for testing model mechanisms, the $u$-constants when available will be equally useful toward this end.

Consider for instance the Monod–Wyman–Changeux model for a tetramer (Fig. 8.6). It contains two protein species devoid of substrate ($R_4$ and $T_4$), and also two protein species with one molecule of substrate ($R_4S$ and $T_4S$):

$$\begin{array}{ccccc} R_4 & \underset{k_{-r}}{\overset{4k_r}{\rightleftharpoons}} & R_4S & \rightleftharpoons & \cdots \\ \Updownarrow & & & & \\ T_4 & \underset{k_{-t}}{\overset{4k_t}{\rightleftharpoons}} & T_4S & \rightleftharpoons & \cdots \end{array}$$

Since the total interconversion between these two stoichiometric classes of protein species must equal the sum of the individual interconversions, the apparent rate constants can be related to the rate constants of the model as follows (the factor **S** appears on both sides of such *uk-relations*, and so can be omitted):

$$u_1(R_4 + T_4) = 4k_r(R_4) + 4k_t(T_4)$$

$$u_{-1}(R_4S + T_4S) = k_{-r}(R_4S) + k_{-t}(T_4S).$$

The relative concentrations for $R_4$, $T_4$, $R_4S$, $T_4S$, etc. are expressible in terms of the model parameters $\bar{K}_R$, $c$ and $L$ under equilibrium conditions (Fig. 8.6). Therefore the $uk$-relations for $u_1$ and $u_{-1}$, and similar $uk$-relations for the other six $u$-constants appropriate to a tetrameric case, are all expressible in these terms. Cancelling $\bar{K}_R$ and **S** when they appear on both sides of such relations, we have:

$$u_1(1 + L) \quad = 4k_r + 4k_tL$$

$$u_{-1}(1 + Lc) = k_{-r} + k_{-t}Lc$$

$$u_2(1 + Lc) \quad = 3k_r + 3k_tLc$$

$$u_{-2}(1 + Lc^2) = 2k_{-r} + 2k_{-t}Lc^2$$
$$u_3(1 + Lc^2) = 2k_r + 2k_tLc^2$$
$$u_{-3}(1 + Lc^3) = 3k_{-r} + 3k_{-t}Lc^3$$
$$u_4(1 + Lc^3) = k_r + k_tLc^3$$
$$u_{-4}(1 + Lc^4) = 4k_{-r} + 4k_{-t}Lc^4 \qquad (8.13)$$

and, in addition,

$$c = \frac{\bar{K}_T}{\bar{K}_R} = \frac{k_t k_{-r}}{k_{-t} k_r}$$

Since Eqn (8.13) relates the eight $u$-constants to the five model parameters $k_r$, $k_{-r}$, $k_t$, $k_{-t}$ and $L$, only five of the $u$-constants are truly independent. The other three can be calculated from Eqn (8.13). Once again, there must be satisfactory agreement between the calculated constants and their experimental values for the model to be valid. Otherwise the model is rejected.

Combining equilibrium and fast-kinetic measurements by means of computer curve fitting, Gibson (1970) has succeeded in deriving experimental estimates for all eight $u$-constants appropriate to the hemoglobin-oxygen system. In principle, any five of these $u$-estimates may be employed in Eqn (8.13) to calculate $k_r$, $k_{-r}$, $k_t$, $k_{-t}$, and $L$, which in turn may be employed to calculate the remaining three $u$-constants. Comparisons between such calculated $u$-constants and the $u$-estimates of Gibson become meaningful if equilibrium conditions are satisfied and the probable errors of the $u$-estimates are firmly defined.

The estimation of all the $u$-constants poses a formidable task for any experimental system, but even with incomplete sets of $u$-constants, $uk$-relations such as those shown in Eqn (8.13) impose useful restrictions on the plausible models. Restrictions of this nature may be combined with stability constants and other kinds of experimental measurements in order to arrive at a meaningful test of the models. The studies by Berger, Antonini, Brunori, Wyman, and Rossi-Fanelli (1967), Hopfield, Shulman and Ogawa (1971), and Gibson (1973) are all fine examples of such a multiple approach.

The relaxation method of Eigen and De Maeyer (1963) is one of the most important methods for estimating actual rate constants, and its application to the yeast glyceraldehyde-3-phosphate dehydrogenase system by Kirschner, Eigen, Bittman and Voigt (1966) has given an unambiguous description of the homotropic mechanism for $NAD^+$-enzyme interactions (see Section 10.2.3). However, in the case of more complex homotropic mechanisms, the spectrum of relaxation times quickly becomes complicated and unique interpretation is difficult to achieve. Nevertheless, Loudon and Koshland (1972) have given a systematic classification of relaxation-time curves that can offer some assistance toward the differentiation of mechanisms.

In the Monod–Wyman–Changeux model, the allosteric constant $L$ is an equilibrium constant. Under favourable conditions, it can be resolved into its constituent rate constants. Such resolution was accomplished by Barber and Bright (1968) for the threonine inhibition of aspartokinase I-homoserine dehydrogenase I. The spectrophotometric curves presented in Fig. 8.11 were obtained by the rapid mixing of two test solutions to initiate aspartatesemialdehyde reduction by NADPH. In curve (A), neither solution contained concentrated threonine, and reaction proceeded uninhibited. In curve (B) the enzyme was pre-equilibrated in the same solution with concentrated threonine, and inhibition was immediate. In

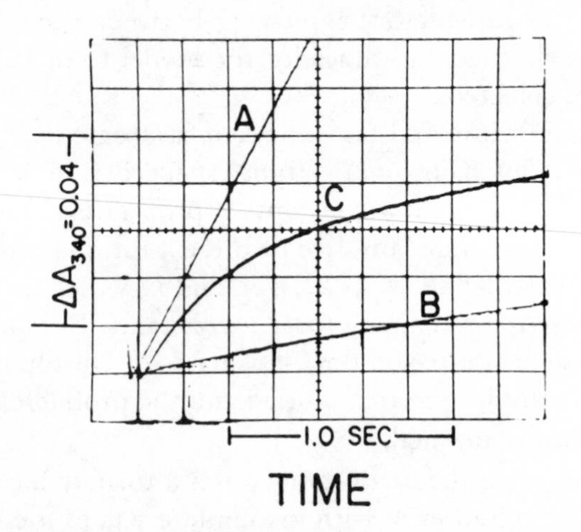


Fig. 8.11. Inhibition of homoserine dehydrogenase I of *E. coli* by threonine. Oscilloscope recorded the oxidation of NADPH. Curve A—enzyme not exposed to concentrated threonine; Curve B—enzyme pre-equilibrated with concentrated threonine; Curve C—concentrated threonine was added at zero time. (From Barber and Bright, 1968.)

curve (C), the enzyme and concentrated threonine came from different solutions; the inhibition process took about one second to become complete, and was first order with respect to the inhibitable enzyme. The rate constant for this onset of inhibition increased hyperbolically with threonine concentration, and it was concluded that the transition from active to inactive protein conformation (i.e. from $R$-state to $T$-state) consisted of one bimolecular step involving threonine plus at least one monomolecular step.

### 8.2.5. Linearized function for concerted transitions

The $y(\text{S})$ function for the Monod–Wyman–Changeux model is

$$y = \frac{\bar{K}_R \text{S}(1 + \bar{K}_R \text{S})^{n-1} + Lc\bar{K}_R \text{S}(1 + c\bar{K}_R \text{S})^{n-1}}{(1 + \bar{K}_R \text{S})^n + L(1 + c\bar{K}_R \text{S})^n}. \tag{8.14}$$

If constant $c$ is zero, so that substrate binds only to the $R$-state but not at all to the $T$-state,

$$y = \frac{\bar{K}_R \text{S}(1 + \bar{K}_R \text{S})^{n-1}}{L + (1 + \bar{K}_R \text{S})^n}. \tag{8.15}$$

Horn and Bornig (1969) suggested that Eqn (8.15) can be *linearized* through inversion, rearrangement, and finally taking logarithms, to give

$$\log\left(\frac{\bar{K}_R \text{S}}{y} - \bar{K}_R \text{S} - 1\right) = \log L - (n - 1)\log(1 + \bar{K}_R \text{S}). \tag{8.16}$$

Plotting $\log(\bar{K}_R \text{S}/y - \bar{K}_R \text{S} - 1)$ versus $\log(1 + \bar{K}_R \text{S})$ yields a straight line with a slope of $-(n - 1)$. Figure 8.12 shows the application by Yon (1972) of this graphical method to rate measurements on wheat-germ aspartate transcarbamylase.

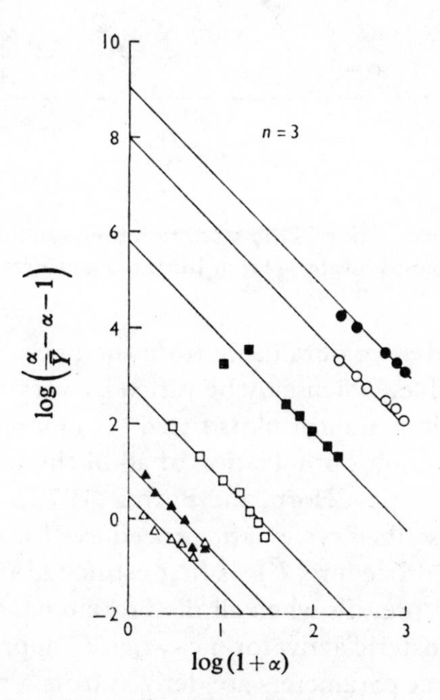


Fig. 8.12. Linearization of the saturation function for wheat-germ aspartate transcarbamoylase: $\alpha = \bar{K}$.(carbamoyl phosphate). (From Yon, 1972.)

The more general Eqn (8.14) also can be linearized to give (Endrenyi, Chan and Wong, 1971):

$$\log \frac{y(1+\bar{K}_R S) - \bar{K}_R S}{c\bar{K}_R S - y(1+c\bar{K}_R S)} = \log L + (n-1)\log \frac{1+c\bar{K}_R S}{1+\bar{K}_R S}. \quad (8.17)$$

Plotting the logarithm on the left-hand side against that on the right-hand side yields a straight line with an intercept equal to $\log L$, and a slope equal to $n - 1$. Figure 8.13 shows the application by Johannes and Hess (1973) of this graphical plot to rate measurements on yeast pyruvate kinase. In using this plot, if estimates of $\bar{K}_R$ and $c$ are already available,

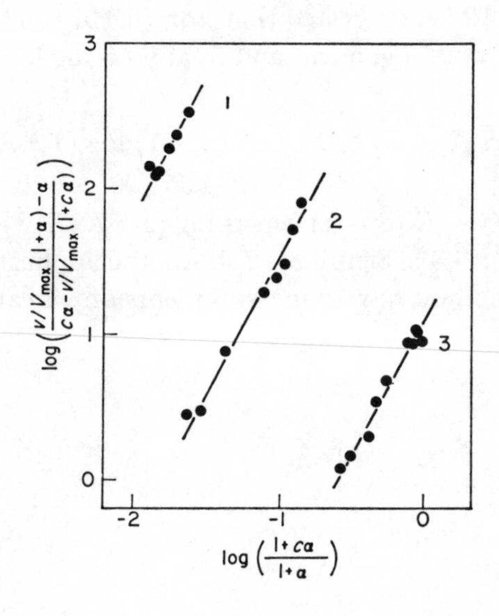


Fig. 8.13. Linearization of the saturation function for yeast pyruvate kinase: $\alpha = \bar{K}$.(phosphoenolpyruvate). (From Johannes and Hess, 1973.)

the parameters $L$ and $n$ are obtainable from the linear plot fitted for a single effective $L'$-value, which may be varied by varying activator or inhibitor concentrations. Linear plots fitted for more than one $L'$-values, on the other hand, permit optimization of all of the four parameters $\bar{K}_R$, $c$, $L$ and $n$. Cumme, Hoppe, Horn, and Bornig (1972), and Johannes and Hess (1973) have described systematic procedures for such optimization. In the Cumme *et al.* procedure, $\bar{K}_R$ is first estimated from the limiting saturation curve that prevails when all the protein has been pulled into the $R$ state by an allosteric activator, i.e. when $L'$ approaches zero. Knowing $\bar{K}_R$ the other three parameters are derived from a reiteration between the saturation curves for any two finite $L'$-values. The Johannes-Hess procedure employs a Fortran computer MONY-program.

### 8.2.6. Measurements of conformation

The conformational state of any protomer within an oligomeric protein must be defined in both ternary and quaternary structural terms. A transition between two conformational states may be largely ternary, largely quaternary, or both ternary and quaternary. Even though a purely quaternary change without some component of ternary change is difficult to conceive, oxygenation of hemoglobin is known to induce greater changes at the quaternary level (Perutz, 1969). Similarly, Markus, McClintock and Bussel (1971) have proposed that the allosteric properties of aspartate transcarbamylase might be explicable mainly in terms of quaternary changes. The finding by Chan and Mort (1973) that an incomplete enzyme missing half of its catalytic subunits exhibits hyperbolic rather than sigmoidal kinetics supports this view.

Changes in protein conformation, ternary together with quaternary ones, can be measured on the basis of changes in the chemical, optical, magnetic, or any other analysable property of the protein molecule. In order to establish the relevance of such measurements to the allosteric process, and furthermore to use the measurements to distinguish between different model mechanisms, quantitative correlations between conformation and activity need to be made. If the transition from an $R$-state to a $T$-state is strictly induced by substrate binding, the fraction of $R$-state will vary with the fractional substrate-saturation.

$$\frac{R}{R+T} = y. \tag{8.18}$$

An allosteric activator may replace the substrate in inducing the $R$-state. An allosteric inhibitor, on the other hand, if it binds only to the $T$-state, will decrease the substrate affinity without affecting Eqn (8.18). In the case of hemoglobin Ogawa and McConnell (1967) used a spin-labelled probe to measure conformational change. The measurements conformed closely to Eqn (8.18), and therefore supported a sequential, substrate-induced mode of change in protomer conformation. A concerted mode of change in protomer conformation would have deviated from Eqn (8.18), except that the expected deviations for this system might be quite small (Edelstein, 1971).

The relative abundance of the $T$ and $R$ states for the concerted-transition model is determined by

$$\frac{T}{R} = L \cdot \frac{(1 + \bar{K}_T \mathrm{S})^n}{(1 + \bar{K}_R \mathrm{S})^n} \tag{8.19}$$

The fraction of the protomers in the $R$-state is in general not proportional to the fractional saturation by substrate. The exact form of the dependence

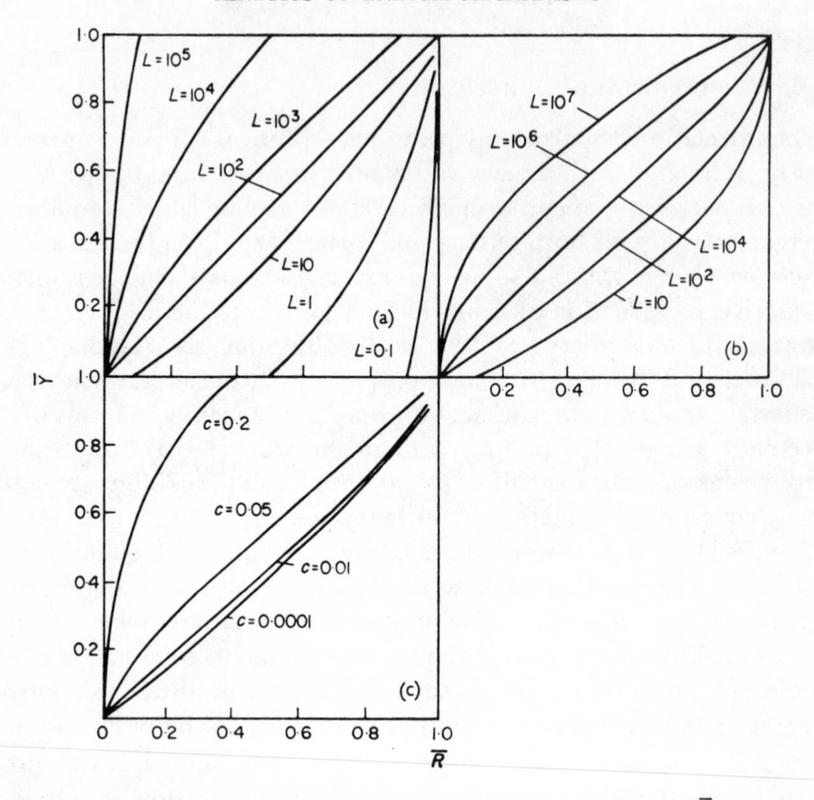


Fig. 8.14. Relationship between the fraction of protein in $R$-state ($\bar{R}$) and the fractional saturation by substrate ($\bar{Y}$), calculated on the basis of the Monod–Wyman–Changeux model: (a) $n = 2$, $c = 0.01$; (b) $n = 4$, $c = 0.01$; (c) $n = 4$, $L = 1000$. (From Will and Damaschun, 1973.)

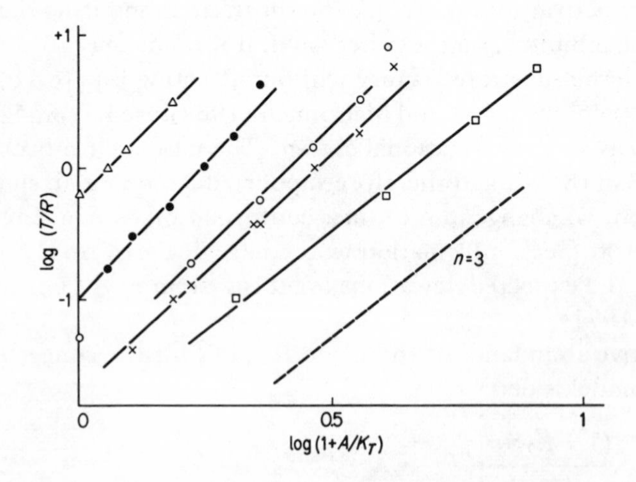


Fig. 8.15. Effects of threonine (A) on the $T/R$ ratio of aspartokinase I-homoserine dehydrogenase I: the different symbols represent measurements at different KCl or aspartate concentrations. (From Janin and Cohen, 1969.)

varies with the model parameters, and Fig. 8.14 shows some of the curves worked out by Will and Damaschun (1973). Allosteric inhibitors favour the less active $T$-state, and allosteric activators the $R$-state. In the case of aspartokinase I-homoserine dehydrogenase I, Janin and Cohen (1969) employed fluorimetry to resolve between $R$ and $T$, and found the effects of threonine and aspartate to conform to

$$\frac{T}{R} = L \cdot \frac{[1 + \bar{K}_{thr}(\text{threonine})]^3}{[1 + \bar{K}_{asp}(\text{aspartate})]^3}.$$

This behaviour (Fig. 8.15) is precisely that expected for a concerted transition of three enzyme protomers between $T$ and $R$, with the $T$-state exclusively binding threonine and the $R$-state exclusively binding aspartate.

## 8.3. *Escherichia coli* Phosphofructokinase

Phosphofructokinase from *Escherichia coli* was found by Atkinson and Walton (1965) to exhibit strong homotropism with respect to the substrate fructose-6-phosphate, and the analysis by Blangy, Buc and Monod (1968) led to the proposal of a two-state concerted mechanism for this tetrameric enzyme:

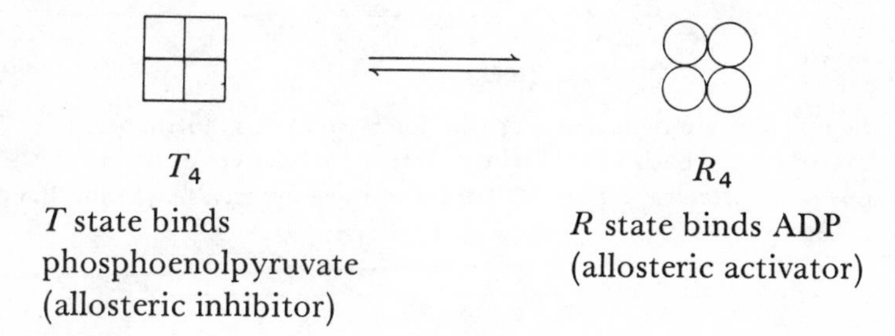


The measurements recorded in Fig. 8.16 were rate rather than binding measurements, but the fractional rate saturation by fructose-6-phosphate was taken to be a measure of the fractional binding saturation, i.e. quasi-equilibrium was assumed. The spectrum of saturation curves approached the limits of pure $R$-state at high concentrations of activator (ADP), and pure $T$-state at high concentrations of inhibitor (phosphoenolpyruvate). When the data were plotted according to the Hill plot, the maximal Hill slope $H_m$ approached unity in the regions near a pure $R$- or $T$-state, but passed through a maximal value of 3.8 between these limiting states (Fig. 8.17). On the basis of a concerted two-state model, the observed behaviour requires that fructose-6-phosphate binds non-exclusively to both the $R$ and $T$ states, differing by a factor of $5 \times 10^{-4}$ in their affinity for fructose-6-phosphate:

$$K_R = 1/\bar{K}_R = 1.25 \times 10^{-5}\text{M}$$

$$K_T = 1/\bar{K}_T = 2500 \times 10^{-5}\text{M}$$

$$c = \frac{\bar{K}_T}{\bar{K}_R} = 5 \times 10^{-4}$$

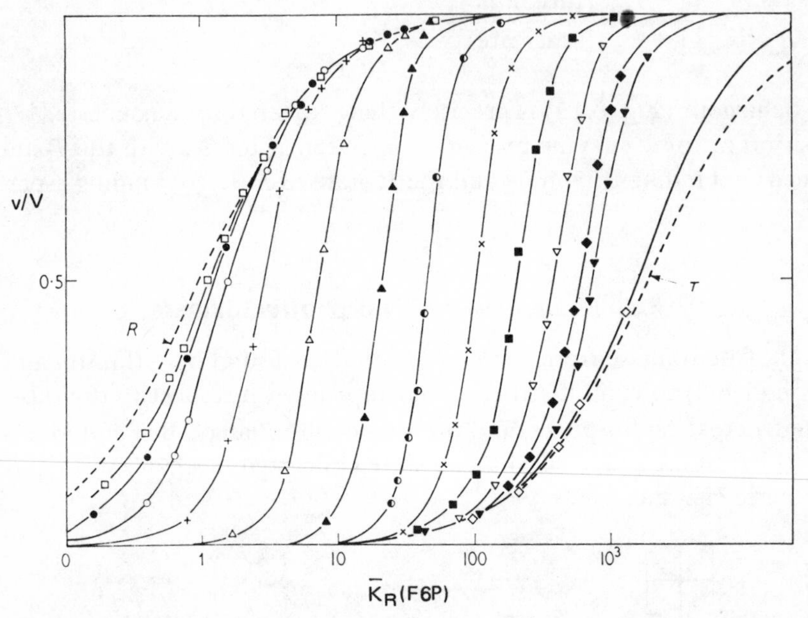


Fig. 8.16. The $v$(fructose-6-phosphate) profile for *E. coli* phosphofructokinase: $y$ was to be estimated on the basis of $v/V$. The different curves were obtained at different concentrations of an effector, either ADP or phosphoenolpyruvate. (From Blangy, Buc and Monod, 1968.)

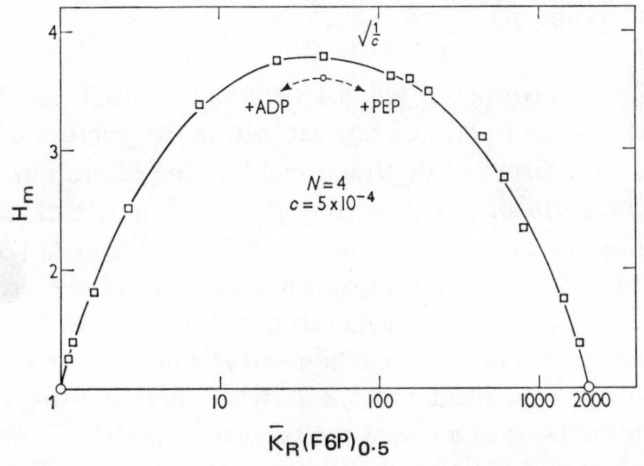


Fig. 8.17. Variation of maximal Hill slope for phosphofructokinase with half-saturation concentration of *F6P*. (From Blangy *et al.*, 1968.)

Substituting the values of $(H_m)_{max}$ and $c$ into Eqn (8.6) yields

$$3.8 - 1 = (n - 1)\left(\frac{1 - 2.24 \times 10^{-2}}{1 + 2.24 \times 10^{-2}}\right)^2$$

which gives a value of 4.06 for $n$. This value, so close to 4, lends strong support to the postulated model. When the data were plotted by the Endrenyi plot of Eqn (8.17), they indicated extensive linearity as well as a slope (= $n - 1$) very close to 3 (Fig. 8.18), again confirming a concerted transition of four protomers.

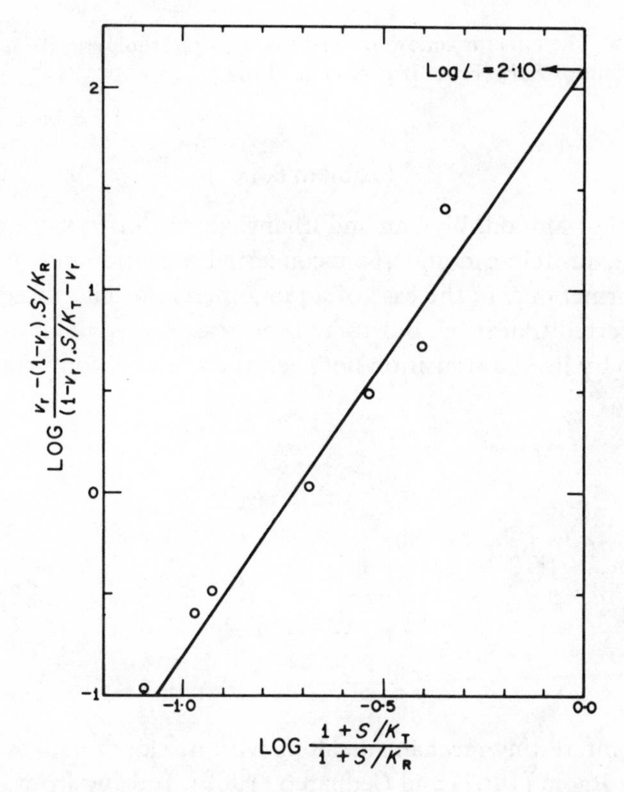


Fig. 8.18. Linearization of a saturation curve from Fig. 8.16. Slope = 2.91. (From Endrenyi *et al.*, 1971.)

Although fructose-6-phosphate binds nonexclusively to both the $R$ and $T$ states, the value of $c$ is so small that, for most concentrations of activator and inhibitor, enzyme activity largely will be due to the $R$ state. The plots of Buc and Buc thus become applicable. Both GDP as activator and phosphoenolpyruvate as inhibitor conformed to these plots, showing linearity as well as convergence on the horizontal axis (Fig. 8.19). These results, entirely in keeping with the model, suggest that GDP binds exclusively to the $R$ state, and phosphoenolpyruvate binds exclusively to the $T$ state.

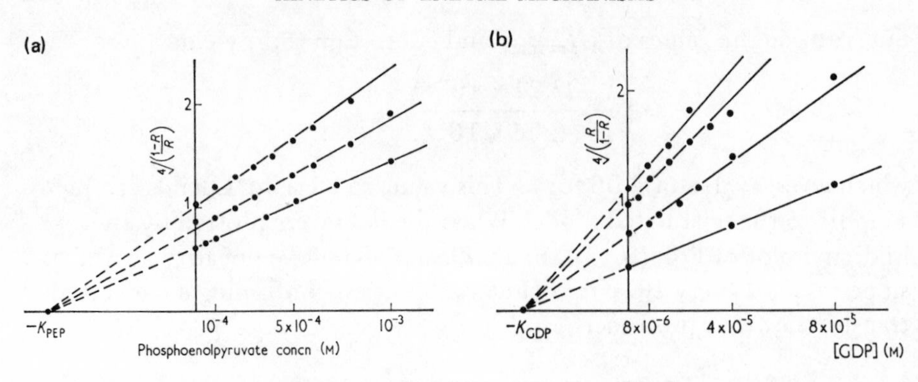


Fig. 8.19. The effector actions of GDP and phosphoenolpyruvate at different concentrations of *F6P*. (From Blangy *et al.*, 1968.)

## Problem 8.1

In the model of Monod, Wyman and Changeux, the different protomers in an oligomeric protein can undergo a concerted transition between the $R$ and $T$ conformations. In the case of a monomeric protein, there cannot be any concerted transition, but there is no reason why the protein cannot undergo by itself a transition between the $R$ and $T$ conformations, e.g. as follows:

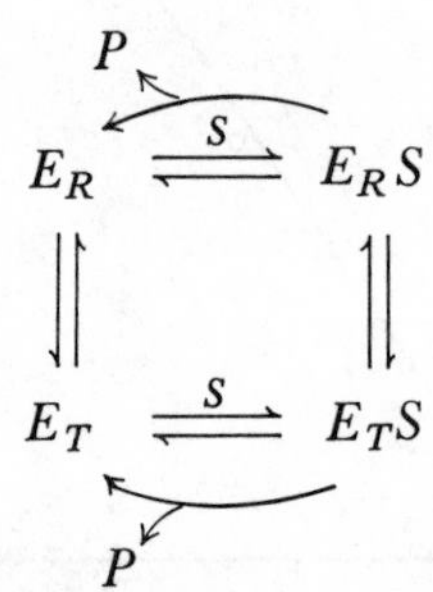


This highly interesting mechanism, along with its close relatives, has been analysed by Rabin (1967) and Cennamo (1969). Judging from its topology, is it capable of predicting nonhyperbolic kinetics? or nonhyperbolic substrate binding? Also, what happens if interconversion is allowed between $E_R$ and $E_T$, but disallowed between $E_RS$ and $E_TS$?

# 9 Isotopic Exchanges

Steady-state, transient-state and relaxation kinetic measurements all require the occurrence of net chemical changes in the reaction system, but isotopic exchanges between reactant molecules can be measured either in the presence or in the absence of such net changes. Substitution of an atom in a substrate molecule by its isotope does not affect the reactivity of the substrate in the case of most atoms, but may do so in the case of light atoms such as hydrogen. The exploitation of reactivity differences between the hydrogen, deuterium, and tritium forms of substrates has been referred to earlier in Section 6.3.2. The present chapter examines those isotopic methods which are not dependent on isotopic rate differences.

## 9.1. Exchanges in Partial Systems

### 9.1.1. Sucrose phosphorylase

Sucrose phosphorylase from *Pseudomonas saccharophila* catalyses a reversible transfer of glucose between phosphate and fructose:

$$\underset{(G\text{–}P)}{\text{glucose-1-phosphate}} + \underset{(F)}{\text{fructose}} \rightleftharpoons \underset{(P)}{\text{phosphate}} + \underset{(G\text{–}F)}{\text{sucrose}}$$

Doudoroff, Barker and Hassid (1947) in their classical study discovered that, when glucose-1-phosphate and $^{32}P$-labelled inorganic phosphate were incubated in the presence of this enzyme, the radioactive phosphorus was incorporated into glucose-1-phosphate even though no net transfer of glucose could occur in this partial system, lacking fructose as well as sucrose. To explain this observation, these workers postulated that the enzyme acted as intermediate glucose acceptor in the normal catalysed reaction, accounting for the $^{32}P$-exchange between glucose-1-phosphate and inorganic phosphate:

$$\text{glucose-phosphate}^* + \text{enzyme} \rightleftharpoons \text{glucose-enzyme} + \text{phosphate}^*$$

where * denotes *isotopic labelling.* More recently, Silverstein, Vogt, Reed and Abeles (1967) succeeded in isolating this long-postulated glucose-enzyme intermediate. It should be pointed out that the isolation of the substituted enzyme in such cases, although of great importance, does not in itself prove that the substituted enzyme forms an essential intermediate in the normal transfer between donor and acceptor substrates, nor that it is kinetically significant. However, in the case of sucrose phosphorylase, independent kinetic studies by Walsh (1959) and by Silverstein *et al.* (1967) established that the normal catalysed reaction indeed proceeds via the glucosyl-enzyme according to the following basic mechanism:

$$E \underset{}{\overset{G-P}{\rightleftharpoons}} E/G\text{–}P \underset{P}{\rightleftharpoons} E\text{–}G \overset{F}{\rightleftharpoons} E/G\text{–}F \underset{G-F}{\rightleftharpoons} E$$

Mechanism 9.I

This is a substituted-enzyme, or ping-pong, type of mechanism; it predicts not only a phosphate exchange between $G\text{–}P$ and $P$ in the absence of $F$ and $G\text{–}F$, but also a fructose exchange between $F$ and $G\text{–}F$ in the absence of $G\text{–}P$ and $P$. Sucrose phosphorylase also catalysed the latter exchange (Wolochow, Putman, Doudoroff, Hassid and Barker, 1949).

A large number of two-substrate enzymes are now known to catalyse partial exchanges similar to those catalysed by sucrose phosphorylase.

### 9.1.2. Three-substrate reactions

The aminoacyl-tRNA synthetases catalyse a representative three-substrate reaction:

amino acid + adenosine triphosphate + tRNA
($S$) ($X\text{–}Y$) ($T$)

= aminoacyl-tRNA + adenosine monophosphate + pyrophosphate
($S\text{–}T$) ($X$) ($Y$)

Mechanisms 9.II-9.V are four simple mechanisms compatible with this reaction:

$$E \overset{S}{\rightleftharpoons} E/S \overset{X-Y}{\rightleftharpoons} E/S/X\text{–}Y \underset{Y}{\rightleftharpoons} E/S\text{–}X \overset{T}{\rightleftharpoons}$$

$$E/S\text{–}X/T \underset{X}{\rightleftharpoons} E/S\text{–}T \underset{S-T}{\rightleftharpoons} E$$

Mechanism 9.II

$$E \overset{X-Y}{\rightleftharpoons} E/X\text{–}Y \underset{Y}{\rightleftharpoons} E\text{–}X \overset{S}{\rightleftharpoons} E\text{–}X/S \overset{T}{\rightleftharpoons}$$

$$E\text{–}X/S/T \underset{X}{\rightleftharpoons} E/S\text{–}T \underset{S-T}{\rightleftharpoons} E$$

Mechanism 9.III

$$E \overset{X-Y}{\rightleftharpoons} E/X-Y \underset{Y}{\rightleftharpoons} E-X \overset{S}{\rightleftharpoons} E-X/S \underset{X}{\rightleftharpoons}$$

$$E-S \overset{T}{\rightleftharpoons} E-S/T \underset{S-T}{\rightleftharpoons} E$$

Mechanism 9.IV

$$E \overset{S}{\rightleftharpoons} E/S \overset{T}{\rightleftharpoons} E/S/T \overset{X-Y}{\rightleftharpoons} E/S/T/X-Y \rightleftharpoons$$

$$E/S-T/X/Y \underset{S-T}{\rightleftharpoons} E/X/Y \underset{Y}{\rightleftharpoons}$$

$$E/X \underset{X}{\rightleftharpoons} E$$

Mechanism 9.V

In Mechanism 9.V, the three substrates $S$, $T$ and $X-Y$ all have to be assembled into a central complex before catalysis occurs, and no isotopic exchange of any kind will occur in a partial system lacking any of the substrates. In Mechanism 9.IV, an $X-Y^* \rightleftharpoons Y^*$ exchange can take place in the absence of the other substrates and products by way of

$$E \overset{X-Y^*}{\rightleftharpoons} E/X-Y^* \underset{Y^*}{\rightleftharpoons} E-X$$

On the other hand, an $X^*-Y \rightleftharpoons X$ exchange will require the presence of $S$ and $Y$:

$$E \overset{X^*-Y}{\rightleftharpoons} E/X^*-Y \underset{Y}{\rightleftharpoons} E-X^* \overset{S}{\rightleftharpoons} E-X^*/S \underset{X^*}{\rightleftharpoons} E-S$$

The different requirements for the various exchanges predicted by these and similar mechanisms are all evident from the mechanism topologies, and provide a useful basis for differentating between three-substrate mechanisms in general (Dixon and Webb, 1958).

In the specific example of the aminoacyl-tRNA synthetases, Hoagland, Keller and Zamecnik (1956) and DeMoss and Novelli (1956) discovered that an $X-Y^*$ (adensoine triphosphate) $\rightleftharpoons Y^*$ (pyrophosphate) exchange could take place in a partial system containing $S$ (amino acid) but lacking $T$ (tRNA). This observation pointed to the formation of an enzyme-bound $S-X$ (aminoacyl-adenosine monophosphate), e.g. as indicated in Mechanism 9.II:

$$E \overset{S}{\rightleftharpoons} E/S \overset{X-Y^*}{\rightleftharpoons} E/S/X-Y^* \underset{Y^*}{\rightleftharpoons} E/S-X$$

This exchange would depend only on the formation of $E/S-X$, and not on the exact sequence in which $S$ and $X-Y$ add to the enzyme. If $X-Y$ should add before $S$ instead, the exchange pathway would simply be:

$$E \overset{X-Y^*}{\rightleftharpoons} E/X-Y^* \overset{S}{\rightleftharpoons} E/S/X-Y^* \underset{Y^*}{\rightleftharpoons} E/S-X$$

## 9.2. Exchanges at Equilibrium

There are no net chemical changes either in a partial reaction system lacking some of the reactants, or in a reaction system that has reached thermodynamic equilibrium. In both cases isotopic measurements permit an estimation of the different exchange fluxes in the system. The theory of equilibrium exchange fluxes began with the fundamental work of Boyer (1959), and was extended by Alberty, Bloomfield, Peller and King (1962), Morales, Horovitz and Botts (1962), and Boyer and Silverstein (1963). Equilibrium flux equations may be derived by a number of methods. Amonst these, that devised by Yagil and Hoberman (1969) is particularly elegant in its simplicity.

### 9.2.1. Flux equations

In the method of Yagil and Hoberman (1969), the rate of an isotopic flux through $n$ parallel reaction steps is simply the sum of the $n$ individual rates:

$$f = \sum_{1}^{n} r_i \tag{9.1}$$

$f$ = flux rate
$r_i$ = rate of each reaction step.

The rate of an isotopic flux through $n$ consecutive reaction steps, on the other hand, is given by the sum of the $n$ reciprocals:

$$\frac{1}{f} = \sum_{1}^{n} \frac{1}{r_i}. \tag{9.2}$$

Equation (9.1) resembles, not by coincidence, the equation for the conductance of $n$ resistors arranged in parallel, and Eqn (9.2) that for $n$ resistors arranged in series. This formal analogy between chemical and electrical fluxes not only reveals the physical meaning of Eqns (9.1) and (9.2), but also enables the derivation of isotopic flux equations for even complicated mechanisms via the construction of analogous electrical circuits (Flossdorf and Kula, 1972).

Use of the Yagil-Hoberman method may be illustrated by considering the model Mechanism 9.VI for a two-substrate group-transfer of the type $G{-}X + Y = X + G{-}Y$:

$$E \underset{k_{-1}}{\overset{k_1(GX)}{\rightleftharpoons}} E/G{-}X \underset{k_{-2}}{\overset{k_2(Y)}{\rightleftharpoons}} E/G{-}X/Y \underset{k_{-0}}{\overset{k_0}{\rightleftharpoons}} E/X/G{-}Y \underset{k_{-3}(GY)}{\overset{k_3}{\rightleftharpoons}}$$

$$E/X \underset{k_{-4}(X)}{\overset{k_4}{\rightleftharpoons}} E \qquad \text{Mechanism 9.VI}$$

Depending on which of the three moieties $G$, $Y$ and $X$ is isotopically labelled, there are altogether three pairs of measurable exchanges:

$$GX^* \rightleftharpoons X^*$$
$$G^*X \rightleftharpoons G^*Y$$
$$Y^* \rightleftharpoons GY^*$$

The principle of microscopic reversibility requires that, at equilibrium, *an exchange flux in one direction must equal the flux in the opposite direction*, so that only three flux equations are needed to describe the six fluxes. The three pathways to be considered are:

*Flux from GX to X:*

$$E \xrightarrow{k_1(GX)} E/GX \xrightarrow{k_2(Y)} E/GX/Y \xrightarrow{k_0} E/X/GY \xrightarrow{k_3} E/X \xrightarrow[X]{k_4} E$$

*Flux from G–X to G–Y:*

$$E \xrightarrow{k_1(GX)} E/GX \xrightarrow{k_2(Y)} E/GX/Y \xrightarrow{k_0} E/X/GY \xrightarrow[GY]{k_3} E/X$$

*Flux from Y to G–Y:*

$$E/GX \xrightarrow{k_2(Y)} E/GX/Y \xrightarrow{k_0} E/X/GY \xrightarrow[GY]{k_3} E/X$$

Since each of the above fluxes proceeds via a single chain of reaction steps, application of Eqn (9.2) defines the flux rates in terms of the reciprocals, or $\omega$s, or the five reaction steps:

$$\frac{1}{f_{GX\rightarrow X}} = \omega_1 + \omega_2 + \omega_0 + \omega_3 + \omega_4$$

$$\frac{1}{f_{GX\rightarrow GY}} = \omega_1 + \omega_2 + \omega_0 + \omega_3 \tag{9.3}$$

$$\frac{1}{f_{Y\rightarrow GY}} = \omega_2 + \omega_0 + \omega_3$$

At equilibrium, the relative concentrations of the five enzyme species are given by:

$$(E):(E/GX):(E/GX/Y):(E/X):(E/X/GY) = 1 + (GX)' + (GX)'(Y)' + (X)' + (X)'(GY)' \tag{9.4}$$

In Eqn (9.4), $(GX)'$ stands for $(GX)/K_{GX}$, where $K_{GX}$ is the dissociation constant of $GX$ from its enzymic complex; similarly, $(Y)'$ stands for $(Y)/K_Y$, $(X)'$ stands for $(X)/K_X$, and $(GY)'$ stands for $(GY)/K_{GY}$. Accordingly the five reciprocals from Eqn (9.3), denoted as $\omega$s for the sake of convenience, can be expressed in terms of the four reactants and total enzyme:

$$\begin{aligned}
\omega_1 &= \frac{1}{k_1(GX)(E)} = \frac{1+(GX)'+(GX)'(Y)'+(X)'+(X)'(GY)'}{k_1(GX)(E)_0} \\
\omega_2 &= \frac{1}{k_2(Y)(E/GX)} = \frac{1+(GX)'+(GX)'(Y)'+(X)'+(X)'(GY)'}{k_2(Y)(GX)'(E)_0} \\
\omega_0 &= \frac{1}{k_0(E/GX/Y)} = \frac{1+(GX)'+(GX)'(Y)'+(X)'+(X)'(GY)'}{k_0(GX)'(Y)'(E)_0} \qquad (9.5) \\
\omega_3 &= \frac{1}{k_3(E/X/GY)} = \frac{1+(GX)'+(GX)'(Y)'+(X)'+(X)'(GY)'}{k_3(X)'(GY)'(E)_0} \\
\omega_4 &= \frac{1}{k_4(E/X)} = \frac{1+(GX)'+(GX)'(Y)'+(X)'+(X)'(GY)'}{k_4(X)'(E)_0}
\end{aligned}$$

At equilibrium, the four reactant concentrations are related through the equilibrium constant $K_{eq}$, and it is impossible to vary just one of the reactants without upsetting the equilibrium. Instead, one (or both) substrate-product pair has to be varied together according to their equilibrium proportions. The effects of saturating the system by a substrate-product pair are revealed by Eqn (9.5) to be:

$$\begin{aligned}
&\text{if } (GX), (X) = \infty: && \omega_1, \omega_2, \omega_0, \omega_3, \omega_4 = \text{finite} \\
&\text{if } (GX), (GY) = \infty: && \omega_1, \omega_2, \omega_0, \omega_3 = \text{finite} \\
& && \omega_4 = \infty \qquad (9.6) \\
&\text{if } (Y), (X) = \infty: && \omega_2, \omega_0, \omega_3, \omega_4 = \text{finite} \\
& && \omega_1 = \infty \\
&\text{if } (Y), (GY) = \infty: && \omega_2, \omega_0, \omega_3 = \text{finite} \\
& && \omega_1, \omega_4 = \infty
\end{aligned}$$

Consequently, if $(GX)$ and $(X)$ become vary large in Eqn (9.3) all three of the flux rates would be finite. If $(GX)$ and $(GY)$ become very large, $\omega_4$ tends to infinity; $f_{GX \to X}$ would become zero but $f_{GX \to GY}$ and $f_{Y \to GY}$ would stay finite. If $(Y)$ and $(X)$ become very large, $\omega_1$ tends to infinity; $f_{GX \to X}$ and $f_{GX \to GY}$ would be prohibited, but $f_{Y \to GY}$ would say finite. If $(Y)$ and $(GY)$ become very large, both $\omega_1$ and $\omega_4$ tend to infinity; again $f_{GX \to X}$ and $f_{GX \to GY}$ would be prohibited, but $f_{Y \to GY}$ would stay finite.

Taken together, Eqns (9.3) and (9.5) define the isotopic flux rates in terms of the reactant concentrations, and predict which of the fluxes will be prohibited when any substrate-product pair is raised to saturation. The predicted prohibition of fluxes varies from mechanism to mechanism, and provides an important tool for distinguishing between the mechanisms not only for multisubstrate systems (Boyer, 1959), but also for modifier systems (Wedler and Boyer, 1973). These equations also provide an important basis for the evaluation of some of the rate constants under limiting concentration conditions (Boyer and Silverstein, 1963; Yagil and Hoberman, 1969).

### 9.2.2. Prohibition patterns

The derivation of flux-prohibition patterns from flux equations such as (9.3) requires the handling of profuse algebra in the case of branching mechanisms. A much simpler way to deduce these patterns can be devised on the basis of *topological reasoning.* The procedure is as follows (Wong and Hanes, 1964):

(a) For example, we shall try to determine for Mechanisms 9.VI–9.XV whether or not high concentrations of $Y$ and $GY$, when varied together at a fixed ratio $q$, will prohibit the $GX \rightleftharpoons X$ exchange fluxes at equilibrium. In any enzyme mechanism, the distribution of total enzyme $(E)_0$ amongst its constituent enzyme species is defined, at equilibrium no less than at steady-state, by the *schematic rule* of King and Altman (Section 2.2). For any enzyme species $E_i$,

$$\frac{(E_i)}{(E)_0} = \frac{\Delta_i}{\Delta_1 + \Delta_2 + \cdots + \Delta_i + \cdots + \Delta_n}. \tag{9.7}$$

The determinant $\Delta_i$ includes all the non-cyclic arrow patterns which converge at $E_i$ (every pattern contains one arrow leaving from each enzyme species other than $E_i$). The form of $\Delta_i$ as a polynomial in $(Y)$ and $(GY)$ is readily apparent from the mechanism topology. Thus inspection of Mechanism 9.VI indicates that, if $(GY)$ is substituted by $(Y)/q$,

$$\begin{aligned} \frac{(E)}{(E)_0}, \frac{(E/GX)}{(E)_0}, \frac{(E/X)}{(E)_0} &= \frac{m_1(Y) + m_0}{d_2(Y)^2 + d_1(Y) + d_0} \\ \frac{(E/GX/Y)}{(E)_0}, \frac{(E/X/GY)}{(E)_0} &= \frac{m_2(Y)^2 + m_1(Y)}{d_2(Y)^2 + d_1(Y) + d_0}. \end{aligned} \tag{9.8}$$

Consequently, if $(Y)$ and $(GY)$ are made very large, $(E)$, $(E/GX)$ and $(EX)$ will be reduced to vanishingly low concentrations, but $(E/GX/Y)$ and $(E/X/GY)$ will persist in finite quantities. This situation is depicted in Fig. 9.1 for this and the other model mechanisms.

$$E \underset{k_{-1}}{\overset{k_1(GX)}{\rightleftharpoons}} E/GX \underset{k_{-2}}{\overset{k_2(Y)}{\rightleftharpoons}} \boxed{E/GX/Y} \underset{k_{-0}}{\overset{k_0}{\rightleftharpoons}} \boxed{E/GY/X} \underset{k_{-3}(GY)}{\overset{k_3}{\rightleftharpoons}}$$

$$E/X \underset{k_{-4}(X)}{\overset{k_4}{\rightleftharpoons}} E \qquad \text{Mechanism 9.VI}$$

$$E \overset{GX}{\rightleftharpoons} E/GX \overset{Y}{\rightleftharpoons} \boxed{E/GX/Y} \rightleftharpoons \boxed{E/GY/X} \underset{X}{\rightleftharpoons}$$

$$\boxed{E/GY} \underset{GY}{\rightleftharpoons} E \qquad \text{Mechanism 9.VII}$$

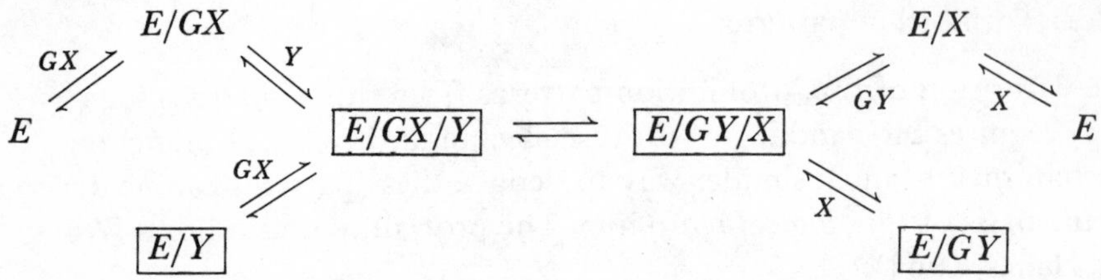

$$\begin{array}{ccccccc} & E/GX & & & & E/X & \\ {\scriptstyle GX}\nearrow\!\!\swarrow & & \searrow\!\!\nwarrow{\scriptstyle Y} & & & \swarrow\!\!\nearrow{\scriptstyle GY} & \searrow\!\!\nwarrow{\scriptstyle X} \\ E & & & \boxed{E/GX/Y} \rightleftharpoons \boxed{E/GY/X} & & & E \\ & & {\scriptstyle GX}\nearrow\!\!\swarrow & & \searrow\!\!\nwarrow{\scriptstyle X} & & \\ & \boxed{E/Y} & & & & \boxed{E/GY} & \end{array}$$

Mechanism 9.VIII

$$E \overset{Y}{\rightleftharpoons} \boxed{E/Y} \overset{GX}{\rightleftharpoons} \boxed{E/GX/Y} \rightleftharpoons \boxed{E/GY/X} \underset{X}{\rightleftharpoons}$$

$$\boxed{E/GY} \underset{GY}{\rightleftharpoons} E \qquad \text{Mechanism 9.IX}$$

$$E \overset{Y}{\rightleftharpoons} \boxed{E/Y} \overset{GX}{\rightleftharpoons} \boxed{E/GX/Y} \rightleftharpoons \boxed{E/GY/X} \underset{GY}{\rightleftharpoons}$$

$$E/X \underset{X}{\rightleftharpoons} E \qquad \text{Mechanism 9.X}$$

$$E \overset{GX}{\rightleftharpoons} E/GX \underset{X}{\rightleftharpoons} E\text{–}G \overset{Y}{\rightleftharpoons} \boxed{E/GY} \underset{GY}{\rightleftharpoons} E$$

Mechanism 9.XI

$$E \overset{GX}{\rightleftharpoons} E/GX \underset{X}{\overset{Y}{\rightleftharpoons}} \boxed{E/GY} \underset{GY}{\rightleftharpoons} E \qquad \text{Mechanism 9.XII}$$

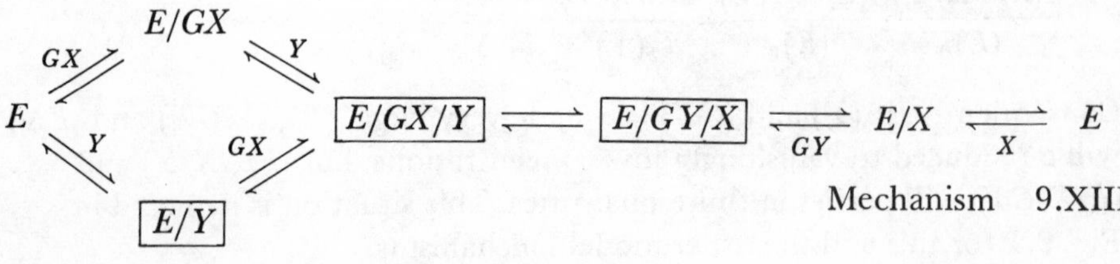

$$\begin{array}{cccc} & E/GX & & \\ {\scriptstyle GX}\nearrow\!\!\swarrow & & \searrow\!\!\nwarrow{\scriptstyle Y} & \\ E & & & \boxed{E/GX/Y} \rightleftharpoons \boxed{E/GY/X} \underset{GY}{\rightleftharpoons} E/X \underset{X}{\rightleftharpoons} E \\ {\scriptstyle Y}\searrow\!\!\nwarrow & & {\scriptstyle GX}\nearrow\!\!\swarrow & \\ & \boxed{E/Y} & & \end{array}$$

Mechanism 9.XIII

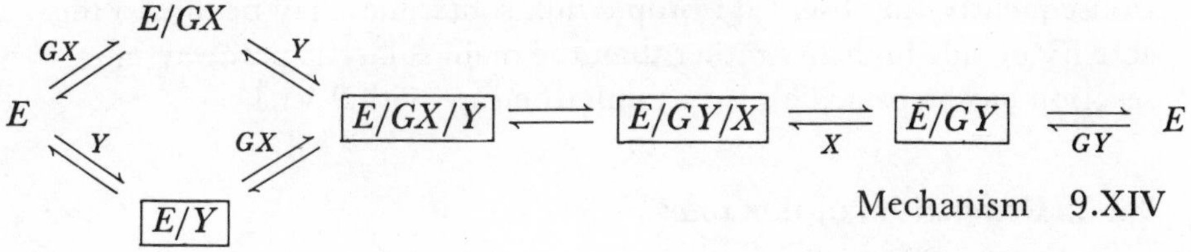

$$E \overset{GX}{\rightleftharpoons} E/GX \overset{Y}{\rightleftharpoons} \boxed{E/GX/Y}$$
$$E \overset{Y}{\rightleftharpoons} \boxed{E/Y} \overset{GX}{\rightleftharpoons} \boxed{E/GX/Y}$$
$$\boxed{E/GX/Y} \rightleftharpoons \boxed{E/GY/X} \underset{X}{\rightleftharpoons} \boxed{E/GY} \underset{GY}{\rightleftharpoons} E$$

Mechanism 9.XIV

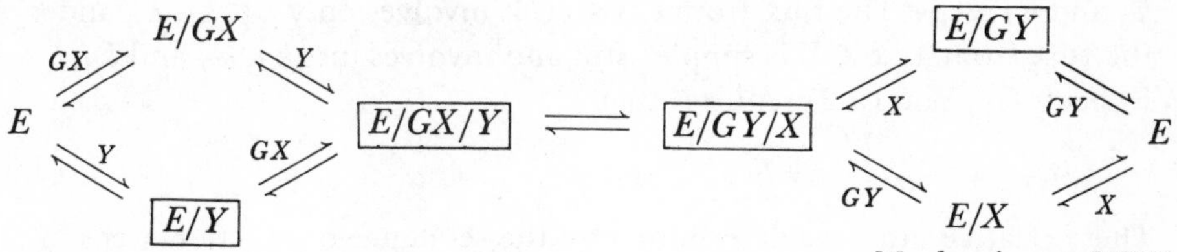

$$E \overset{GX}{\rightleftharpoons} E/GX \overset{Y}{\rightleftharpoons} \boxed{E/GX/Y}$$
$$E \overset{Y}{\rightleftharpoons} \boxed{E/Y} \overset{GX}{\rightleftharpoons} \boxed{E/GX/Y}$$
$$\boxed{E/GX/Y} \rightleftharpoons \boxed{E/GY/X}$$
$$\boxed{E/GY/X} \overset{X}{\rightleftharpoons} \boxed{E/GY} \overset{GY}{\rightleftharpoons} E$$
$$\boxed{E/GY/X} \overset{GY}{\rightleftharpoons} E/X \overset{X}{\rightleftharpoons} E$$

Mechanism 9.XV

Fig. 9.1. Some two-substrate group-transfer mechanisms. The boxed-in enzyme species are those which will persist at finite concentrations when both $Y$ and $GY$ become saturating.

(b) All reaction steps originating from a persisting enzyme species proceed at significant rates. Furthermore, since the system is at equilibrium, microscopic reversibility dictates that all reaction steps leading to a persisting species must also proceed at significant rates†. As a result, the interconversion of any two species will continue at a significant rate if one or both species persist in the face of saturating $(Y)$ and $(GY)$. In contrast, it will be insignificant if both species are reduced to vanishing concentrations.

(c) The exchange of the X-moiety between $GX$ and $X$ will not be prohibited if an uninterrupted chain of significant interconversions can be drawn to provide a pathway for the exchange. Otherwise the exchange will be prohibited. Therefore this procedure predicts that:

> The $GX^* \rightleftharpoons X^*$ exchange will be prohibited by saturating concentrations of $Y$ and $GY$ in Mechanisms 9.VI, 9.VII, 9.X, 9.XI, 9.XII, 9.XIII, but not in Mechanisms 9.VIII, 9.IX, 9.XIV and 9.XV.

The different behaviour of Mechanisms 9.VI and 9.VIII are especially instructive. In 9.VI, high $(Y)$ and $(GY)$ force all the enzyme to accumulate in the two ternary complexes and thereby prohibit the $GX^* \rightleftharpoons X^*$ exchange. In 9.VIII, on the other hand, prohibition is avoided, because the exchange can proceed through the two dead-end side-branches by the following pathway:

$$E/Y \overset{GX}{\rightleftharpoons} E/GX/Y \rightleftharpoons E/X/GY \underset{X}{\rightleftharpoons} E/GY$$

† Microscopic reversibility requires a pair of directly-opposing reaction steps to have equal rates at equilibrium, so that true reversibility is established at every microscopic region of the reaction mechanism.

Consequently, an observed isotopic flux sometimes may be due to the activity of side-branches rather than the main catalytic pathway in a reaction mechanism. This is certainly the case with 9.VIII.

### 9.2.3. Comparison of flux rates

In Mechanism 9.VI the flux from $GX$ to $X$ passes through the $k_1$, $k_2$, $k_0$, $k_3$ and $k_4$ steps. The flux from $GX$ to $GY$ involves only $k_1$, $k_2$, $k_0$ and $k_3$; the flux from $Y$ to $GY$ is simpler still and involves just $k_2$, $k_0$ and $k_3$. Expectedly, Eqn (9.3) requires that

$$f_{GX \to X} \leqslant f_{GX \to GY} \leqslant f_{Y \to GY}. \tag{9.9}$$

These relative rates are determined by the sequence of reaction steps in the linear mechanism, and serve as a diagnostic reflection of the given sequence.

According to Eqn (9.9), $f_{GX \to X}$ must not exceed $f_{GX \to GY}$, which in turn must not exceed $f_{Y \to GY}$. However, these fluxes may all become equal under limiting conditions, the foremost being a rate-limiting isomerization between the ternary complexes $E/GX/Y$ and $E/X/GY$. When the isomerization constants $k_0$ and $k_{-0}$ are very small and therefore rate-limiting, in Eqn (9.3) the reciprocal of the $k_0$-step will be so much greater than all other reciprocals that all three fluxes will approach the same asymptotic value:

$$f_{GX \to X} \simeq f_{GX \to GY} \simeq f_{Y \to GY} \simeq k_0(E/GX/Y) \tag{9.10}$$

The very slow $k_0$-step thus imposes the same numerical restriction on all three fluxes.

Another consequence of the slow isomerization postulated is that the binding of substrates and products to the enzyme will conform to the simpler quasi-equilibrium state rather than the more general steady-state even when the reaction system is away from the thermodynamic equilibrium. Since the equality of different equilibrium flux-rates and the prevalence of quasi-equilibrium stem from a common basis, Boyer and Silverstein (1963) suggested that the validity of quasi-equilibrium can be confirmed by equal, or rejected by unequal, equilibrium flux-rates. If quasi-equilibrium could be confirmed in this manner, branching mechanisms with random pathways for substrate additions also should yield hyperbolic (first-degree) instead of nonhyperbolic (higher-degree) velocity-substrate profiles.

On the basis of the latter expectation, Engers, Bridger and Madsen (1970) have designed a powerful method for distinguishing between two classes of model mechanisms for nonhyperbolic rate behaviour. For any oligomeric enzyme, a nonhyperbolic $v_0(S)$ function might arise from allosteric interactions between the enzyme subunits. It also might arise, even when the enzyme subunits are entirely independent, from the operation of random pathways for substrate addition. The random-pathway explanation is tenable under steady-state conditions, when the degree of the $v_0(S)$ function depends on the number of enzyme species in the mechanism reacting with the substrate, but not under quasi-equilibrium conditions, when the degree of the $v_0(S)$ function depends solely on the number of interacting substrate sites. Consequently, if quasi-equilibrium conditions were confirmed on the strength of equality between the different equilibrium flux rates, steady-state random-pathway models would have to be rejected in favour of allosteric ones. This method has been applied to phosphorylase *a*. In the absence of glucose, the flux rates for the [glucose-1-phosphate ⇌ inorganic phosphate] and [glucose-1-phosphate ⇌ glycogen] exchanges were close to equal; both rates varied hyperbolically with glucose-1-phosphate and inorganic phosphate concentrations. In the presence of glucose, the two flux rates were again close to equal, but both varied sigmoidally with glucose-1-phosphate and inorganic phosphate concentrations. These results, shown in Fig. 9.2, suggested that the reaction system was under quasi-equilibrium, and therefore the sigmoidal rate profiles were due to allosterism rather than steady-state random pathways.

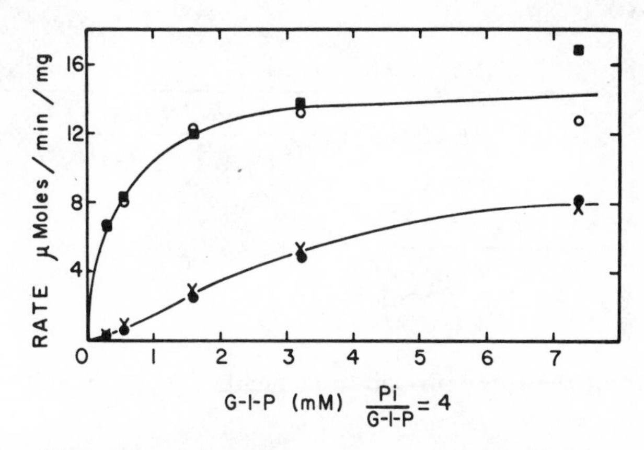


Fig. 9.2. Equilibrium isotopic exchanges catalysed by phosphorylase *a* in the absence (upper curve) or presence (lower curve) of glucose. (■, ●: $P_i \rightleftharpoons G1P$ exchange; ○, x; glycogen ⇌ $G1P$ exchange). (From Engers, Bridger and Madsen, 1970.)

## 9.3. Exchanges at Steady-State

In studying isotopic exchanges at equilibrium, the need to maintain equilibrium calls for the joint variation of one or more substrate-product pairs, and precludes the variation of individual concentrations one at a time. This restriction could be avoided by examining the isotopic fluxes under steady-state rather than equilibrium conditions. The steady-state flux equations needed for this purpose, which are in general more complex than their equilibrium counterparts, were first derived by Britton (1964, 1966).

Alternative derivations also have been presented by Cleland (1967a) and by Schachter (1972).

The method of Britton may be illustrated by considering once again the $Y \rightarrow GY$ flux for Mechanism 9.VI, which involves six reaction steps between four enzyme species:†

$$E/GX \underset{k_{-2}}{\overset{k_2(Y)}{\rightleftharpoons}} E/GX/Y \underset{k_{-0}}{\overset{k_0}{\rightleftharpoons}} E/X/GY \underset{k_{-3}(GY)}{\overset{k_3}{\rightleftharpoons}} E/X$$

The rate of this flux depends on the probabilities of three successive stages of *isotopic transfer*;

(a) Probability of transfer from $Y$ to $E/GX/Y$ is $k_2(Y)(E/GX)$

(b) Probability of transfer from $E/GX/Y$ on to $E/X/GY$ instead of back to $Y$ is

$$\frac{k_0(E/GX/Y)}{k_0(E/GX/Y)+k_{-2}(E/GX/Y)} = \frac{k_0}{k_0+k_{-2}}$$

(c) Probability of transfer from $E/X/GY$ on to $GY$ instead of all the way back to $Y$ is

$$\frac{k_3(E/X/GY)}{k_3(E/X/GY)+k_{-0}(E/X/GY)\cdot\dfrac{k_{-2}(E/GX/Y)}{k_0(E/GX/Y)+k_{-2}(E/GX/Y)}}$$

$$= \frac{k_3}{k_3+k_{-0}\left(\dfrac{k_{-2}}{k_0+k_{-2}}\right)}$$

Combination of the three probabilities yields the steady-state flux equation:

$$f_{Y\rightarrow GY} = k_2(Y)(E/GX)\cdot\frac{k_0}{k_0+k_{-2}}\cdot\frac{k_3}{k_3+k_{-0}\left(\dfrac{k_{-2}}{k_0+k_{-2}}\right)} \qquad (9.11)$$

At steady-state, the expression for $(E/GX)$, as for any other enzyme species in the mechanism, is given by the schematic rule. Inspection of Mechanism

† In the equilibrium case for the same flux treated in Eqn (9.3), only the $k_2$, $k_0$ and $k_3$ steps and not the $k_{-2}$, $k_{-0}$ and $k_{-3}$ steps enter into calculation. The reason is that at equilibrium, microscopic reversibility demands that the $k_2$ and $k_{-2}$ steps, the $k_0$ and $k_{-0}$ steps, or the $k_3$ and $k_{-3}$ steps must be equal. At steady-state, however, microscopic reversibility does not hold, and all six reaction steps must enter separately into calculation.

9.VI indicates that $(Y)$ will influence $(E/GX)$ in accordance with a first-degree inhibitor function:

$$(E/GX) = \frac{m_1}{d_1(Y) + d_0}(E)_0.$$

Equation (9.11) requires the flux rate to vary with $(Y)(E/GX)$, which means that the flux rate will vary hyperbolically with $(Y)$:

$$f_{Y \to GY} = \frac{m_1(Y)}{d_1(Y) + d_0}.$$

Inspection of the mechanism further reveals that $(GX)$ will influence $(E/GX)$, and therefore the flux rate, hyperbolically in the absence of reaction product(s)

$$f_{Y \to GY} = \frac{m_1(GX)}{d_1(GX) + d_0}.$$

Adding $GY$ to the system influences $(E/GX)$ and the flux rate in accordance with a first-degree modifier function:

$$f_{Y \to GY} = \frac{m_1(GY) + m_2}{d_1(GY) + d_2}.$$

On the other hand, adding $X$ to the system influences $(E/GX)$ and the flux rate in accordance to a first-degree inhibitor function in the absence of $GY$, or a first-degree modifier function in the presence of $GY$, because the $k_{-4}(X)$ arrow can join in a sequence terminating at $E/GX$ only when the $k_{-3}(GY)$ arrow is significant:

$$f_{Y \to GY} = \frac{m_1}{d_1(X) + d_0} \qquad \text{if } GY \text{ is absent}$$

$$f_{Y \to GY} = \frac{m_1(X) + m_0}{d_1(X) + d_0} \qquad \text{if } GY \text{ is present.}$$

In terms of enzyme distribution, high concentrations of $X$ causes all the enzyme to accumulate in $E/X$ in the absence of $GY$. This counter-productive accumulation does not take place in the presence of $GY$, which allows $E/X$ to be converted to $E/GY/X$ and in turn the other enzyme species.

The feasibility of studying the rate effects of individual concentrations is a distinct advantage of the steady-state flux method. Furthermore, Britton (1966) has discovered a simple way to apply the method based on the analysis of *flux ratios*. For any mechanism, the ratio between two different steady-state fluxes can be obtained from the flux equations. The expressions for these flux-ratios are as a rule simpler, and therefore more useful, than those for the fluxes themselves. Figure 9.3 portrays the different

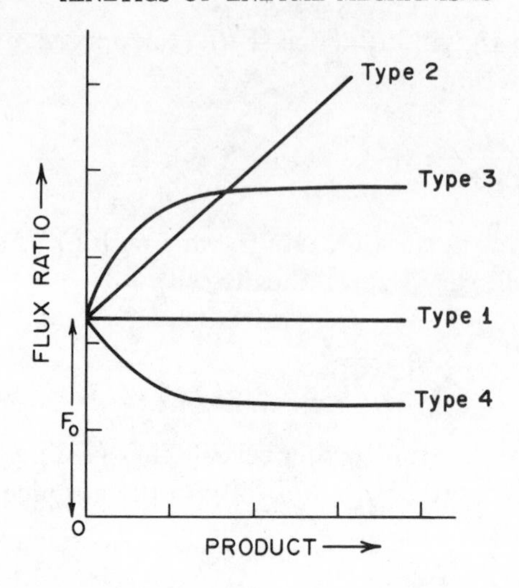


Fig. 9.3. Curve types for the variation of steady-state isotopic flux-ratio with product concentration.

curve types relating the ratio between the $GX \rightarrow X$ and $GX \rightarrow GY$ fluxes to a product concentration. From the equations derived by Britton, several straightforward conclusions can be drawn:

(a) Let $F = \dfrac{f_{GX \rightarrow X}}{f_{GX \rightarrow GY}}$ .

(b) If $X$ and $GY$ dissociate randomly from the enzyme, e.g., in Mechanism 9.XV, a plot of $F$ vs. $(X)$ will give a Type 3 curve; $F_0$ (the vertical intercept) = if $GY$ is not added to the reaction system, but $F_0 < 1$ if a finite concentration of $GY$ is added. A plot of $F$ vs. $(GY)$ will give a Type 4 curve with $F_0 = 1$ if $X$ is not added, but $F_0 > 1$ if a finite concentration of $X$ is added.

(c) If $X$ dissociates before $GY$, e.g., in Mechanisms 9.VII, 9.IX, 9.XI, and 9.XIV, a plot of $F$ vs. $(X)$ will give a Type 2 curve with $F_0 = 1$, whether or not $GY$ is added. A plot of $F$ vs. $(GY)$ will give a Type 1 curve with $F_0 = 1$ if $X$ is not added, but $F_0 > 1$ if a finite concentration of $X$ is added.

(d) If $GY$ dissociates before $X$, e.g., in Mechanisms 9.VI, 9.X and 9.XIII, the predictions would be comparable to those outlined in (c), except with the effects of $X$ and $GY$ now being reversed. In such cases it would be diagnostically more revealing to plot instead $1/F$ vs. $(GY)$ to obtain a Type 2 curve, and $1/F$ vs. $(X)$ to obtain a Type 1 curve.

Since the flux-ratio expressions in such instances are uncomplicated, information concerning the magnitudes of the various rate constants in the mechanism also may be obtained from a quantitative analysis of the slopes and intercepts of these plots.

## Problem 9.1

In a group-transfer reaction of the type $GX + Y = X + GY$, isolation of a substituted enzyme does not in itself prove that the substituted enzyme $E{-}G$ is a significant enzyme species in the kinetic mechanism of the enzyme. Why?

# 10 Rate Constants

The kinetic description of an enzyme mechanism demands not only the establishment of the topology of interconversions between enzyme species, but also the evaluation of the elementary rate constants for these interconversions. Without complete knowledge of the elementary rate constants, the reaction mechanism can be discussed in terms of the various enzyme species and their protein chemistry, but never quantitatively. In this chapter we shall examine the three major experimental approaches for evaluating elementary rate constants, viz. transient-state kinetics, relaxation kinetics, and steady-state kinetics. The importance of knowledge of rate constants to our understanding of enzymic catalysis is greatly enhanced by the fact that they are not truly immutable constants. Instead, they vary with changes in environmental factors, especially temperature and pH. We shall consider also the influence of such factors on the rate constants, and the insight to be gained from an analysis of this influence.

## 10.1. From Transient-State Kinetics

### 10.1.1. Flow methods

The usual method of measuring the rate of a chemical reaction is to mix separate solutions of the reactants, and monitor the progress of the reaction by means of discrete or continuous sampling of the reaction mixture. Inasmuch that slow mixing (relative to the half-time of the reaction) distorts the time-course of the reaction, the speed at which mixing can be accomplished sets an upper limit to the rate of reaction steps that can be isolated and measured in this manner. The mixing of two solutions by manual methods as a rule takes one second or more, which is much too slow for the direct measurement of most enzymic reaction steps. Accordingly, such measurements became practicable only when Hartridge and Roughton (1923) introduced the use of flow techniques for analysing fast reactions.

There are two types of flow techniques. In the original *continuous-flow* technique, the enzyme and substrate solutions are forced into a mixing chamber at high speed, and the reaction mixture flows from the chamber into an observation tube. The time required by the mixture to reach any point along the tube depends on the distance from the mixing chamber, and on the flow velocity. The progress of the reaction over an interval of milliseconds can be monitored either by observation at different points along the tube at a single flow velocity, or by observation at a single point at different flow velocities. In the *stopped-flow* technique, developed by Chance (1954) and by Gibson (1969), the reaction mixture is rapidly injected into the observation tube and just as rapidly brought to a full stop. The progress of reaction is monitored at a single observation point after the flow is stopped; the fast recording of optical or other physical properties of the system is made feasible by the use of a cathode-ray oscilloscope. Continuous-flow, because it calls for a relatively prolonged flow of the reaction mixture, usually containing a high concentration of enzyme, is more demanding than stopped-flow with respect to the consumption of enzymes and substrates. In both cases advances in the engineering design of the flow apparatus have permitted an increasingly faster and less turbulent mixing and flow, and the resolution time now approaches $10^{-4}$ second.

### 10.1.2. Formation of enzyme-substrate complex

The basic mechanism for one-substrate reactions consists of three elementary steps characterized by the rate constants $k_1$, $k_{-1}$, and $k_2$;

$$E \underset{k_{-1}}{\overset{k_1 S}{\rightleftharpoons}} ES \xrightarrow{k_2} E + \text{products} \qquad \text{Mechanism 10.I}$$

When the substrate/enzyme ratio is high, the system will be in steady-state during the greater part of the time-course of the reaction. However, since all the enzyme occurs as $E$ and none as $ES$ prior to mixing with substrate, the system after mixing first goes through a transient-phase during which the $(ES)/(E)$ ratio rises from zero to a steady-state value. At the beginning of the transient phase, before significant amounts of $ES$ have formed, the $k_{-1}$ and $k_2$ steps are negligible and only the $k_1$ step is significant; in effect the system consists of only a single second-order step:

$$E + S \xrightarrow{k_1} ES$$

If a physical property of free enzyme or free substrate should be sufficiently different from that of the enzyme-substrate complex to allow the single-step reaction to be monitored by flow techniques, $k_1$ could be estimated directly without any interference from the $k_{-1}$- and $k_2$-steps. Once $k_1$ is obtained, the other two constants may be estimated from steady-state measurements. The steady-state equation is:

$$v = \frac{k_2(E)_0 \cdot S}{S + K_m^S}$$

$$K_m^S = \frac{k_{-1} + k_2}{k_1} \qquad (10.1)$$

Thus $k_2$ may be calculated from the saturation velocity $k_2(E)_0$. Knowing both $k_1$ and $k_2$, $k_{-1}$ may be calculated from $K_m^S$.

The classical study by Chance (1943) of the peroxidase mechanism 10.II illustrates this approach for measuring rate constants:

$$E \underset{k_{-1}}{\overset{k_1 H_2O_2}{\rightleftharpoons}} E/H_2O_2 \xrightarrow{k_2 (= k_2' \cdot A)} E + \text{products} \quad \text{Mechanism} \quad 10.II$$

Peroxidase is a hematin enzyme. Its absorption spectrum is drastically shifted upon combination with $H_2O_2$, and $k_1$ could be estimated from rapid observation of this single-step combination. Moreover, the $E/H_2O_2$ complex being measurable throughout the reaction, $k_2$ (or, more precisely, $k_2'A$ since $k_2$ in this instance was a pseudo-first order constant that included the concentration of some hydrogen donor $A$ such as leucomalachite green) could be estimated directly from the relation

$$\frac{d(\text{product})}{dt} = k_2(E/H_2O_2).$$

The verification of this relation for this system, and the agreement between the transient-state and steady-state estimates of $k_2$, provided the strongest evidence for the general theory that the enzyme-substrate complex is an *obligatory* intermediate in enzymic catalysis.

### 10.1.3. Formation of reaction product

The formation of reaction product also can be monitored during the transient state by means of flow techniques. For Mechanism 10.I, the rate of change of $(ES)$ is,

$$\begin{aligned}\frac{d(ES)}{dt} &= k_1 S(E) - (k_{-1} + k_2)(ES) \\ &= k_1 S[(E)_0 - (ES)] - (k_{-1} + k_2)(ES) \\ &= k_1 S(E)_0 - (k_1 S + k_{-1} + k_2)(ES) \qquad (10.2)\end{aligned}$$

The rate of formation of reaction product $P$ is:

$$\frac{dP}{dt} = k_2(ES) \qquad (10.3)$$

and so,

$$\frac{d^2\mathbf{P}}{dt^2} = k_2 \frac{d(ES)}{dt}. \tag{10.4}$$

By substituting Eqns (10.3) and (10.4) into (10.2), $(ES)$ can be eliminated to give

$$\frac{d^2\mathbf{P}}{dt^2} + (k_1\mathbf{S} + k_{-1} + k_2)\frac{d\mathbf{P}}{dt} - k_2 k_1 \mathbf{S}(E)_0 = 0. \tag{10.5}$$

If a high $\mathbf{S}/(E)_0$ ratio is employed, relatively little substrate would be consumed throughout the transient-state, so that **S** would remain very close to its zero-time value of $\mathbf{S}_0$. Given this provision, Eqn (10.5) becomes a second-order differential equation with constant coefficients. Its solution, obtained by Gutfreund and Roughton (in Roughton, 1954) is:

$$\mathbf{P} = \mathbf{P}_0 + \frac{k_2 k_1 \mathbf{S}_0 (E)_0 t}{k_1 \mathbf{S}_0 + k_{-1} + k_2} + \frac{k_2 k_1 \mathbf{S}_0 (E)_0}{(k_1 \mathbf{S}_0 + k_{-1} + k_2)^2}\,[e^{-(k_1\mathbf{S}_0 + k_{-1} + k_2)t} - 1]. \tag{10.6}$$

If no reaction product was pre-added to the system, $\mathbf{P}_0$ is equal to zero in Eqn (10.6). Also, the exponential term can be expanded into its standard power-series form of

$$e^x = 1 + x + \frac{x^2}{2!} + \frac{x^3}{3!} + \cdots \frac{x^n}{n!} + \cdots \tag{10.7}$$

This power-series will be a rapidly diminishing one very early in the transient-state when $t$ will be small. Substituting only its first three terms into Eqn (10.6) reduces the latter to

$$\mathbf{P} = \frac{k_2 k_1 \mathbf{S}_0 (E)_0 t^2}{2}. \tag{10.8}$$

Therefore reaction product accumulates exponentially with $t$ early in the transient-state (Fig. 10.1), and plotting **P** versus $t^2$ gives a linear plot the slope of which is equal to $k_2 k_1 \mathbf{S}_0 (E)_0/2$. With $k_2$ obtainable from the steady-state saturation velocity, $k_1$ can be calculated from this slope; and once $k_1$ and $k_2$ are known, $k_{-1}$ is obtainable from the steady-state $K_m^S$ (Eqn 10.1).

Later, as the value of $t$ increases, the exponential term in Eqn (10.6) becomes negligible, and the system formally passes from the transient-state into the steady-state. Equation (10.6) is reduced to (assuming that $\mathbf{P}_0 = 0$):

$$\mathbf{P} = \frac{k_2 k_1 \mathbf{S}_0 (E)_0 t}{k_1 \mathbf{S}_0 + k_{-1} + k_2} - \frac{k_2 k_1 \mathbf{S}_0 (E)_0}{(k_1 \mathbf{S}_0 + k_{-1} + k_2)^2}. \tag{10.9}$$

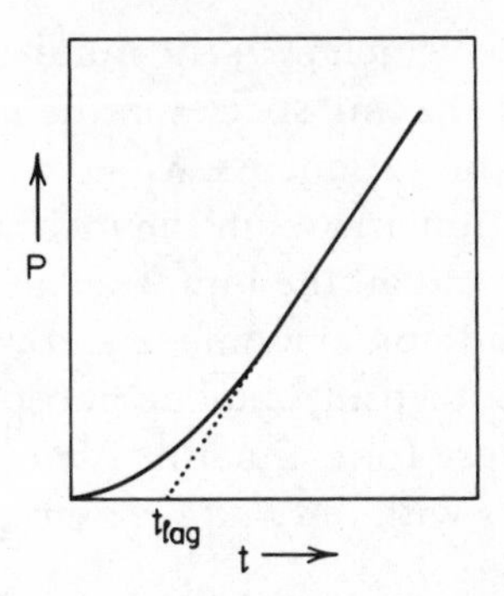


Fig. 10.1. Formation of reaction product during the transient phase.

Accordingly, P now accumulates linearly with time (Fig. 10.1). Extrapolation of this linear region intersects the $t$-axis at $t_{lag}$. Since product concentration is zero on the $t$-axis, the value of $t_{lag}$ satisfies Eqn (10.10):

$$\frac{k_2 k_1 S_0 (E)_0 t_{lag}}{k_1 S_0 + k_{-1} + k_2} - \frac{k_2 k_1 S_0 (E)_0}{(k_1 S_0 + k_{-1} + k_2)^2} = 0 \tag{10.10}$$

which yields

$$t_{lag} = \frac{1}{k_1 S_0 + k_{-1} + k_2} . \tag{10.11}$$

If substrate concentration is raised to saturating levels, this is reduced to

$$t_{lag} = \frac{1}{k_1 S_0} . \tag{10.12}$$

Accordingly, as suggested by Gutfreund (1955), $t_{lag}$ obtained at saturating levels of substrate provides yet another route for the estimation of $k_1$ during the transient-state.

The transient-state equations for more complex mechanisms have been treated generally by Darvey (1968) and Hijazi and Laidler (1973b). Darvey showed that $t_{lag}$ is independent of enzyme concentration, and is expressible by a ratio between two concentration polynomials:

$$t_{lag} = \frac{m_z S^z + m_{z-1} S^{z-1} + \cdots + m_1 S + m_0}{d_z S^z + d_{z-1} S^{z-1} + \cdots + d_1 S + d_0} .$$

The $d$-coefficients are all of the same sign, but the sign of the $m$-coefficients varies. Therefore $t_{lag}$ may be either negative or positive, as had been observed by Ouellet and Stewart (1959).

Hijazi and Laidler showed that the equations for a variety of one-substrate, two-substrate, and inhibition mechanisms all obey the general form of,

$$P = vt + \sum_{i=1}^{n} \beta_i e^{-\lambda_i t} - \sum_{i=1}^{n} \beta_i$$

where $v$ is the steady-state velocity; $n$, the number of exponential terms, is equal to the number of enzyme species in the mechanism other than the free enzyme; the sum of the exponents, $\lambda_1 + \lambda_2 + \lambda_3 + \cdots + \lambda_n$, is equal to the sum of all the reaction arrows in the mechanism (viz. all the monomolecular rate constants, and all the bimolecular rate constants each multiplied by its associated non-enzymic ligand). This equation, which has been applied to chymotrypsin, alkaline phosphatase and myosin, provides a general procedure for estimating bimolecular rate constants. Whenever a ligand $S$ reacts with *only one* enzyme species in the mechanism in a bimolecular $k_s$-step, i.e.

$$S + E_s \xrightarrow{k_s} E_sS$$

the rate constant $k_s$ will be obtainable as the slope of the predicted linear plot of $(\lambda_1 + \lambda_2 + \lambda_3 + \cdots + \lambda_n)$ versus S.

## 10.2. From Relaxation Kinetics

The equilibrium position of any chemical reaction is determined by the standard free-energy change of the reaction:

$$\Delta G^0 = -RT \ln K_{eq}$$
$$R = \text{gas constant} \qquad (10.13)$$
$$T = \text{absolute temperature}$$

A perturbation in the temperature or any physical factor that affects $\Delta G^0$ invariably alters $K_{eq}$. Consequently, if such a perturbation is imposed on a reaction system that has been at equilibrium, a net chemical change must take place in the system in order to bring about an appropriately new equilibrium position. If the perturbation could be very rapidly imposed, it would be possible to follow the chemical response, or *relaxation,* of the system. On this basis Eigen (1954) has developed an all-powerful tool for studying elementary rate processes, with a resolution time ranging from minutes down to fractions of a nanosecond.

A number of physical perturbations are useful in chemical relaxation studies. In order of increasing rapidity, these are pressure shock, temperature jump, sound absorption, and electric field change. In the technique of temperature jump, which has been most widely applied to enzyme system a high-voltage current is discharged through a reaction mixture having a high electrolyte conductance. A jump of several degrees in temperature can be induced within microseconds, permitting, usually by optical means, the observation of relaxation responses down to $10^{-4}$–$10^{-5}$ second. The response again can be recorded with the aid of an oscilloscope.

### 10.2.1. Relaxation times

*(a) First-order reaction*

A chemical relaxation which can be treated simply is that of a reversible interconversion between two enzyme conformations, $E_R$ and $E_T$:

$$E_R \underset{k_{-1}}{\overset{k_1}{\rightleftharpoons}} E_T$$

Initially, the system is at equilibrium at temperature $T$. After the temperature is jumped to $T'$, the concentrations of the two enzyme species at various times can be denoted as:

| *Enzyme conformation* | *Initial concentration at T* | *Final concentration at T′* | *Concentration during relaxation* |
|---|---|---|---|
| $R$ | $E_R^0$ | $E_R'$ | $E_R' + x$ |
| $T$ | $E_T^0$ | $E_T'$ | $E_T' - x$ |

During the relaxation response, the enzyme is redistributed between the two conformations so that $x$, which is the distance from the new equilibrium position, is finally diminished to zero. At any instant during the relaxation, the law of mass action requires that

$$\frac{d(E_T' - x)}{dt} = k_1(E_R' + x) - k_{-1}(E_T' - x)$$

or, upon rearrangement,

$$\frac{dE_T'}{dt} - \frac{dx}{dt} = k_1 E_R' - k_{-1} E_T' + (k_1 + k_{-1})x. \tag{10.14}$$

Eventually, after the equilibrium has been reestablished at the new temperature, the reaction system will become once again stationary, at which point we can write:

$$\frac{dE_T'}{dt} = k_1 E_R' - k_{-1} E_T' = 0.$$

Accordingly, these terms can be removed without consequence from Eqn (10.14), leaving only

$$-\frac{dx}{dt} = (k_1 + k_{-1})x$$

which yields upon integration:

$$x = x^0\, e^{-(k_1 + k_{-1})t}. \tag{10.15}$$

In this expression $x^0$ respresents the amplitude of $x$ at the start of the response, namely the distance between $E_R^0$ and $E_R'$. With this in mind, Eqn (10.15) can be rewritten as:

$$x = (E_R^0 - E_R')\, e^{-t/\tau}$$
$$\frac{1}{\tau} = k_1 + k_{-1}. \tag{10.16}$$

The *relaxation time* $\tau$ is the time required for $x$ to be reduced to $1/e$ of its original amplitude (Fig. 10.2). Its reciprocal i.e. the *reciprocal relaxation*

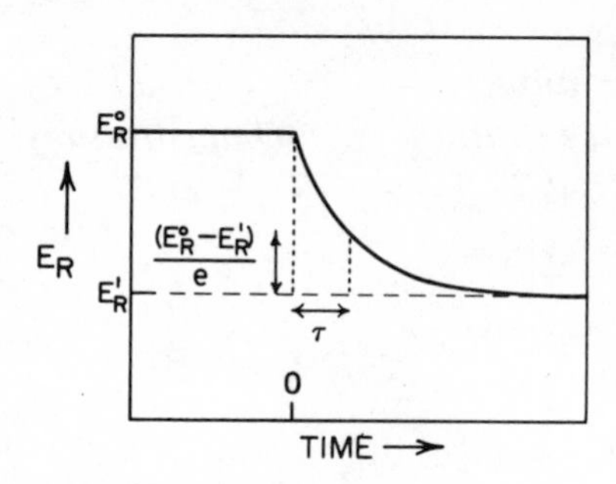


Fig. 10.2. The relaxation process.

*time*, gives the sum $(k_1 + k_{-1})$. The ratio of $k_1/k_{-1}$ is of course just the equilibrium constant at $T'$:

$$K_{eq} = \frac{E_T'}{E_R'} = \frac{k_1}{k_{-1}}.$$

Knowing both the sum and the ratio between $k_1$ and $k_{-1}$, these two rate constants are readily estimated.

(*b*) *Second-order reaction*

We may next consider a reversible combination between substrate and enzyme:

$$E + S \underset{k_{-1}}{\overset{k_1}{\rightleftharpoons}} ES$$

At any instant during the relaxation, the law of mass action requires that

$$\frac{d[(ES)' - x]}{dt} = k_1\,(\mathrm{E}' + x)\,(\mathrm{S}' + x) - k_{-1}[(ES)' - x]$$

or, upon rearrangement,

$$\frac{d(ES)'}{dt} - \frac{dx}{dt} = k_1\mathrm{E}'\mathrm{S}' - k_{-1}(ES)' + k_1(\mathrm{E}' + \mathrm{S}')x + k_{-1}x + k_1x^2 \tag{10.17}$$

Again looking ahead to the stationary equilibrium at the new temperature $T'$, we can write

$$\frac{d(ES)'}{dt} = k_1 \mathrm{E}'\mathrm{S}' - k_{-1}(ES)' = 0.$$

Removal of these terms from Eqn (10.17) leaves

$$-\frac{dx}{dt} = [k_1(\mathrm{E}' + \mathrm{S}') + k_{-1}]x + k_1 x^2. \tag{10.18}$$

If $x$ is very small, the $x^2$ term can be neglected, and integration yields:

$$x = (\mathrm{E}^0 - \mathrm{E}')\, e^{-t/\tau}$$

$$\frac{1}{\tau} = k_1(\mathrm{E}' + \mathrm{S}') + k_{-1}. \tag{10.19}$$

In this case the reciprocal relaxation time is predicted to vary linearly with the sum of free enzyme and free substrate. The bimolecular constant $k_1$ is given by the slope of this linear plot, and the monomolecular constant $k_{-1}$ is given by the intercept of this linear plot on the $1/\tau$ axis.

The derivation of Eqn (10.19) requires the assumption that $x$ is very small and therefore $x^2$ is negligible in Eqn (10.18). In general, the valid application of chemical relaxations depends on working with concentration perturbations that are small relative to the total concentrations. Mechanisms with multiple reversible steps generate multiple relaxation times. The various reaction steps being coupled to one another, the spectrum of relaxation times are in effect the eigen values of the system of rate equations. The general problem of extracting rate constants from relaxation times has been treated at length by Hammes and Schimmel (1967), and Czerlinski (1968). On the experimental side, it might be noted that the extent to which any two relaxation times can be resolved depends on how different they are in magnitude. Since the possibility of poor resolution always exists, the detection of an apparently single relaxation time is not conclusive proof that the mechanism has only a single pair of reversible steps, whereas detection of more than one relaxation time would be conclusive proof that there are multiple steps.

### 10.2.2. Isomerization of enzyme-substrate complex

Steady-state kinetics provide a powerful tool capable of delineating the topology of different ligand reactions in a mechanism, but is incapable of detecting isomerization of central enzyme-substrate complexes that do not react with any ligand. Either transient-state or relaxation kinetics must be employed to study such isomerizations. The simplest model for the isomerization of an enzyme-substrate complex is:

$$E \underset{k_{-1}}{\overset{k_1 S}{\rightleftharpoons}} (ES) \underset{k_{-2}}{\overset{k_2}{\rightleftharpoons}} (ES)'$$

Its two relaxation times $\tau_1$ and $\tau_2$ are the two roots of the quadratic Eqn (10.20) (Hammes and Schimmel, 1967):

$$(1/\tau)^2 - [k_1(\mathrm{E} + \mathrm{S}) + k_{-1} + k_2 + k_{-2}](1/\tau) + [k_1(\mathrm{E} + \mathrm{S}) + k_{-1}] \times (k_2 + k_{-2}) - k_{-1}k_2 = 0 \quad (10.20)$$

If the initial formation of $(ES)$ is much more rapid than its isomerization to $(ES)'$, the solutions to this equation would be:

$$\frac{1}{\tau_1} = k_{-1} + k_1(\mathrm{E} + \mathrm{S})$$

$$\frac{1}{\tau_2} = \frac{k_2 k_1(\mathrm{E} + \mathrm{S})}{k_1(\mathrm{E} + \mathrm{S}) + k_{-1}} + k_{-2} \quad (10.21)$$

Of the two, $1/\tau_1$ varies linearly with (E + S), whereas $1/\tau_2$ varies hyperbolically with (E + S). This latter type of hyperbolic behaviour is

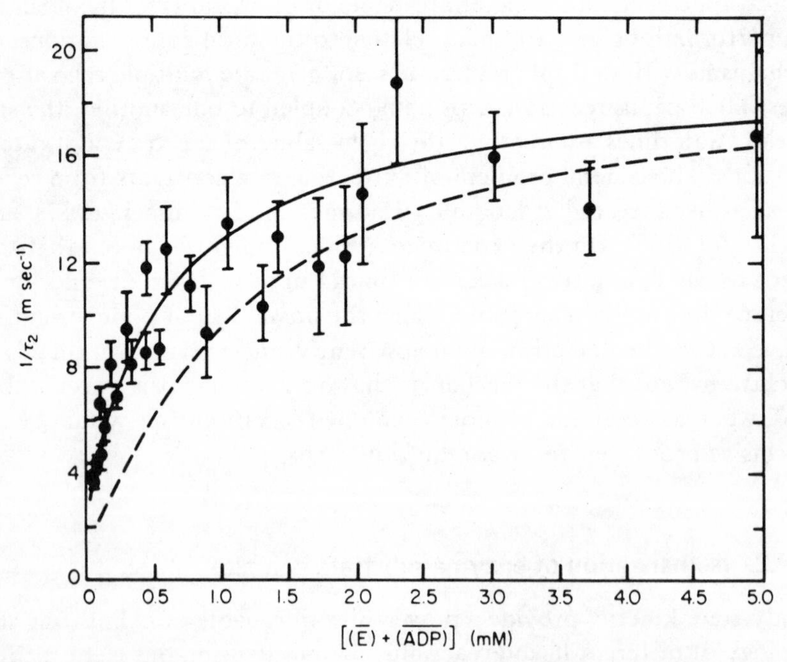


Fig. 10.3. Reciprocal relaxation time for creatine kinase-ADP interactions. The solid curve is the theoretical line for the isomerization of enzyme-substrate complex; the dashed curve is obtained after correction for the coupling of reaction steps. (From Hammes and Hurst, 1969.)

exemplified by the observation of Hammes and Hurst (1969) on creatine kinase-ADP interactions (Fig. 10.3).

The *elementary* step in enzyme kinetics is often elementary only until it can be resolved further into components which are even more elementary. In such cases, the original estimate for the rate constant would require reinterpretation in terms of the constants for the new steps. This process will continue until all meaningful conformations of every enzyme species in the mechanism have been described. Thus even the $(ES) \rightarrow (ES)'$ step of isomerization might eventually prove to be composite. In the case of the transamination between erythro-β-hydroxyaspartate and aspartate aminotransferase, Hammes and Haslam (1969) were able to resolve as many as eight relaxation processes.

### 10.2.3. Yeast glyceraldehyde-3-phosphate dehydrogenase

The study by Kirschner, Eigen, Bittman and Voigt (1966) on the binding of $NAD^+$ to the tetrameric glyceraldehyde-3-phosphate dehydrogenase from yeast is an elegant application of relaxation kinetics. Altogether three relaxation processes were detected from oscilloscope traces recorded at different scanning speeds (Fig. 10.4). This observation conformed to the two-state allosteric model of Monod. Wyman and Changeux, which postulates that the four enzyme subunits undergo a concerted transition between the $R$ and $T$ conformations:

| | | |
|---|---|---|
| $R$-state | ○○ / ○○ | association rate constant $= k_R$<br>dissociation rate constant $= k_D$ |
| | $k_0 \downarrow\uparrow k_0'$ | |
| $T$-state | ⊞ | association rate constant $= k_R'$<br>dissociation rate constant $= k_D'$ |

Three relaxation processes are predicted by this model: one for $NAD^+$ binding to $R$, one for $NAD^+$ binding to $T$, and one for the allosteric transition between $R$ and $T$.

As shown in Fig. 10.5, the fastest relaxation time $\tau_1$ conformed to the rate equation for a simple bimolecular step between $D$ (namely $NAD^+$) and $R$, which is entirely analogous to Eqn (10.19):

$$\frac{1}{\tau_1} = k_D + k_R(D + \phi R) \qquad (10.22)$$

where $D$ represents free $NAD^+$, and $\phi R$ the sum of free enzyme sites in

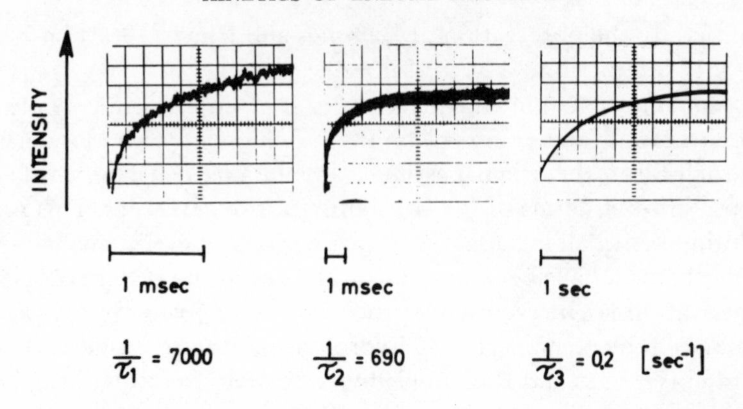


Fig. 10.4. Resolution between three relaxation times for yeast glyceraldehyde-3-phosphate dehydrogenase. (From Kirschner, Eigen, Bittman and Voigt, 1966.)

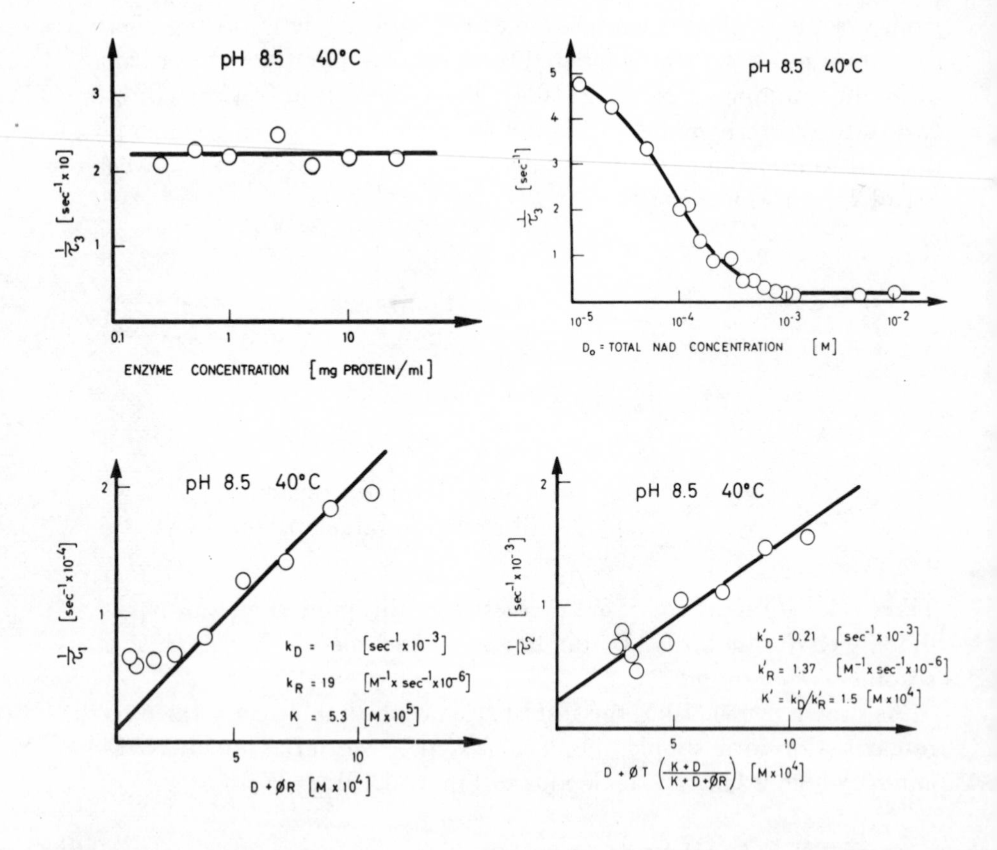


Fig. 10.5. Dependence of glyceraldehyde-3-phosphate dehydrogenase relaxation times on $NAD^+$ and enzyme concentrations. $D$ stands for free $NAD^+$, $\phi R$ total protein sites in $R$-state, and $\phi T$ total protein sites in $T$-state. (From Kirschner *et al.*, 1966.)

the $R$-conformation. The second relaxation time $\tau_2$ conformed to the rate equation for the slower of two bimolecular steps which are coupled through the common ligand $D$:

$$\frac{1}{\tau_2} = k'_D + k'_R \left(D + \phi T \left[\frac{K + D}{K + D + \phi R}\right]\right) \quad (10.23)$$

$\phi T$ being the sum of all free enzyme sites in the $T$-conformation. The third and slowest relaxation time $\tau_3$ conformed to Eqn (10.24), which could be obtained on the basis of the simplifying assumptions that concentration changes in $D$ during relaxation were negligible, and that the transition between the $T$ and $R$ states were unaffected by coenzyme binding:

$$\frac{1}{\tau_3} = k'_0 + k_0 \left[\frac{1 + \frac{D}{k'}}{1 + \frac{D}{k}}\right]^4 \quad (10.24)$$

According to this equation, $1/\tau_3$ should be independent of enzyme concentration (for the same reason that $1/\tau$ in Eqn 10.16 is independent of enzyme concentration, viz. the transition between the $R$ and $T$ states is strictly intramolecular), but should vary with $D$ in the form of a fourth-degree function. Both of these predictions were confirmed by the observations presented in Fig. 10.5. Taken together, Eqn (10.22–24) made possible the estimation of all six rate constants:

$$\left.\begin{aligned} k_R &= 1.9 \times 10^7\ \text{M}^{-1}\ \text{sec}^{-1} \\ k_D &= 1 \times 10^3\ \text{sec}^{-1} \end{aligned}\right\} k_D/k_R = 5.3 \times 10^{-5}\ \text{M}$$

$$\left.\begin{aligned} k'_R &= 1.37 \times 10^6\ \text{M}^{-1}\ \text{sec}^{-1} \\ k'_D &= 210\ \text{sec}^{-1} \end{aligned}\right\} k'_D/k'_R = 1.5 \times 10^{-4}\ \text{M}$$

$$\left.\begin{aligned} k_0 &= 5.5\ \text{sec}^{-1} \\ k'_0 &= 0.18\ \text{sec}^{-1} \end{aligned}\right\} L = k_0/k'_0 = 30.5$$

The $T$-state is thirty-fold favoured over the $R$-state in the allosteric equilibrium ($L = 30.5$), but about only one third as active as the $R$-state in binding $NAD^+$. Furthermore, stopped-flow measurements indicated that the $T$-state, although it binds $NAD^+$, is catalytically inactive (Kirschner, 1971a). Since the $R$ and $T$ states differ with respect to $NAD^+$ binding and even more so with respect to catalytic activity, the allosteric system is a *KV-system.*

### 10.3. From Steady-State Kinetics

Rate law (10.1) for the one-substrate Mechanism 10.I may be rewritten to give

$$k_2 = V_S, \text{saturation velocity per unit enzyme} \tag{10.25}$$

$$\frac{K_m^S}{V_S} = \frac{k_{-1}}{k_1 k_2} + \frac{1}{k_1}. \tag{10.26}$$

Equation (10.26) leads to (10.27) and therefore (10.28):

$$\frac{K_m^S}{V_S} > \frac{1}{k_1} \tag{10.27}$$

$$\frac{V_S}{K_m^S} < k_1. \tag{10.28}$$

Equation (10.25) yields $k_2$ when the concentration of enzyme is known, and Eqn (10.28) usefully provides a *minimal* estimate of $k_1$. On this basis Peller and Alberty (1959) calculated the minimum estimates of $k_1$ for various reversible as well as irreversible reactions (Table 10.1). In the $k_1$-step, the enzyme and substrate molecules collide in solution, and a fraction of the collisions will produce an enzyme-substrate complex. If this productive fraction approaches unity, the complex would be formed as fast as the enzyme and substrate molecules could diffuse together. Under these conditions reaction is said to be *diffusion-controlled.* For a small substrate molecule with a diffusion coefficient of about $10^{-5}$ cm$^2$ sec$^{-1}$, and a larger enzyme molecule with a diffusion coefficient of about $10^{-7}$ cm$^2$ sec$^{-1}$, the diffusion-controlled $k_1$ is expected to be about $10^8$–$10^{10}$ M$^{-1}$ sec$^{-1}$, which represents the absolute ceiling for $k_1$. Table 10.1 shows that the $k_1$-values for some of the enzymes are clearly approaching this diffusion-controlled region.

Steady-state measurements give only $k_2$ and a minimal estimate of $k_1$ in the case of Mechanism 10.I, but give all of $k_1$, $k_{-1}$, and $k_2$ in the case of the peroxidase-type Mechanism 10.II (Slater and Bonner, 1952), where $k_2$ is a pseudo-first order constant. The steady-state saturation velocity and Michaelis constant for this mechanism are:

$$V_{H_2O_2} = k_2' A$$

$$K_m^{H_2O_2} = \frac{k_2' A + k_{-1}}{k_1}.$$

First, $k_2'$ can be determined from the saturation velocity per unit enzyme.

TABLE 10.1

Minimal estimate of $k_1$, the bimolecular combination rate constant between substrate and enzyme. (From Peller and Alberty, 1959.)

| Enzyme | Substrate | Minimal estimate of $k_1$ ($M^{-1}$ $sec^{-1}$) | Reference |
|---|---|---|---|
| β-Amylase | Amylose | $5.8 \times 10^7$ | Bailey and French (1957) |
| Urease | Urea | $5.0 \times 10^6$ | Wall and Laidler (1953) |
| Adenosine Triphosphatase | Adenosine triphosphate | $8 \times 10^6$ | Ouellet, Laidler and Morales (1952) |
| Cytochrome *c* reductase | Cytochrome *c* | $9 \times 10^7$ | Hogness (1942) |
| Acetylcholin-esterase | Acetylcholine | $\sim 10^9$ | Nachmansohn and Wilson (1951) |
| Fumarase | | | |
| (1) Phosphate | Fumarate | $4.2 \times 10^7$ | Frieden, Wolfe and Alberty (1957) |
| pH = 7.0 | L-malate | $1.6 \times 10^7$ | |
| (2) "Tris" acetate | Fumarate | $6.5 \times 10^8$ | |
| pH = 7.0 | L-malate | $1.6 \times 10^8$ | |

Since $K_m^{H_2O_2}$ increases linearly with $A$ with a slope equal to $k_2'/k_1$, once $k_2'$ is known $k_1$ also can be obtained. Finally, having both $k_2'$ and $k_1$, $k_{-1}$ can be calculated from $K_m^{H_2O_2}$.

The more fruitful analysis of Mechanism 10.II as compared to Mechanism 10.I underlines the advantage of having multiple concentration factors that can be varied experimentally. This advantage had been put to full use in the three enzyme systems next to be considered. With chymotrypsin, the useful factors were a series of structurally related substrates; comparison of their kinetic properties helped to identify the rate-limiting steps in the reaction mechanism. With carbamate kinase and liver alcohol dehydrogenase, they were the two substrates and two products of, respectively, the phosphate-transfer and hydrogen-transfer reactions.

## 10.3.1. Chymotrypsin

Mechanism 10.III has been proposed for the chymotryptic hydrolysis of acyl-esters (Hartley and Kilby, 1954) and of acyl-amides (Gutfreund and Sturtevant, 1956).

$$E \underset{k_{-1}}{\overset{k_1 S}{\rightleftharpoons}} ES \xrightarrow[P]{k_2} E\text{—}Q \xrightarrow[Q]{k_3} E \qquad \text{Mechanism 10.III}$$

$P$ = alcohol or amine

$Q$ = organic acid

The steady-state rate equation for this mechanism is:

$$v_0 = \frac{V_S \cdot S}{S + K_m^S}$$

$$V_S = k_{cat} = \frac{k_2 k_3}{k_2 + k_3} \qquad (10.29)$$

$$K_m^S = \frac{k_3(k_{-1} + k_2)}{k_1(k_2 + k_3)}$$

In the $k_3$-step, the acyl-enzyme $E$–$Q$ is hydrolysed to form the free organic acid. If hydroxylamine is present in the aqueous medium, it can compete against water for reaction with the acyl-enzyme to form instead the hydroxamic acid. Since the nature of $P$ has no influence on reactions involving $E$–$Q$, the mechanism predicts that the partition of $E$–$Q$ between water and hydroxylamine should not vary with the alcoholic moiety of the ester. In the case of the hippuric acyl-enzyme, this partition can be represented as:

$$\text{hippuric acyl-}E \begin{cases} \xrightarrow{k_3} E + \text{hippuric acid} \\ \xrightarrow{k_3'(NH_2OH)} E + \text{hippuryl hydroxamic acid} \end{cases}$$

Epand and Wilson (1963) found that the partition ratio between hippuric acid and hippuryl hydroxamic acid was the same for the methyl ester as

TABLE 10.2

Reaction of esters of hippuric acid with water and hydroxylamine catalysed by α-chymotrypsin. (From Epand and Wilson, 1963.)

| | Ester | Fraction of ester converted to hydroxamic acid |
|---|---|---|
| In 0.1 M hydroxylamine: | Methyl | 0.374–0.380 |
| | Ethyl | 0.358, 0.375 |
| | Isopropyl | 0.378 |
| | Isobutyl | 0.363 |
| | Choline bromide | 0.378, 0.369 |
| | Homocholine bromide | 0.366–0.379 |
| | 4-Pyridinemethyl | 0.367, 0.369 |
| In 0.1 M hydroxylamine, 12% isopropanol: | Methyl | 0.273, 0.281 |
| | Isoamyl | 0.262, 0.271 |
| | Benzyl | 0.277, 0.285 |
| | Glycerol | 0.275 |

for nine other esters (Table 10.2). This striking lack of influence by the alcoholic moiety is expected for the acyl-enzyme mechanism 10.III, but otherwise difficult to explain.

If $k_3 \ll k_2$ in Eqn (10.29), the expression for $k_{cat}$ is reduced to,

$$k_{cat} = \frac{k_2 k_3}{k_2 + k_3} \simeq \frac{k_2 k_3}{k_2} = k_3$$

and $k_{cat}$ becomes a measure of $k_3$. Accordingly, $k_{cat}$ also should not vary with the alcoholic moiety of the ester. This appears to be the case in the chymotryptic hydrolyses of the ethyl, methyl and *p*-nitrophenyl esters of *N*-acetylphenylalanine. The $k_{cat}$s for these hydrolyses were about the same, even though *p*-nitrophenol is a far better leaving group than ethanol or methanol (Zerner, Bond and Bender, 1964).

Contrary to the case of esters, there is little evidence that $k_3 \ll k_2$ for the hydrolysis of amides. Brandt, Himoe and Hess (1967) in fact found that the steady-state Michaelis constant $K_m^S$ (31 mM) for *N*-acetyl-L-phenylalanine amide was close to the enzyme-substrate dissociation constant $K_S$ (28 mM). This finding suggests that $k_2 < k_3$ or $k_{-1}$, so that

$$K_m^S \simeq \frac{k_3 k_{-1}}{k_1 k_3} = \frac{k_{-1}}{k_1}, \text{ namely } K_S$$

Consequently, the $k_{cat}$ estimated from steady-state velocities is a satisfactory estimate for $k_3$ in the case of the ester substrates under favorable conditions, but not in the case of the amide substrates.

### 10.3.2. Carbamate kinase

The mechanism for carbamate kinase was identified by Marshall and Cohen (1966) to be:

$$E \underset{k_{-1}}{\overset{k_1(\text{MgATP})}{\rightleftharpoons}} E/\text{MgATP} \underset{k_{-2}}{\overset{k_2\,(\text{C})}{\rightleftharpoons}} \begin{matrix}\text{Central}\\ \text{complex}\end{matrix} \underset{k_{-3}(\text{CP})}{\overset{k_3}{\rightleftharpoons}} E/\text{MgADP}$$

$$\underset{k_{-4}(\text{MgADP})}{\overset{k_4}{\rightleftharpoons}} E \qquad\qquad \text{Mechanism} \quad 10.\text{IV}$$

The forward and reverse initial velocities for this enzyme are characterized by eight kinetic parameters, all of which could be obtained from primary and secondary linear plots (see Section 6.5). Table 10.3 shows the compositions of these parameters in terms of the rate constants, together with their experimental values. This matching between the compositions and the experimental values of the parameters is sufficient basis for calculating in successive stages all eight of the elementary rate constants. Each stage of calculation in Table 10.3 utilizes the rate constants already calculated in the preceding stages. Thus, knowing $k_1$ from Stage 1, the relation $k_{-1}/k_1 = 1.2 \times 10^{-4}$ M is sufficient basis for calculating $k_{-1}$ in Stage 2, and so on. The steady-state calculation of all eight rate constants permitted by this mechanism stands in striking constrast to the calculation of only one of the three rate constants permitted by Mechanism 10.I. This contrast confirms that, for the purpose of rate constant estimation, the advantage of having four instead of one controllable concentration factors in the mechanism can far outweigh the disadvantage of having to treat a more complex mechanism.

TABLE 10.3

Calculation of rate constants for carbamate kinase (Modified from Marshall and Cohen, 1966.)†

| Stage | Composition and experimental value of parameter | Calculated rate constant |
|---|---|---|
| 1 | $\phi_{MgATP} = \frac{1}{k_1} = 9.0 \times 10^{-8}$ M sec | $k_1 = 1.1 \times 10^7\ M^{-1}\ sec^{-1}$ |
| 2 | $K_{iATP} = \frac{\phi_{MgATP \cdot C}}{\phi_C} = \frac{k_{-1}}{k_1} = 1.2 \times 10^{-4}$ M | $k_{-1} = 1.3 \times 10^3\ sec^{-1}$ |
| 3 | $\frac{1}{V_r} = \frac{1}{k_{-1}} + \frac{1}{k_{-2}} = 1.37 \times 10^{-3}$ sec | $k_{-2} = 1.7 \times 10^3\ sec^{-1}$ |
| 4 | $\phi_{MgADP} = \frac{1}{k_{-4}} = 7.0 \times 10^{-8}$ M sec | $k_{-4} = 1.4 \times 10^7\ M^{-1}\ sec^{-1}$ |
| 5 | $K_{iADP} = \frac{\phi_{MgADP \cdot CP}}{\phi_{CP}} = \frac{k_4}{k_{-4}} = 7.6 \times 10^{-6}$ M | $k_4 = 106\ sec^{-1}$ |
| 6 | $\frac{1}{V_f} = \frac{1}{k_3} + \frac{1}{k_4} = 1.09 \times 10^{-2}$ sec | $k_3 = 680\ sec^{-1}$ |
| 7 | $\phi_C = \frac{k_3 + k_{-2}}{k_2 k_3} = 8.4 \times 10^{-7}$ M sec | $k_2 = 4.2 \times 10^6\ M^{-1}\ sec^{-1}$ |
| 8 | $\phi_{CP} = \frac{k_{-2} + k_3}{k_{-3} k_{-2}} = 1.34 \times 10^{-7}$ M sec | $k_{-3} = 1.0 \times 10^7\ M^{-1}\ sec^{-1}$ |

† It might be noted that Marshall and Cohen obtained two different values for each of $k_{-1}/k_1$ and $k_4/k_{-4}$ from two different experimental routes of estimation. Correspondingly, different routes of calculation on these bases would yield nonidentical sets of rate constants. The particular set presented in this table is chosen only because of its very plain sequence of deduction. The original study by Marshall and Cohen should be consulted for judicious comments regarding the reliability of the various estimates.

### 10.3.3. Liver alcohol dehydrogenase

The mechanism for ethanol oxidation by the alcohol dehydrogenase of horse liver, at non-excess concentrations of ethanol and in the absence of reaction products, was established by Hanes *et al.* (1972) to be:

$$E/-/- \underset{k_{-1}}{\overset{k_1\,\mathrm{eth}}{\rightleftharpoons}} E/-/\mathrm{eth} \underset{k_{-3}}{\overset{k_3\,\mathrm{O}}{\rightleftharpoons}} \boxed{E/O/\mathrm{eth} \rightleftharpoons E/R/\mathrm{ald}}$$

$$E/-/- \underset{k_{-2}}{\overset{k_2\,\mathrm{O}}{\rightleftharpoons}} E/O/- \underset{k_{-4}}{\overset{k_4\,\mathrm{eth}}{\rightleftharpoons}} \boxed{E/O/\mathrm{eth} \rightleftharpoons E/R/\mathrm{ald}}$$

$$E/R/\mathrm{ald} \xrightarrow[k_5]{\rightarrow \mathrm{ald}} E/R/- \xrightarrow[k_6]{\rightarrow R} E/-/-$$

Mechanism 10.V

where (eth) stands for ethanol, (ald) for acetaldehyde, ($O$) for $NAD^+$ and ($R$) for NADH (see Section 5.3). The initial rate equation for this mechanism is (cf. solution to Problem 2.1):

$$v_f = \frac{\text{NUMERATOR}}{\text{DENOMINATOR}}$$

$$\text{NUMERATOR} = k_1k_3k_4k_5k_6(\mathrm{eth})^2(\mathrm{NAD}^+) + k_2k_3k_4k_5k_6(\mathrm{eth})(\mathrm{NAD}^+)^2 + (k_1k_{-2}k_3 + k_{-1}k_2k_4)k_5k_6(\mathrm{eth})(\mathrm{NAD}^+)$$

$$\begin{aligned}
\text{DENOMINATOR} = {} & k_1k_3k_4(k_5+k_6)(\mathrm{eth})^2(\mathrm{NAD}^+) + k_1k_4k_6(k_{-3}+k_5)(\mathrm{eth})^2 \\
& + k_2k_3k_4(k_5+k_6)(\mathrm{eth})(\mathrm{NAD}^+)^2 + [k_1k_3(k_{-2}k_5 + k_{-2}k_6 + k_{-4}k_6) \\
& + k_{-1}k_2k_4(k_5+k_6) + k_4k_6(k_2k_{-3} + k_3k_5)]\,(\mathrm{eth})(\mathrm{NAD}^+) \\
& + [k_1k_{-2}k_6(k_{-3}+k_{-4}+k_5) + k_{-1}k_4k_6(k_{-3}+k_5)]\,(\mathrm{eth}) \\
& + k_2k_3k_6(k_{-4}+k_5)(\mathrm{NAD}^+)^2 + [k_{-1}k_2k_6(k_{-3}+k_{-4}+k_5) \\
& + k_{-2}k_3k_6(k_{-4}+k_5)](\mathrm{NAD}^+) + k_{-1}k_{-2}k_6(k_{-3}+k_{-4}+k_5)
\end{aligned} \qquad (10.30)$$

Unlike the linear carbamate-kinase mechanism, for which all eight rate constants are obtainable from algebraic analysis, the ten rate constants in Mechanism 10.V could not be extracted solely by algebraic analysis. Instead, this branching system usefully illustrates how algebraic analysis could be supplemented by computer curve-fitting and statistical analysis to serve the purpose of estimation of rate constants.

First, numerical relationships between rate constants, as well as lower or upper limits for some of the constants, were established through algebraic analysis. As summarized in Table 10.4, such analysis placed narrow limits on $k_6$ (Eqn 10.36), required $k_5$ to be greater than 32.5 $sec^{-1}$ (Eqn 10.37), $k_4$ to be greater than 8.9 $sec^{-1}$ $mM^{-1}$ (Eqn 10.38), and $k_3$ to be greater than 302 $sec^{-1}$ $mM^{-1}$ (Eqn 10.39). Moreover, $k_{-4}$, $k_{-3}$ and $k_{-2}$ could be calculated from the other constants on the basis of, respectively, Eqns (10.38), (10.39) and (10.40). Therefore, of the ten rate constants, only $k_1$, $k_{-1}$ and $k_2$ can vary without restriction.

To estimate precisely the ten rate constants, the results of the algebraic analysis were used as a starting point. Direct search methods of nonlinear regression (see Section 11.3) were employed to trace the influences of each rate constant on the velocity predictions, in order to arrive at a best fit of Eqn (10.30) to the data represented by curve B and Sets 20–24 in Fig. 5.3. The best-fit rate constants are summarized in Fig. 10.6. As Fig.

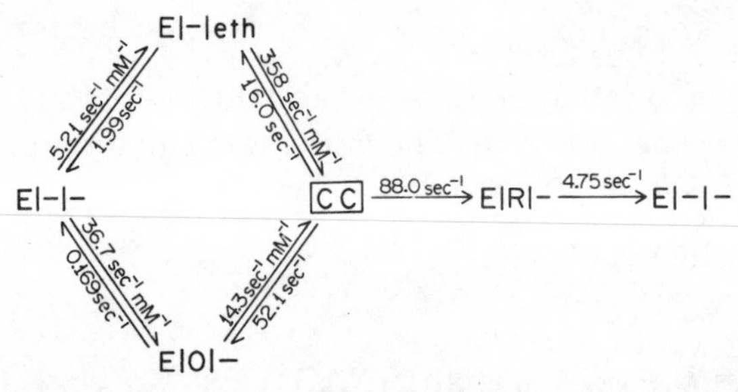


**Fig. 10.6.** Rate constants of the liver alcohol dehydrogenase mechanism. (From Wong and Hanes, 1973.)

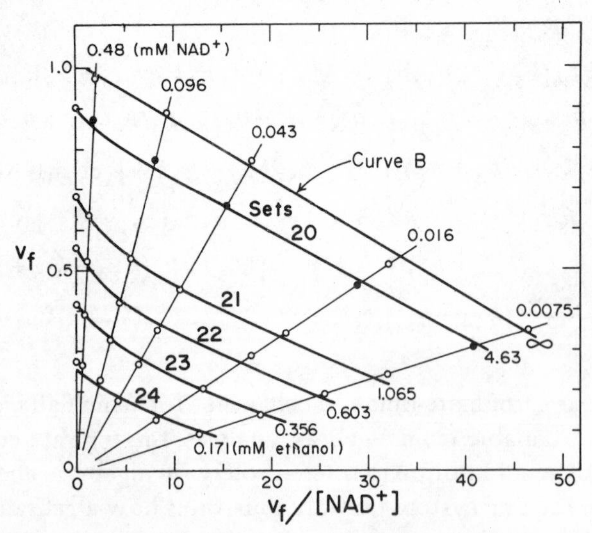


**Fig. 10.7.** Fit between theoretical curves and data points of liver alcohol dehydrogenase. (From Wong and Hanes, 1973.)

10.7 shows, these rate constants gave a satisfactory fit to the placement and curvature of the six sets of data points. The sensitivity of this best fit will be examined in Section 11.3.

Taniguchi, Theorell and Akeson (1967) have studied $NAD^+$ binding to the enzyme, and the variation of the enzyme-substrate dissociation constant with pH. At pH 8.6, the direct estimate of this constant would be about 11–12 $\mu$M. From Fig. 10.6, the kinetic estimate of this constant at the same pH would be

$$K_{NAD^+} = \frac{k_{-2}}{k_2} = 4.6\ \mu\text{M}.$$

The agreement between these two independent estimates provides strong confirmation for this set of rate constants.

The rate constants provide a precise description of the heterotropic interaction between the two cosubstrates, ethanol and $NAD^+$, which emerges from a comparison of steps 1 and 4, and of steps 2 and 3:

$$k_4/k_1 = 2.7$$

$$k_3/k_2 = 9.7$$

$$k_{-4}/k_{-1} = 26$$

$$k_{-3}/k_{-2} = 95$$

Thus the presence of $NAD^+$ on the enzyme promotes the combination of ethanol with the enzyme by 2.7-fold, but promotes the dissociation of ethanol from the enzyme by 26-fold. Analogously, the presence of ethanol on the enzyme promotes the combination of $NAD^+$ with the enzyme by 9.7-fold, but promotes the dissociation of $NAD^+$ from the enzyme by 95-fold. Consequently, the presence of each substrate opens up the binding site for its cosubstrate so that *both combination and dissociation are facilitated.* In each case, dissociation is promoted more than combination, so that the net effect is one of destabilized binding of the cosubstrate.

Although ligand-binding studies are commonly employed to characterize homotropic and heterotropic interactions, they convey far less information than rate studies. For instance, ethanol can destabilize the binding of $NAD^+$ to an enzyme site by any one of three kinds of effects:

(1) Ethanol opens up the $NAD^+$-site: both $k_{-3}/k_{-2}$ and $k_3/k_2$ are greater than unity, but the former exceeds the latter.

(2) Ethanol tightens the $NAD^+$-site: both $k_{-3}/k_{-2}$ and $k_3/k_2$ are less than unity, but the former exceeds the latter.

(3) Ethanol causes $k_{-3}/k_{-2}$ to be greater than unity, but $k_3/k_2$ to be less than unity.

TABLE 10.4

Analysis of initial velocities of liver alcohol dehydrogenase. (From Wong, Gurr, Bronskill and Hanes, 1972.)

---

(a) Figure 5.3 describes the variation of initial velocity with (eth) and ($NAD^+$). It yields:

$$V_{\text{eth,NAD}} = \frac{k_5 k_6}{k_5 + k_6} = 4.44 \text{ sec}^{-1} \tag{10.31}$$

$$K_m^{\text{eth}} = \frac{k_{-4} k_6 + k_5 k_6}{k_4(k_5 + k_6)} = 0.50 \text{ mM}. \tag{10.32}$$

$$K_m^{\text{NAD}} = \frac{k_{-3} k_6 + k_5 k_6}{k_3(k_5 + k_6)} = 0.0147 \text{ mM}. \tag{10.33}$$

---

(b) Figure 5.5 describes the action of (acetaldehyde) on the forward reaction. Its effects on $K_{(m)}^{\text{ethanol}}$ is described by:

$$K_{(m)}^{\text{ethanol}} = \frac{k_{-4} k_6 + k_5 k_6 + k_{-4} k_{-5} \text{(acetaldehyde)}}{k_4 k_5 + k_4 k_6 + k_4 k_{-5} \text{(acetaldehyde)}}$$

This Michaelis constant was close to 4 when (acetaldehyde) became very large; therefore,

$$\frac{k_{-4}}{k_4} \simeq 4 \text{ mM}. \tag{10.34}$$

---

(c) Equation (10.32) can be split into two terms, and each part must be smaller than the whole:

$$\frac{k_{-4} k_6}{k_4(k_5 + k_6)} < 0.50 \text{ mM}.$$

Substitution of (10.34) into this equation gives

$$\frac{k_6}{k_5 + k_6} < 0.12$$

which in turn implies that

$$\frac{k_5}{k_5 + k_6} > 0.88 \tag{10.35}$$

Equation (10.31) by itself requires each of $k_5$ and $k_6$ to exceed 4.44 $\text{sec}^{-1}$. Substitution of (10.35) into (10.31) further requires $k_6$ to be less than 5.05 $\text{sec}^{-1}$. Consequently,

$$4.44 \text{ sec}^{-1} < k_6 < 5.05 \text{ sec}^{-1} \tag{10.36}$$

and combination of (10.35) and (10.36) yields

$$k_5 > 32.5 \text{ sec}^{-1} \tag{10.37}$$

---

Table 4—*cont.*

(d) Division of (10.31) by (10.32) gives:

$$\frac{k_4 k_5}{k_{-4} + k_5} = 8.9 \text{ sec}^{-1} \text{ mM}^{-1} \tag{10.38}$$

Division of (10.31) by (10.33) gives,

$$\frac{k_3 k_5}{k_{-3} + k_5} = 302 \text{ sec}^{-1} \text{ mM}^{-1}. \tag{10.39}$$

Also, microscopic reversibility stipulates that the equilibrium constants around a closed reaction loop should be equal to unity, i.e.

$$k_1 k_{-2} k_3 k_{-4} = k_{-1} k_2 k_{-3} k_4 \tag{10.40}$$

---

The end result of all three effects is a destablized binding of $NAD^+$, viz.

$$\frac{k_{-3}}{k_3} = K'_{\text{NAD}} > K_{\text{NAD}} = \frac{k_{-2}}{k_2}$$

Measurements of $NAD^+$-binding could not distinguish between the three kinds of effects. Rate studies could, and the values of $k_3/k_2$ and $k_{-3}/k_{-2}$ clearly indicate that Effect-1 operates for liver alcohol dehydrogenase.

The use of computer curve-fitting to estimate rate constants is by no means restricted to the treatment of initial velocities. Hess, Chance, Busse and Wurster (1972) have applied curve-fitting to the whole progress curves of lactate dehydrogenase, and estimated all eight rate constants in Mechanism 10.VI:

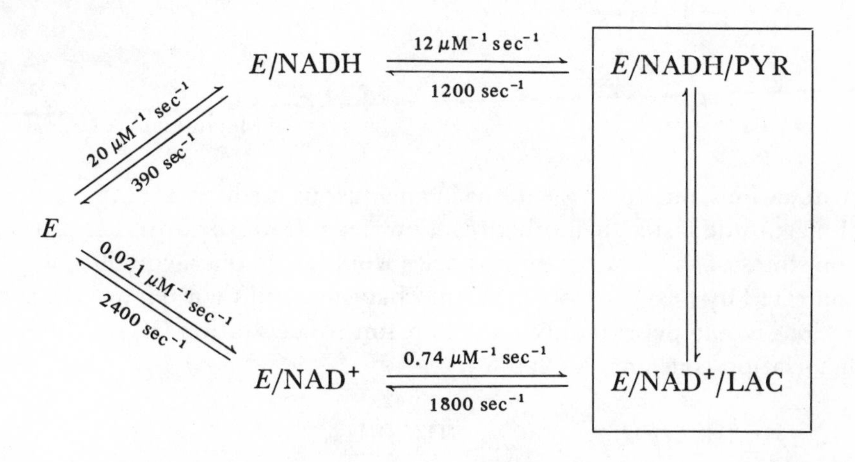


Mechanism 10.VI

where PYR stands for pyruvate, and LAC for lactate.

## 10.4. Effects of pH

Enzymes are polyelectrolytes with both positive and negative ionizations of amino acid residues. Although the distribution of charges on the enzyme molecule will vary continuously with the pH of the medium, enzyme activity often shows little change over some ranges of the pH scale, but sharp changes over other ranges. This gives rise to the concept that out of the entire constellation of charged groups, sometimes only a small number might influence the catalytic activity over a defined range of pH (Michaelis and Davidsohn, 1911). Accordingly, analysis of enzyme activity as a function of pH might reveal the identity of these essential groups.

### 10.4.1. Effects on rate constants

In general terms we may consider the conversion of any enzyme species $E_i$ to another species $E_j$:

$$\begin{aligned} &E_i \xrightarrow{k_i} E_j \\ &\text{rate} = k_i(E_i) \end{aligned} \tag{10.41}$$

Both $E_i$ and $E_j$ can occur in many states of protonation. Mechanism 10.VII represents the case where only three of these states are important:

$$\begin{array}{ccc} H_2^+E_i & \xrightarrow{k_{i2}} & H_2^+E_j \\ K_1 \Updownarrow & & \Updownarrow \\ H_1^+E_i & \xrightarrow{k_{i1}} & H_1^+E_j \\ K_2 \Updownarrow & & \Updownarrow \\ H_0^+E_i & \xrightarrow{k_{i0}} & H_0^+E_j \end{array} \qquad \text{Mechanism 10.VII}$$

Protonations and deprotonations in an aqueous medium are often an order of magnitude faster than other reaction steps. *If so,* the different protonation states of the same enzyme species would be in quasi-equilibrium (indicated by heavy arrows in the mechanism), and their relative concentrations would depend only on the proton concentration ($H^+$) and the two dissociation constants $K_1$ and $K_2$:

$$K_1 = \frac{(H_1^+E_i)(H^+)}{(H_2^+E_i)}, \qquad K_2 = \frac{(H_0^+E_i)(H^+)}{(H_1^+E_i)}$$

$$(H_0^+E_i) = \frac{K_1K_2}{(H^+)^2 + (H^+)K_1 + K_1K_2} \cdot \Sigma(E_i) = \psi_0\Sigma(E_i)$$

$$(H_1^+E_i) = \frac{(H^+)K_1}{(H^+)^2 + (H^+)K_1 + K_1K_2} \cdot \Sigma(E_i) = \psi_1 \Sigma(E_i)$$

$$(H_2^+E_i) = \frac{(H^+)^2}{(H^+)^2 + (H^+)K_1 + K_1K_2} \cdot \Sigma(E_i) = \psi_2 \Sigma(E_i) \quad (10.42)$$

The three expressions $\psi_0$, $\psi_1$ and $\psi_2$ so defined specify the fraction of total $E_i$ occurring in each of the three protonation states. The *macroscopic* rate given by Eqn (10.41) can be resolved into its three *microscopic* components:

$$\begin{aligned} \text{rate} &= k_{i0}(H_0^+E_i) + k_{i1}(H_1^+E_i) + k_{i2}(H_2^+E_i) \\ &= (k_{i0}\psi_0 + k_{i1}\psi_1 + k_{i2}\psi_2)(\Sigma E_i). \end{aligned} \quad (10.43)$$

Comparison of Eqns (10.41) and (10.43) reveals the relationship between the macroscopic rate constant and the three microscopic rate constants:

$$k_i = k_{i0}\psi_0 + k_{i1}\psi_1 + k_{i2}\psi_2. \quad (10.44)$$

More generally, if there are $n$ different protonation states, the macroscopic rate constant may be resolved into a sum of $n$ terms:

$$k_i = k_{i0}\psi_0 + k_{i1}\psi_1 + k_{i2}\psi_2 + \cdots + k_{in}\psi_n. \quad (10.45)$$

Equation (10.45) defines the *pH-profile* of the rate constant $k_i$. A complete description of the pH effects on any enzyme mechanism may be obtained by *replacing every rate constant in the rate law by its pH-profile.*

The three simplest pH-profiles for $k_i$ are in Fig. 10.8:

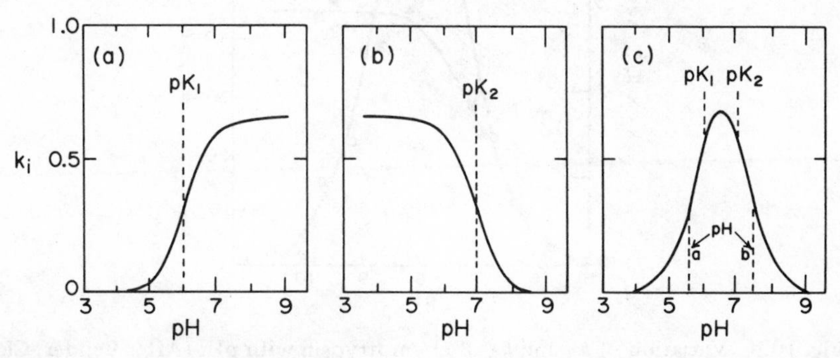


Fig. 10.8. $k_i$-pH profiles. (a) left-sigmoid; (b) right-sigmoid; (c) bell-shaped (note that $pH_a \neq pK_1$, and $pH_b \neq pK_2$). (After Alberty and Massey, 1954.)

(a) *Left-sigmoid*: $H_2^+E_i$ is inactive ($k_{i2} = 0$), whereas $H_1^+E_i$ and $H_0^+E_i$ are equally active ($k_{i1} = k_{i0}$). Consequently the enzyme is inactive at lower pH but becomes active at higher pH. The rise of enzyme activity with pH is sigmoidal, with pH $= pK_1$ at the midpoint of the rise.

(b) *Right-sigmoid*: $H_0^+E_i$ is inactive ($k_{i0} = 0$), whereas $H_1^+E_i$ and $H_2^+E_i$ are equally active ($k_{i1} = k_{i2}$). Consequently the enzyme is active at lower pH but becomes inactive at higher pH. The fall of enzyme activity with pH is sigmoidal, with pH = $pK_2$ at the midpoint of the fall.

(c) *Bell-shaped:* Both $H_2^+E_i$ and $H_0^+E_i$ are inactive ($k_{i2}$ and $k_{i0} = 0$). Consequently, the enzyme is active at intermediate pH, but exhibits a sigmoidal fall on both sides. $(H^+)_a$ and $(H^+)_b$, the midpoints of the fall on the two sides, are related to $K_1$ and $K_2$ by Eqn (10.46) in the event that $\Sigma(E_i)$ does not vary with pH:

$$K_1K_2 = (H^+)_a(H^+)_b \qquad (10.46)$$
$$K_1 = (H^+)_a + (H^+)_b - 4\sqrt{(H^+)_a(H^+)_b}$$

$K_1$ and $K_2$ are not the same as $(H^+)_a$ and $(H^+)_b$. However, if the two sigmoids are widely separated from one another, $(H^+)_b$ and $4\cdot\sqrt{(H^+)_a(H^+)_b}$ will be much smaller than $(H^+)_a$, and the latter will be very close to $K_1$. It also follows that $(H^+)_b$ will be very close to $K_2$. Equation (10.46) was first derived by Alberty and Massey (1954).

Experimental examples of these elementary pH profiles are well known. Fig. 10.9 shows the pH-profiles of $k_2$ and $k_3$ evaluated by Bender, Clement,

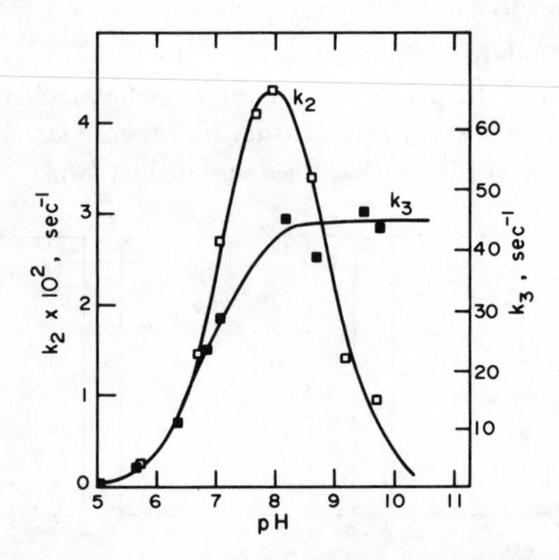


Fig. 10.9. Variation of $k_2$ and $k_3$ of chymotrypsin with pH. (After Bender, Clement, Kezdy and Heck, 1964.)

Kezdy and Heck (1964) for chymotrypsin on the basis of Mechanism 10.III. The $k_3$-profile was left-sigmoid with a $pK$ close to 7. The $k_2$-profile was largely bell-shaped, with p$K_1$ close to 7 and p$K_2$ close to 9.

When the three protonation states of the enzyme are all active, the pH-profile can be analysed on the basis of Eqn (10.44) to yield not only $K_1$ and $K_2$, but also $k_{i0}$, $k_{i1}$ and $k_{i2}$. This kind of analysis was exemplified by the studies of Winer and Schwert (1958) on lactic dehydrogenase, and

Raval and Wolfe (1962) on malic dehydrogenase, which yielded the p$K$s and the microscopic rate constants of NADH binding for these enzymes. All three protonation states of both enzymes were active in binding NADH, and p$K_1$ and p$K_2$ (more especially p$K_2$) were enhanced in the enzyme-NADH complex relative to the free enzyme.

### 10.4.2. One-substrate reactions

The preceding analysis of pH effects on rate constants is general to all classes of enzyme mechanisms. The estimates of any rate constant obtained at different pHs will define its pH-profile, which in turn will serve as a basis for evaluating the p$K$s and the microscopic rate constants. In this connection the one-substrate Mechanism 10.I poses a special problem insofar that only one of its three rate constants is obtainable from steady-state kinetics for analysis.

The steady-state behaviour of Mechanism 10.I at any constant pH is described by Eqn (10.47):

$$v_0 = \frac{V_S \cdot \mathrm{S}}{\mathrm{S} + K_m^S}$$

$$V_S = k_2 \tag{10.47}$$

$$K_m^S = \frac{k_{-1} + k_2}{k_1}.$$

As the pH is varied within limits of full reversibility, the relationships between the three macroscopic rate constants and the microscopic protonation-states become all important. Waley (1953) has formulated a fundamental model for delineating these relationships. In this model the free enzyme and the enzyme-substrate complex each have three protonation states which are in quasi-equilibrium relative to one another, but only the middle states are functionally active:

$$\begin{array}{ccccc}
\mathrm{H}_2^+E & & \mathrm{H}_2^+ES & & \\
K_1 \updownarrow & & \updownarrow K_1' & & \\
\mathrm{H}_1^+E & \underset{k_{-11}}{\overset{k_{11}\mathrm{S}}{\rightleftharpoons}} & \mathrm{H}_1^+ES & \xrightarrow{k_{21}} & \mathrm{H}_1^+E + \text{Products} \\
K_2 \updownarrow & & \updownarrow K_2' & & \\
\mathrm{H}_0^+E & & \mathrm{H}_0^+ES & &
\end{array}$$

Mechanism 10.VIII

Since $\mathrm{H}_1^+E$ is the only active form of the free enzyme, and $\mathrm{H}_1^+ES$ the only active form of the enzyme-substrate complex, the pH-profiles for the three macroscopic rate constants $k_1, k_{-1}$ and $k_2$ are each described by just a single term:

$$k_1 = k_{11}\psi_{\mathrm{H_1}E}$$
$$k_{-1} = k_{-11}\psi_{\mathrm{H_1}ES} \qquad (10.48)$$
$$k_2 = k_{21}\psi_{\mathrm{H_1}ES}$$

where $\psi_{\mathrm{H_1}E}$ or $\psi_{\mathrm{H_1}ES}$ is the fraction of $\Sigma(E)$ or $\Sigma(ES)$ found in the active form:

$$\psi_{\mathrm{H_1}E} = \frac{(\mathrm{H^+})K_1}{(\mathrm{H^+})^2 + (\mathrm{H^+})K_1 + K_1K_2}$$
$$\psi_{\mathrm{H_1}ES} = \frac{(\mathrm{H^+})K'_1}{(\mathrm{H^+})^2 + (\mathrm{H^+})K'_1 + K'_1K'_2}. \qquad (10.49)$$

Substitution of Eqns (10.48) into (10.47) yields the pH-dependence of the *saturation velocity* and of the *Michaelis constant:*

$$V_S = k_{21}\cdot\psi_{\mathrm{H_1}ES}$$
$$K_m^S = \frac{k_{-11} + k_{21}}{k_{11}}\cdot\frac{\psi_{\mathrm{H_1}ES}}{\psi_{\mathrm{H_1}E}} \qquad (10.50)$$
$$\frac{V_S}{K_m^S} = \frac{k_{21}\cdot k_{11}}{k_{-11} + k_{21}}\cdot\psi_{\mathrm{H_1}E}$$

Since $V_S$ is proportional to $\psi_{\mathrm{H_1}ES}$, the dissociation constants $K'_1$ and $K'_2$ can be determined from the bell-shaped variation of $V_S$ with pH. And, since $V_S/K_m^S$ is proportional to $\psi_{\mathrm{H_1}E}$, the dissociation constants $K_1$ and $K_2$ can be estimated from the bell-shaped variation of $V_S/K_m^S$ with pH. Both of these variations are bell-shaped because only $\mathrm{H_1^+}ES$ and $\mathrm{H_1^+}E$ are active. If $\mathrm{H_0^+}ES$ or $\mathrm{H_2^+}ES$ is as active as $\mathrm{H_1^+}ES$, the pH profile for $V_S$ would be correspondingly left or right-sigmoid. Similarly, if $\mathrm{H_0^+}E$ or $\mathrm{H_2^+}E$ is as active as $\mathrm{H_1^+}E$, the pH-profile for $V_S/K_m^S$ would be correspondingly left or right sigmoid.

The convenient fact that $K_1$ and $K_2$ can be obtained from the pH-profile of $V_S/K_m^S$ is due to the same pH-function $\psi_{\mathrm{H_1}ES}$ being applicable to both $k_2$ and $k_{-1}$. Consequently this function is cancelled out from the expression for $V_S/K_m^S$, leaving behind only $\psi_{\mathrm{H_1}E}$ in the expression. If $k_{-1}$ and $k_2$ have different pH profiles, this advantage would be lost, and interpretation would have to rely on a more sophisticated analysis (Bloomfield and Alberty, 1962).

The Michaelis constant $K_m^S$, unlike either $V_S$ or $V_S/K_m^S$, is related to both $\psi_{\mathrm{H_1}ES}$ and $\psi_{\mathrm{H_1}E}$. Nevertheless, Walker and Schmidt (1944) and Dixon (1953b) have been able to develop an interpretation of its more complex dependence on pH. Transforming $K_m^S$ from Eqn (10.50) into negative logarithms gives (denoting "−log" by "$p$"):

$$pK_m^S = pK_{m1} + p(\psi_{H_1ES}) - p(\psi_{H_1E}) \qquad (10.51)$$

where $K_{m1}$ is $(k_{-11} + k_{21})/k_{11}$, the Michaelis constant specifically for the middle protonation state. In the event that the substrate also undergoes ionization, Eqn (10.51) may be further expanded to (10.52):

$$pK_m^S = pK_{m1} + p(\psi_{H_1ES}) - p(\psi_{H_1E}) - p(\psi_S) \qquad (10.52)$$

where $\psi_S$ describes the fraction of substrate in its active protonation state. If the various pKs involved are well separated from one another, the variation of $pK_m^S$ with pH can be interpreted by means of a set of rules first formulated for *quasi-equilibrium* kinetics by Dixon (1953b) and later extended to *steady-state* kinetics by Dixon and Webb (1958):

(a) If the p*K*s are well separated, the plot of $pK_m^S$ versus pH will consist of straight-line sections joined by short curved bends.

(b) The straight sections have an integral slope equal to −2, −1, 0, +1, or +2.

(c) Each bend indicates the p*K* of an ionizing group. The straight sections on the two sides of the bend when extrapolated will intersect at a pH equal to this p*K*.

(d) Each p*K* produces a unit change in the slope of the straight sections.

(e) The change in slope is positive (i.e. a concave bend) for each p*K* in the *ES* complex, but negative (i.e. a convex bend) for each p*K* in the free *E* or free *S*.

(f) The curved bend misses the intersection point of the straight sections by a distance equal to log 2 if one p*K* group is involved in the bend, or by a distance equal to log 3 if two p*K* groups are involved.

(g) The integral slope value of any straight-line section is equal to the change of charge occurring in that pH range when the enzyme-substrate complex dissociates into free enzyme and free substrate, e.g.

$ES^{-n} \rightarrow E^{-n} + S$ slope = 0

$ES^{-n} \rightarrow E^{-(n+1)} + S^{+1}$ slope = 0

$ES^{-n} \rightarrow E^{-(n-1)} + S$ slope = +1

These rules, the valid range of which has been rigorously defined by Reiner (1959), do not distinguish between the protonations of the free enzyme and the free substrate, but the latter could be determined separately. The observations by Dodgson, Spencer and Williams (1955) on the arylsulphatase of *Alcaligenes metalcaligenes* provide an elegant illustration of the variation of $pK_m^S$ with pH. The plots in Fig. 10.10 for three different substrates each showed two convex bends, indicating two important proton dissociations for the free enzyme, with $pK_1$ about 8.1 and $pK_2$ about 9.4; there was only one concave bend, indicating one important proton dissociation for the enzyme-substrate complex, with p*K* about 7.5–7.8. The substrate nitrocatechol sulphate had a phenolic group; it showed an additional convex bend, p*K* about 6.5, for the free substrate, and an additional concave bend, p*K* about 6.9, for the enzyme-substrate complex.

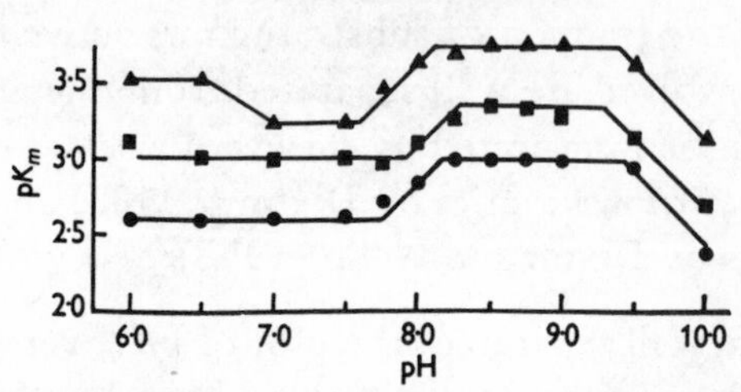


Fig. 10.10. Variation of $pK_m$ of arylsulphatase for three different substrates with pH: ▲—nitrocatechol sulphate; ■—*p*-nitrophenyl sulphate, ●—*p*-acetylphenylsulphate. (From Dodgson, Spencer and Williams, 1955.)

### 10.4.3. Role of proton dissociation

Edsall (1943) has given the p*K* and $\Delta H$ values for the ionizable groups of protein molecules. These values, summarized in Table 10.5, provide a basis for identifying the key ionizations in the enzyme mechanism. First of all, the important p*K* values extracted from the velocity-pH profiles would implicate some of the ionizable groups more strongly than others, e.g. a p*K* of about 3 would suggest a carboxyl decidedly more than an amino

TABLE 10.5

p*K* values and heats of ionization of amino acid. (From Edsall, 1943.)

| Group | p*K* (25°) | $\Delta H$ (Kcal/mole) |
|---|---|---|
| Carboxyl ($\alpha$) | 3.0–3.2 | ±1.5 |
| Carboxyl (aspartyl) | 3.0–4.7 | ±1.5 |
| Carboxyl (glutamyl) | ca. 4.4 | ±1.5 |
| Phenolic-hyroxyl (tyrosine) | 9.8–10.4 | 6 |
| Sulphydryl | 9.1–10.8 | |
| Imidazolium (histidine) | 5.6–7.0 | 6.9–7.5 |
| Ammonium ($\alpha$) | 7.6–8.4 | 10–13 |
| Ammonium ($\alpha$, cystine) | 6.5–8.5 | |
| Ammonium ($\epsilon$, lysine) | 9.4–10.6 | 10–12 |
| Guanidinium (arginine) | 11.6–12.6 | 12–13 |

group. Secondly, the $\Delta H$ of ionization might be calculated from p$K$ values obtained at different temperatures, using the thermodynamic relationship given below in Eqn (10.56). Since different ionizable groups have different $\Delta H$-values, the experimental $\Delta H$ is a useful clue for identifying the nature of any p$K$-group. The effectiveness of this diagnostic approach is illustrated by Wyman's (1939) analysis (from titrations carried out at different temperatures) of the apparent $\Delta H$ of ionization of oxyhemoglobin. As Fig. 10.11 shows, the apparent $\Delta H$ was small at low p$H$s, in keeping with the small $\Delta H$s of carboxyl ionizations. Over the middle p$H$s, the apparent $\Delta H$ rose to about 6 Kcals/mole which is consistent with imidazole. Finally, at pH 9 and above, the high $\Delta H$s attributable to the basic amino acids became apparent. From these data Wyman was able to define broadly the proton dissociations in hemoglobin which are affected by the binding of oxygen to the heme.

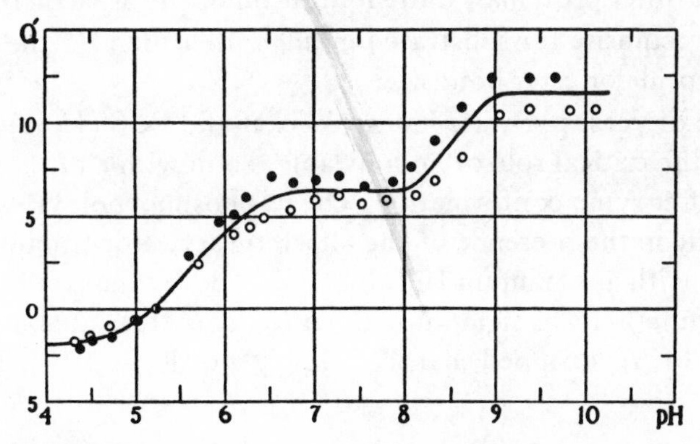


Fig. 10.11. Dependence of apparent heat of dissociation (Kcal/eq) of oxyhemoglobin on pH. Different symbols refer to experiments over different temperature ranges. (From Wyman, 1948.)

The exact p$K$ value of any amino acid residue depends on its microenvironment (Tanford, 1962). For example, in egg-white lysozyme, tyrosine-53 exhibited a p$K$ as high as 12–12.5, glutamic-35 exhibited a p$K$ as high as 6–6.5, and aspartic-66 exhibited a p$K$ as low as 1.5–2.5. Such abnormal p$K$s reflect the unusual electrostatic and hydrophobic settings in which these amino acids are situated within the tertiary enzyme structure (Hess and Rupley, 1971).

No less important a question than the identities of the prominent p$K$ groups are the roles played by these groups in the enzyme mechanism. Broadly, the function of any ionizable group is classifiable into one of three general types, even though all schemes to sub-divide an integral protein molecule into distinct *zones of function* must be ultimately more arbitrary than real, These three types of functions are (a) catalysis, (b) substrate-binding, and (c) enzyme-conformation. Examples of all three are known.

Cunningham (1957), on the basis of the velocity-pH profile of chymotrypsin, suggested that a deprotonated imidazole nitrogen serves as hydrogen-bond acceptor for the hydroxyl of serine-195, rendering the hydroxyl oxygen a stronger nucleophile. This suggestion was graphically confirmed when X-ray diffraction revealed a close contact between histidine-57 and both aspartic-102 and serine-195 (Blow, Birktoft and Hartley, 1969). The latter workers proposed that a relay system of hydrogen bonds extends from aspartic-102 to histidine-57 to serine-195, and activates the serine-195 hydroxyl preparatory to its acylation by substrate.

For trypsin, Smith and Shaw (1969) suggested that a negatively charged aspartic-177 carboxyl might be responsible for the well-known tryptic specificity for positively-charged substrates. Since this position in the amino acid sequence is occupied also by aspartic acid in proteinases with a similar propensity for positively-charged substrates, but by a neutral amino acid in other proteinases, the ionization of the aspartic residue appears to be requisite for substrate binding rather than for the catalytic scission of peptide or ester bonds.

In the case of yeast pyruvate kinase, Wieker and Hess (1971) presented evidence for the critical role of an ionizable group with a p$K$ about 5.4 in the control of enzyme conformation. The $v$ (phosphoenolpyruvate) function was hyperbolic in the presence of the allosteric activator fructose diphosphate (FDP), with a maximum Hill slope, $H_m$, close to unity. In the absence of FDP, the function was sigmoid with an $H_m$ close to 3 at above pH 7, but the value of $H_m$ dropped sharply below pH 6 (Fig. 10.12). The difference

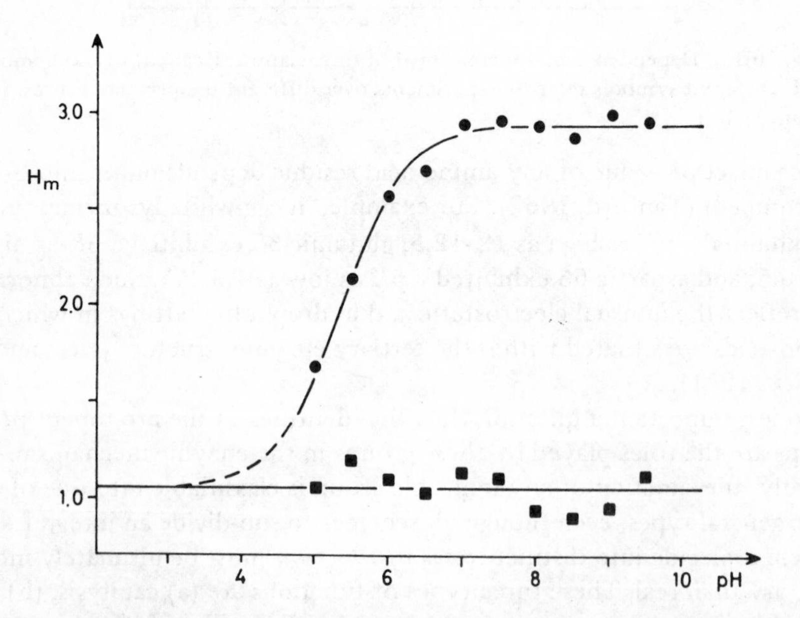


Fig. 10.12. Variation of maximal Hill slope for yeast pyruvate kinase with pH in the absence (●) or presence (■) of fructose diphosphate. (From Wieker and Hess, 1971.)

in $K_m^{PEP}$ between the FDP-activated and the non-activated forms of the enzyme also became greatly diminished below pH 6. Therefore Wieker and Hess proposed that the protonation of the p$K$ 5.4 group gave the same allosteric activation of the enzyme as accomplished by FDP, i.e. a shift from the less active $E_T$ conformation into a more active $E_R$ conformation:

$$E_R \underset{\text{FDP or H}^+}{\rightleftharpoons} E_T$$

The identical roles assigned to FDP and $H^+$ thus confer on the p$K$ 5.4 proton the mantle of an *allosteric effector* controlling enzyme conformation.

## 10.5. Effects of Temperature

### 10.5.1. Chemical equilibrium

The equilibrium constant $K$ for any chemical reaction is determined by Eqn (10.53):

$$\Delta G^0 = -RT \ln K \qquad (10.53)$$

$R$ = gas constant = 1.987 cal/degree/mole

$T$ = absolute temperature = degrees in centigrade + 273

ln = 2.303 log

$\Delta G^0$ is the *free energy change* between the product(s) and the reactant(s) in their standard states, and consists of one temperature-independent component and one temperature-dependent component:

$$\Delta G^0 = \Delta H^0 - T\Delta S^0. \qquad (10.54)$$

The symbol $H$ stands for *enthalpy* (or heat content), and $\Delta H^0$ is equal to the negative of the heat of reaction at constant pressure. The symbol $S$ stands for *entropy*, which is a measure of molecular chaos in the sense that a perfect crystal of a pure substance at absolute zero temperature, when all is still and ordered, would have zero entropy. Combination of Eqns (10.53) and (10.54) furnishes an explicit expression for $K$:

$$K = e^{-\Delta G^0/RT} = e^{-\Delta H^0/RT} \cdot e^{\Delta S^0/R} \qquad (10.55)$$

Therefore the variation of $K$ with temperature is

$$\frac{d \ln K}{dT} = \frac{\Delta H^0}{RT^2} \cdot \qquad (10.56)$$

When the reaction is *endothermic*, $\Delta H^0$ is positive, and $K$ increases with increasing temperature. When the reaction is *exothermic*, $\Delta H^0$ is negative, and $K$ decreases with increasing temperature. These prescriptions are entirely in keeping with the principle of Le Chatelier.

### 10.5.2. Activation energy and frequency factor

Higher temperatures, although they may increase or decrease the equilibrium constant, increase the rates of the elementary steps in both directions of the chemical reaction. Arrhenius in 1889 proposed that the effects of temperature on rate constants are adequately described by a simple equation, now known as the *Arrhenius equation,* which can be written in several equivalent forms:

$$\text{differential form: } \frac{d \ln k}{dT} = \frac{E}{RT^2} \tag{10.57a}$$

$$\text{logarithmic form: } \ln k = -\frac{E}{RT} + \ln A \tag{10.57b}$$

$$\text{exponential form: } k = Ae^{-E/RT} \tag{10.57c}$$

where $E$ is the *activation energy,* and $A$ is an integration constant called the *frequency factor.* Equation (10.57a) bears a striking resemblance to Eqn (10.56). In the latter, $\Delta H$ is a heat difference betweeen the reactants and the products. Analogously, $E$ may be regarded as an energy difference between the reactants of a reaction step and an *activated complex* that is an essential intermediate in the reaction step. According to the activated-complex theory of reaction rates, an equilibrium exists between the reactants and the activated complex, and the overall rate constant is proportional to this equilibrium constant $K^\ddagger$:

$$\text{Reactants} \overset{K^\ddagger}{\rightleftharpoons} \text{Activated complex} \rightarrow \text{Products}$$

$$k = \frac{RT}{Nh} \cdot (\text{transmission coef}) \cdot K^\ddagger \tag{10.58}$$

where $N$ is Avogadro's number ($6.023 \times 10^{23}$/mole), and $h$ is Planck's constant ($1.5834 \times 10^{-34}$ cal sec). The transmission coefficient expresses the efficiency of crossing the free-energy barrier and will be treated as equal to unity, even though for some enzymic reactions it might be less than unity. On the basis of Eqn (10.55), which applies to all forms of chemical equilibria, the equilibrium constant $K^\ddagger$ is expressible in terms of $\Delta G^\ddagger$, $\Delta H^\ddagger$ and $\Delta S^\ddagger$, which are respectively the free-energy, enthalpy, and entropy difference between the reactants and the activated complex:

$$K^\ddagger = e^{-\Delta G^\ddagger/RT} = e^{-\Delta H^\ddagger/RT} \cdot e^{\Delta S^\ddagger/R} \tag{10.59}$$

Substitution of this into Eqn (10.58) yields

$$k = \frac{RT}{Nh} \cdot e^{-\Delta G^\ddagger/RT} = \frac{RT}{Nh} \cdot e^{-\Delta H^\ddagger/RT} \cdot e^{\Delta S^\ddagger/R} \tag{10.60a}$$

which in turn yields,

$$\ln k = \ln \frac{RT}{Nh} - \frac{\Delta G^{\ddagger}}{RT} \tag{10.60b}$$

and,

$$\frac{d \ln k}{dT} = \frac{1}{T} + \frac{\Delta H^{\ddagger}}{RT^2}. \tag{10.60c}$$

Equations (10.57) and (10.60) represent alternate expressions for the rate constant: the former came from Arrhenius, and the latter from the absolute reaction-rate treatment of the activated-complex theory (Eyring, 1935). Lining up (10.57a) and (10.60c) leads to (10.61), and lining up (10.57c) and (10.60a) leads to (10.62) if $\Delta H^{\ddagger}$ is taken to be close to $E$:

$$E = RT + \Delta H^{\ddagger} \tag{10.61}$$

$$A = \frac{RT}{Nh} e^{\Delta S^{\ddagger}/R}. \tag{10.62}$$

When $k$ is estimated at any one temperature, Eqn (10.60b) permits the calculation of $\Delta G^{\ddagger}$. When $k$ is estimated at a series of different temperatures, it becomes possible to estimate the activation energy $E$ and the frequency factor $A$, since according to Eqn (10.57b) plotting ln $k$ versus $1/T$ would yield a linear *Arrhenius plot* with a slope equal to $-E/R$ and a vertical intercept equal to ln $A$ (Fig. 10.13). Having obtained the activation

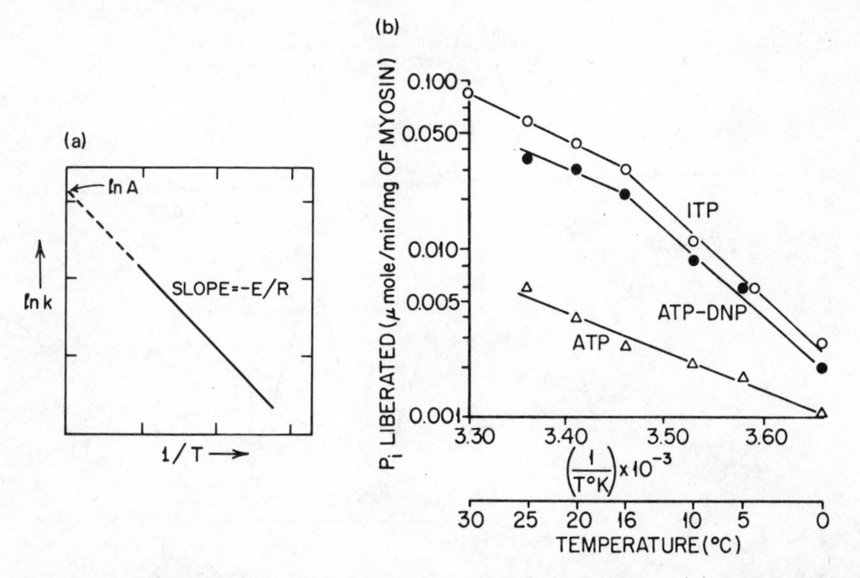


Fig. 10.13. (a) Arrhenius plot for rate constant is usually linear. (b) Arrhenius plot for reaction velocity may be straight, e.g. for myosin-catalysed hydrolysis of ATP; or segmented, e.g. for myosin-catalysed hydrolysis of ATP in the presence of dinitrophenol (DNP) or of ITP. (From Koshland, 1959.)

energy $E$ and the frequency factor $A$ by this means, Eqn (10.61) permits the calculation of the enthalpy of activation $\Delta H^{\ddagger}$, and Eqn (10.62) permits the calculation of the entropy of activation $\Delta S^{\ddagger}$. Alternatively $\Delta S^{\ddagger}$ can be calculated on the basis of $\Delta G^{\ddagger} = \Delta H^{\ddagger} - T\Delta S^{\ddagger}$. The complete $\Delta G^{\ddagger}$ profile for any enzyme mechanism is obtained by calculating the $\Delta G^{\ddagger}$ for all its elementary steps. For this purpose, Eqn (10.60b) may be rearranged into a more numerical form:

$$\Delta G^{\ddagger} = RT\left(\ln \frac{RT}{Nh} - \ln k\right)$$

$$= 1.987 \times T \times 2.303\,(\log T + \log 2.083 \times 10^{10} - \log k)\ \text{cal/mole}$$

For example, in the case of the triosephosphate isomerase mechanism (Section 6.3.2), we have $k_{-1} = 2 \times 10^4\ \text{sec}^{-1}$. At 27°,

$$\Delta G^{\ddagger} = 1.987 \times 300 \times 2.303\ (\log 300 + \log 2.083 + 10 - \log k_{-1})$$

$$= 11607\ \text{cals/mole}$$

A similar calculation can be carried out for $k_2$, $k_{-2}$, and $k_3$, as well as $k_1$(DHAP) and $k_{-3}$(GAP) if the concentrations of dihydroxyacetone phosphate (DHAP) and glyceraldehyde phosphate (GAP) are known. Such calculations made by Albery and Knowles (1974), on the basis that the

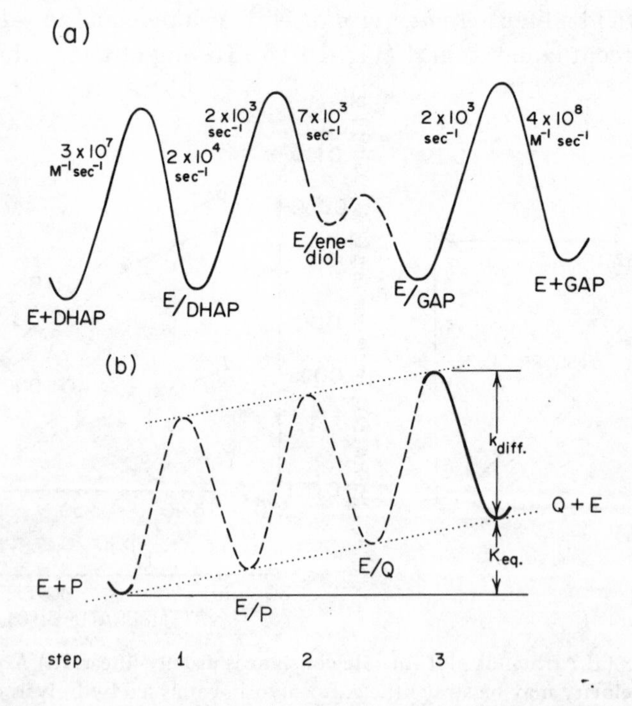


Fig. 10.14. Free-energy profile for (a) triosephosphate isomerase, and (b) for a "perfect" enzyme. (From Albery and Knowles, 1974.)

levels of DHAP and GAP are about 40 μM in muscle, gave the free-energy profile for this enzyme shown in Fig. 10.14a. Knowles (1974) noted that in this profile the other free-energy "bumps" are not much lower than the one for the $E$ + GAP reaction, nor are the "dips" much higher than the $E$ + DHAP state, and suggested that the evolution of triose phosphate isomerase has achieved catalytic *perfection.*

Figure 10.14b shows the free energy profile for what Knowles considers a perfected enzyme catalysing the interconversion of $P$ and $Q$. If the combination of $E$ and $Q$ in step 3 is already *diffusion-controlled*, then there will be little more the enzyme, or the continued evolution of the enzyme can do once it has (a) lowered the free energies of all transition states below that of step 3, and (b) raised the free energies of all intermediate states in the mechanism above that of the $E + P$ state. The profile for triosephosphate isomerase undoubtedly bears a striking resemblance to this idealized state.

The Arrhenius plot of ln $k$ versus $1/T$ is expected to be linear for any rate constant in an enzyme mechanism unless deviations due to temperature-linked phase and ionic changes in the solvent or in the enzyme molecule become important. In contrast, an Arrhenius plot of ln $v$ versus $1/T$ constructed for the overall reaction velocity may become nonlinear for a number of reasons. After all, the overall velocity depends on all the rate constants in the enzyme mechanism, and its temperature profile reflects the totality of temperature effects on all the different rate constants. If it is dominated by a single rate-limiting step throughout the temperature-range under observation, its Arrhenius plot would remain linear; otherwise the plot would be segmented. Figure 10.13b shows both linear and segmented Arrhenius velocity profiles for myosin-catalysed hydrolyses. Dixon and Webb (1958) and Levy, Sharon and Koshland (1959) have given a detailed analysis of the significance of segmented profiles.

### 10.5.3. Catalytic enhancement

The uncatalysed and enzyme-catalysed rates for the transformation of a substrate $S$ into reaction products are:

uncatalysed: $v_u = k_u \mathrm{S}$

enzyme-catalysed: $$\frac{v_e}{(E)_0} = \frac{k_2\mathrm{S}}{\mathrm{S} + \dfrac{k_2 + k_{-1}}{k_1}} = \frac{k_2\mathrm{S}}{\mathrm{S} + K_m^S}$$

where $k_1$, $k_{-1}$ and $k_2$ are the three rate constants of Mechanism 10.I, and $K_m^S$ is the Michaelis constant. The catalytic enhancement per unit enzyme is measurable by the ratio $v_e/v_u(E)_0$:

$$\frac{v_e}{v_u(E)_0} = \frac{k_2/k_u}{\mathrm{S} + K_m^S} \,. \tag{10.63}$$

Decreasing S improves catalytic efficiency, but a *high* $k_2$ and a *low* $K_m^S$ are the real roots of catalytic enhancement from the viewpoint of enzyme design. If $k_2$ is too small or $K_m^S$ too large, satisfactory enhancement would be out of reach.

What are some of the effects that contribute to a high $k_2$? Bruice (1970) has separated such effects into three major types:

(a) *Propinquity effect*
When two interacting groups (two substrates, or one substrate and one catalytic amino acid residue) are brought into contact on the enzyme surface, the rate advantage achieved is roughly 5- to 10-fold over random collisions between the same two groups in free solution. If the two groups are not only brought into mere contact, but into contact with proper orientation as well as exclusion of solvent, the rate advantage would be of the order of $10^3$-fold. If the enzyme in fact aligns a constellation of more than two, say five interacting groups (e.g. two substrates and three catalytic amino acid residues), the rate advantage achieved over a corresponding fifth-order collision in solution would be of the order of $10^{25}$-fold. Such propinquity calculations developed by Koshland (1962) and Bruice (1970) usefully define a conceivable order of magnitude of the propinquity factor. Also, Reuben (1971) has suggested that, since the mean life-time of an *ES* complex may be between $10^{-7}$–$10^{-4}$ sec, in contrast to a dissociation time constant of about $10^{-13}$ sec for small-molecule complexes in solution, this extremely long *duration* of propinquity achieved by the enzyme may give an additional rate enhancement of $10^6$–$10^9$ fold.

(b) *Milieu effect*
Within the *ES* complex, if the enzyme can present charged or polar groups to the substrate at positions where formal or partial charges are to occur in the activated complex, the milieu discrepancy between *ES* and its activated complex would be minimized. Accordingly $\Delta G^\ddagger$ would be minimized, and $k_2$ maximized.

(c) *Conformational-strain effect*
If the enzyme can alter the conformation of substrate within *ES*, so as to decrease resonance stabilization of its ground state or increase resonance stabilization of its activated state, again $\Delta G^\ddagger$ would be reduced and $k_2$ enhanced. For esterase and chymotrypsin, Hofstee (1954) found that the $\Delta G^\ddagger$ for the hydrolysis of long-chain fatty-acid esters was more favourable than short-chain ones by 550–650 cal/mole/$CH_2$: this suggests that every extra $CH_2$-group in the substrate molecule created extra strain. Also, Rupley *et al.* (1967) could demonstrate a distortion of the substrate $C_1$—O bond to be scissioned in the *ES* complex of lysozyme. Such catalytic effects constitute enthalpic or entropic compensations made by the enzyme molecule

to minimize the enthalpic or entropic discrepancy between the ground-state and the activated complex of $ES$ (Westheimer, 1962; Hammes, 1964). Although an unambiguous delineation between enthalpic and entropic contributions is always difficult, the propinquity effect might be largely entropic, and the milieu and conformational-strain effects both enthalpic and entropic. It is instructive that in the examples of rhodanase analysed by Leininger and Westley (1968), and $\Delta^5$-3-keto-isomerase analysed by Jones and Wigfield (1967), where a comparison between the enzyme-catalysed and uncatalysed rate constants could be achieved, both $\Delta H^\ddagger$ and $\Delta S^\ddagger$ catalytic effects were observed.

What are some of the elements that contribute to a low $K_m^S$? The expression for $K_m^S$ is $(k_2 + k_{-1})/k_1$. Since $k_2$ should not be too small for the sake of catalytic enhancement, it becomes essential to have a large $k_1$ and/or a small $k_{-1}$. Therefore the dissociation constant $k_{-1}/k_1$ should be small, which means that the $\Delta G$ of $ES$ formation should be as low as possible. In general, the major sources of free energy that can act to reduce $\Delta G$ are the electrostatic, hydrophobic and hydrogen bonding forces between enzyme and substrate, but the potential contribution from a *substrate-induced stabilization* of enzyme structure in some systems also should not be completely overlooked.

Changes in enzyme structure induced by the binding of substrates are well known (Koshland, 1958). Jencks (1966) had proposed that the substrate-induced structure of the enzyme must be energetically less favourable relative to the native structure, because the enzyme exists in the native structure in the absence of substrate. Although this proposal may be valid perhaps for most enzymes, it is not inherently infallible. In the case of some enzymes, the native structure might represent a local rather than a global minimum on the potential-energy surface, the pathway from the local to the global minimum being open only in the presence of substrate. Such *substrate-induced stabilization* can be represented as follows:

$$E_R \overset{S}{\rightleftharpoons} E_RS \rightleftharpoons E_TS \rightleftharpoons E_TP \rightleftharpoons E_RP \underset{P}{\rightleftharpoons} E_R$$

where $E_R$ is the native structure which is active in the binding of substrate, and $E_T$ the substrate-induced structure which is energetically more stable than $E_R$. Such a phenomenon of induced stabilization would contribute to the increased resistance to thermal denaturation shown by so many enzymes in the presence of their substrates. Furthermore, it would be reminiscent of the phenomenon of allosteric-transition, which can be represented, if $E_T$ cannot bind substrate directly, as:

$$\begin{array}{ccc} E_R & \overset{S}{\rightleftharpoons} & E_RS \\ \updownarrow & & \updownarrow \\ E_T & & E_TS \end{array}$$

where $E_R$ is the active structure, and $E_T$ very often the more stable structure. Thus basically the phenomena of induced-stabilization and allosteric-transition would both involve the active (in substrate binding) enzyme structure being the less stable structure, and in this light allosteric transitions might represent an evolutionary adaptation of induced-stabilization to fulfill regulatory purposes.

## Problem 10.1

Derive Eqn (10.46) which relates the mid-point pHs of a bell-shaped curve to the p$K$s of two ionizable groups on the enzyme. It might be noted that the derivation given by Alberty and Massey (1954) begins by obtaining the expression for the pH at the peak of the bell-shaped curve.

# 11 Statistical Methods

Other branches of mathematics might be the embodiment of perfection, but the development of statistics stems from a plain recognition of human incapacity to attain perfect knowledge. It is indeed a sobering lesson to learn that the concept of probability itself is beset with fundamental uncertainties! However, limitations of space, and even more importantly of the comprehension of the author, will guide this chapter away from the deeper reaches of statistical theory into safer channels oriented more toward kinetic studies on enzymes.

## 11.1. Method of Least Squares

In a typical kinetic experiment, a series of determinations of initial velocity, or $v_i$, are made at a series of concentrations of some ligand, $X_i$. Three important questions then arise:

1. What form of a mathematical model (function) can be fitted to the data?

2. For any given model, how might its parameters be optimized to result in the best fit to the data?

3. How may the relative merits of any two models be compared on the basis of their respective ability to fit the data?

In general, the answer to the first question is guided to some extent by *a priori* considerations. For example, our knowledge of the steady-state kinetic theory informs us that enzymic rate functions are most likely to take the form of a ratio between two polynomials in ligand concentrations. Other mathematical forms such as exponential or geometric functions are less likely to be relevant. The third question (treated in Section 11.6) can be examined only after the second question has been examined, because two different models can be meaningfully compared only if both of them have been *optimally* fitted to the data. Therefore, the second question needs to be resolved at the outset, so that we may estimate from the data the optimized parameters for any model under consideration. The method of least squares is one of the most important methods for this purpose of parameter estimation.

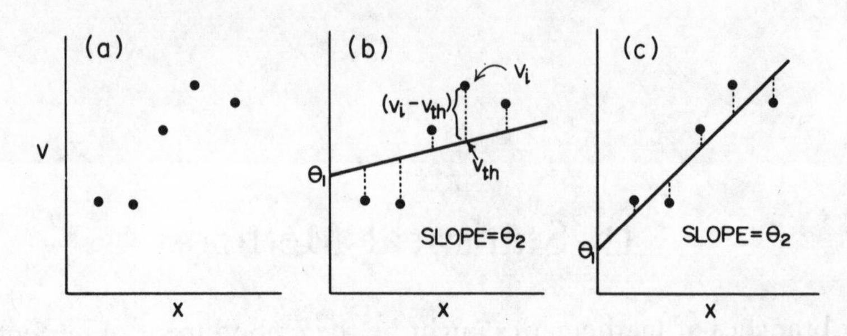


**Fig. 11.1. Fit of straight line to data.**

Take for instance the set of velocity measurements displayed in Fig. 11.1a. Any mathematical model to be fitted to these measurements can be written as

$$v_{th} = f(X, \theta_1, \theta_2, \ldots, \theta_j). \tag{11.1}$$

Namely, the model supposes that the theoretical velocity, $v_{th}$, is a function of the concentration of a ligand, **X**, and a set of parameters. Three candidate models might be:

$$v_{th} = \theta_1 + \theta_2 X \tag{11.2}$$

$$v_{th} = \theta_1 + \theta_2 X + \theta_3 X^2 \tag{11.3}$$

$$v_{th} = \frac{\theta_1 X}{X + \theta_2}. \tag{11.4}$$

Equation (11.2) is a *linear model*, because $v_{th}$ is linearly related to the concentration variable **X**. Equations (11.3) and (11.4) are *nonlinear models*, because $v_{th}$ is not linearly related to **X**. The parameters for a linear model are obtainable by *linear regression* procedures, and those for a nonlinear model by *nonlinear regression* procedures. In either case, to estimate by the method of least squares is to estimate those values of parameters which minimize the sum of squares of deviations between theory and experiment. For example, in Fig. 11.1, parts (b) and (c) show two straight lines fitted to the data points from part (a). Both straight lines obey Eqn (11.2), but the numerical estimates for the parameters $\theta_1$ and $\theta_2$ are not the same for the two lines. The residual deviation between the theoretical line and each experimental point, indicated by a vertical dotted line, might be represented as

$$(v_i - v_{th})$$

When each residual is squared, the sum of these squares is a useful criterion for judging the goodness of fit between theory and experiment. In Fig. 11.1, the sum of squares is smaller for line (c) than for line (b), and line (c) is said to fit the data better than line (b). Best fit is obtained when numerical estimates of the $\theta$-parameters are found which minimize the sum of squares of residuals, namely,

$$\text{Best fit} = \text{Min } \Sigma(v_i - v_{\text{th}})^2. \tag{11.5}$$

The least-squares criterion as stated in this equation is justifiable if and only if four assumptions regarding the nature of experimental errors are valid (Endrenyi and Kwong, 1972):

1. When a measurement is repeated infinitely, the errors from the mean should average to zero.

2. The errors should be random; there should be no correlation between the errors of different experimental points.

3. The variance should be constant for all observations, the random error being unrelated to the magnitude of the independent or dependent variables.

4. The errors of the independent variables are either zero, or at least negligible relative to the errors of the dependent variables.

Assumption 1 usually poses little problem. Assumption 2 becomes invalid whenever there are systematic deviations in the experimental procedure, e.g. when the enzyme undergoes denaturation in the course of the experiment, so that measurements taken at different stages of the experiment do not have comparable errors. Such weaknesses often could be overcome by careful planning, e.g. by randomizing the sequence in which different velocity measurements are to be made. Assumption 4 is often, but not always, valid. Independent variables, which are usually concentration measurements, are often more accurately obtained than dependent variables, which are usually velocity or ligand-binding measurements. Assumption 3 poses the greatest problem. It is quite often invalid. If so, the weakness has to be compensated for by assigning more weight to the reliable measurements, and less weight to the unreliable ones (Section 11.4).

## 11.2. Linear Regression

In order to derive the regression formulae for calculating the least-squares parameters of the linear Eqn (11.2), we may begin by placing this equation into Eqn (11.5):

$$\text{Best fit} = \text{Min } \Sigma(v_i - v_{\text{th}})^2 = \text{Min } \Sigma(v_i - \hat{\theta}_1 - \hat{\theta}_2 \text{X})^2 \tag{11.6}$$

$\hat{\theta}_1$ stands for any numerical estimate of the parameter $\theta_1$, and $\hat{\theta}_2$ stands for any numerical estimate of the parameter $\theta_2$. The sum of squares will vary when the two estimates are varied. At the minimum, the partial derivative of the sum of squares with respect to either estimate will be equal to zero:

$$\frac{\partial \Sigma (v_i - \hat{\theta}_1 - \hat{\theta}_2 \mathbf{X})^2}{\partial \hat{\theta}_1} = -2(\Sigma v_i - N\hat{\theta}_1 - \hat{\theta}_2 \Sigma \mathbf{X}) = 0 \tag{11.7}$$

$$\frac{\partial \Sigma (v_i - \hat{\theta}_1 - \hat{\theta}_2 \mathbf{X})^2}{\partial \hat{\theta}_2} = -2(\Sigma \mathbf{X} v_i - \hat{\theta}_1 \Sigma \mathbf{X} - \hat{\theta}_2 \Sigma \mathbf{X}^2) = 0 \tag{11.8}$$

where $N$ is the total number of experimental points in the summation. Equation (11.7) may be rearranged into

$$\hat{\theta}_1 = \frac{\Sigma v_i - \hat{\theta}_2 \Sigma \mathbf{X}}{N}. \tag{11.9}$$

Substitution of Eqn (11.9) into (11.8) yields the least-squares estimate of $\hat{\theta}_2$:

$$\hat{\theta}_2 = \frac{\Sigma(\mathbf{X} v_i) - \dfrac{(\Sigma v_i)(\Sigma \mathbf{X})}{N}}{\Sigma \mathbf{X}^2 - \dfrac{(\Sigma \mathbf{X})^2}{N}}. \tag{11.10}$$

Placing this least-squares estimate of $\hat{\theta}_2$ back into Eqn (11.9) will give the least-squares estimate of $\hat{\theta}_1$ as well. Equations (11.9) and (11.10) are of course not restricted to a linear variation of $v_i$ with $\mathbf{X}$. They are general for any pair of linearly related variables.

## 11.3. Nonlinear Regression

Again, in nonlinear regression, estimates of the model parameters are varied until the sum of squares of residuals is minimized. Take the nonlinear model of Eqn (11.4). As $\theta_1$ and $\theta_2$ are varied, $v_{th}$ will vary, causing the sum of squares also to vary. Therefore the sum of squares has the form of a surface on the $\theta_1$-$\theta_2$ plane. There are many algorithms, or search methods, for seeking out the minimum on this surface. The excellent review by Box, Davies and Swann (1969) should be consulted to gain a working acquaintance with these methods, but even a most cursory survey will convey some of their fascinating variety and ingenuity. None of the methods gives any assurance that the minimum located is the *global minimum* in the entire parameter space, and not just some *local minimum*. However, if the fit between theory and experiment at the minimum turns out to be satisfactory, free of bias, and if the same minimum is located from different starting points and via different search methods, it may be assumed at least tentatively that the global minimum has been located. The assumption derives further support when the experimental measurements are precise, and when the parameter space is smooth and free from sharp singularities.

*Direct search methods*

(a) *Tabulation methods*: A plausible region of the $\theta_1$-$\theta_2$ surface is divided into either a regular or a random grid, and the sum of squares is calculated for each node in order to locate the minimum node. In a Fibonacci search, the spacing of the grid points is aided by the use of the Fibonacci numbers. These numbers are generated by the formula $F_n = F_{n-1} + F_{n-2}$; their beginning members are: 1, 1, 2, 3, 5, 8, 13, 21, 34, 55, . . .

(b) *Linear methods*: Search proceeds along a set of directional vectors. In the alternating-variable method, each of the $\theta$-axis is explored in turn (Fig. 11.2). In the powerful method of Powell (1964), a set of conjugate directions is generated to ensure, for quadratic or approximately quadratic functions, convergence to the minimum.

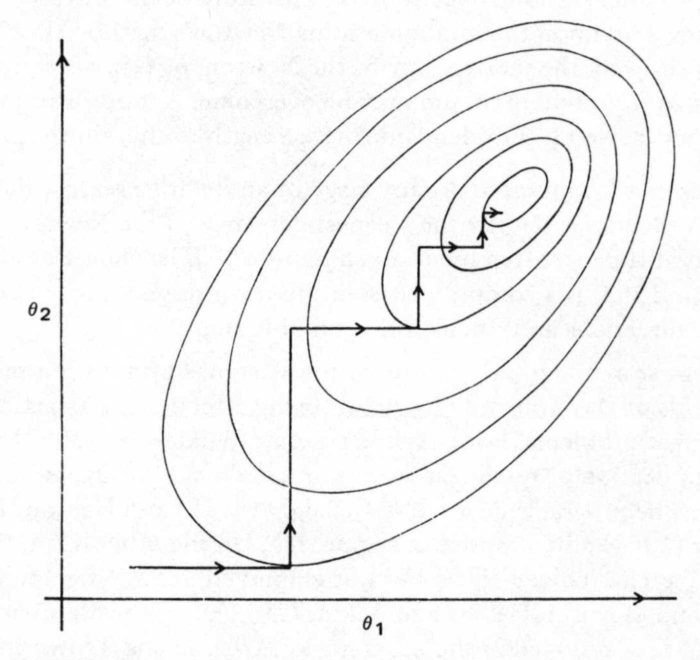


Fig. 11.2. Search for minimum in sum of squares of residuals on the $\theta_1$-$\theta_2$ surface by the alternating variable method. Each arrow indicates the direction in one step of the search. (From Box, Davies, and Swann, 1969.)

(c) *Sequential methods*: These methods are originated from the *evolutionary operation* principle of Box (1957) and adopt geometric patterns for scanning the sum of squares surface. In the simplex method of Spendley, Hext and Himsworth (1962), the sum of squares is calculated at the three vertices of a triangle, and the vertex that scores the highest is inverted to form a new triangle. The process is then repeated. (Fig. 11.3).

*Gradient methods*

(a) *Steepest-descent methods*: In this approach which can be traced back to Cauchy, at any point during the search the gradient of the surface is calculated and a step is taken along the direction of steepest downward gradient. The gradient is then determined again at the new locale, and so on (Fig. 11.4).

(b) *Newton's method*: A second-order approximation of the Taylor series of the sum of squares function guides the descent. Descent is efficient in the vicinity of the minimum but not at a distance from the minimum. Accordingly, good starting estimates of the parameters are needed for the procedure.

(c) *Davidon's method*: The search begins with steepest descent, gradually accumulates information concerning the curvature of the surface, and finally converges upon the minimum using Newton's method (Davidon, 1959). In this way the inefficiency of the Newton, or Gauss-Newton, method at a distance from the minimum may be overcome. A modification by Fletcher and Powell (1963) lends further strength to this approach.

(d) *Marquardt's method*: At any stage of an iterative search, the directions of descent recommended by the steepest-descent and the Newton approaches are usually different, often by as much as 80–90°. The algorithm of Marquardt (1963) is a compromise solution for interpolating between these two directions and calculating a suitable step-size.

Different search methods are suited to different optimization problems, the methods of Davidon and Marquardt being effective for a particularly wide range of problems. For enzymic systems, Wilkinson's (1961) trail-blazing application of nonlinear regression employed the Gauss-Newton approach. The programs devised by Cleland (1967b) and Hanson, Ling and Havir (1967) also adopted this approach. On the other hand, the Davidon-Fletcher-Powell algorithm was employed in the programs of Kowalik and Morrison (1968) and Atkins (1971), and the Marquardt algorithm was employed in the programs of Arihood and Trowbridge (1970), and of Reich, Wangermann, Falck and Rohde (1972). The analysis of phosphorylase *b* kinetics illustrates the utility of the Reich program. When it was postulated for this system that the subunit interactions in the enzyme affected only substrate binding and not catalytic activity. i.e. a *K*-effects model, the best fit obtained by nonlinear regression exhibited systematic deviations between theoretical curve and experimental points. However, when it was postulated that the subunit interactions affected substrate binding as well as catalytic activity, i.e. a *KV*-effects model, the best fit

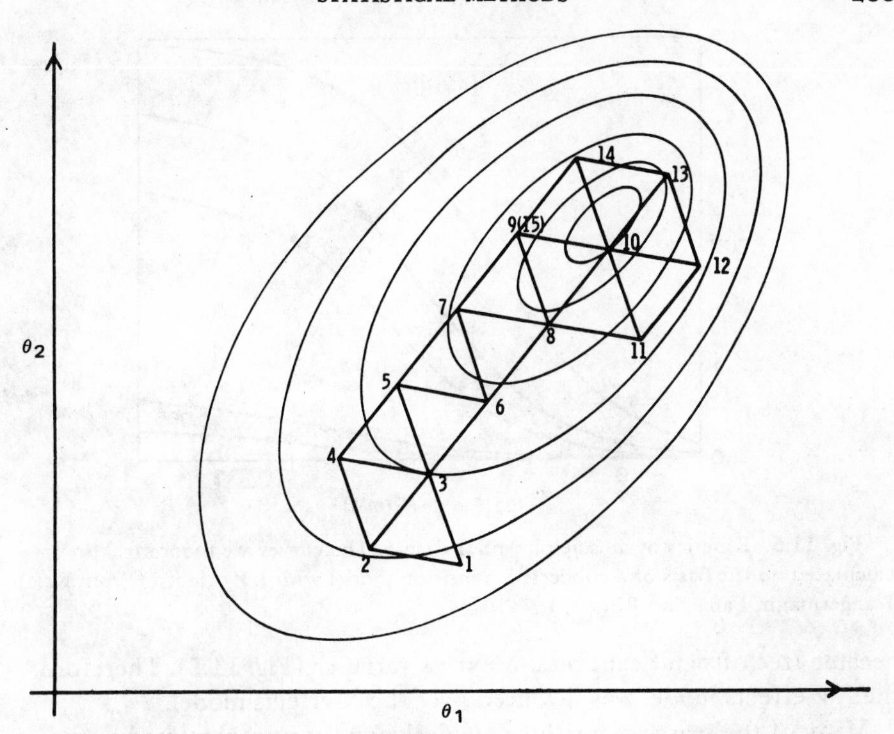


Fig. 11.3. Search by the simplex method. (From Box, Davies and Swann, 1969.)

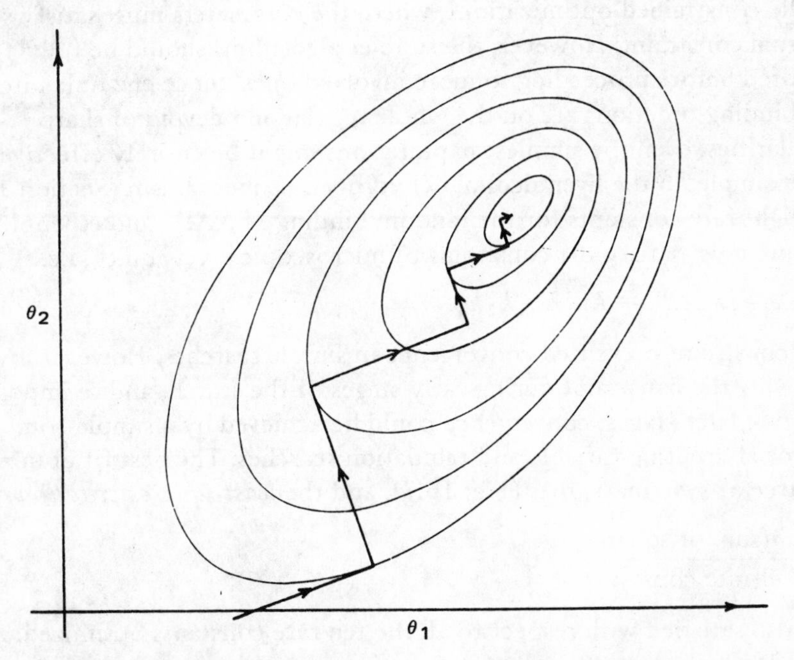


Fig. 11.4. Search by the steepest-descent method. (From Box, Davies and Swann, 1969.)

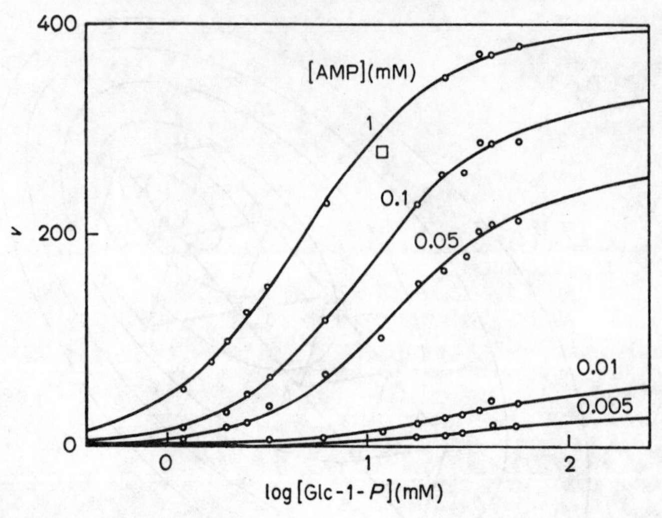


Fig. 11.5. Kinetics of muscle phosphorylase *b*. The curves are theoretical lines calculated on the basis of a concerted-transition model with *KV*-effects. (From Reich, Wangermann, Falck and Rohde, 1972.)

became free of significant bias or excess variance (Fig. 11.5). Therefore the *KV*-effects model was favoured over the *K*-effects model.

Many of the search algorithms are designed for unconstrained optimizations, where the parameters to be optimized can be varied independently and free of constraints. More involved algorithms are often needed to handle constrained optimizations, where the parameters must satisfy some external constraint. However, the simpler algorithms should be fully explored before proceeding to more involved ones. Since enzymic rate and binding functions are on the whole regular and devoid of sharp singularities, even the simplest explorations might be entirely effective. For example, in the liver alcohol dehydrogenase mechanism (Section 10.3.3) the eight rate constants for the random binding of $NAD^+$ and ethanol to the enzyme must satisfy the constraint of microscopic reversibility, i.e.

$$k_1 k_3 k_{-2} k_{-4} = k_{-1} k_{-3} k_2 k_4.$$

This constraint prevented convergence in simple searches. However, by removing the constraint during early stages of the search, and re-imposing it during later stages, convergence could be achieved by a simple combination of alternating-variable and tabulation searches. The best fit obtained was free of systematic bias (Fig. 10.7), and the *least-squares criterion* of

$$\frac{\partial(\text{sum of squares})}{\partial(\text{rate constant})} = 0$$

was also satisfied with respect to all the ten rate constants optimized. As Fig. 11.7 indicates, the sum of squares increases whenever one of the ten rate constants is either increased or decreased from its optimized estimate.

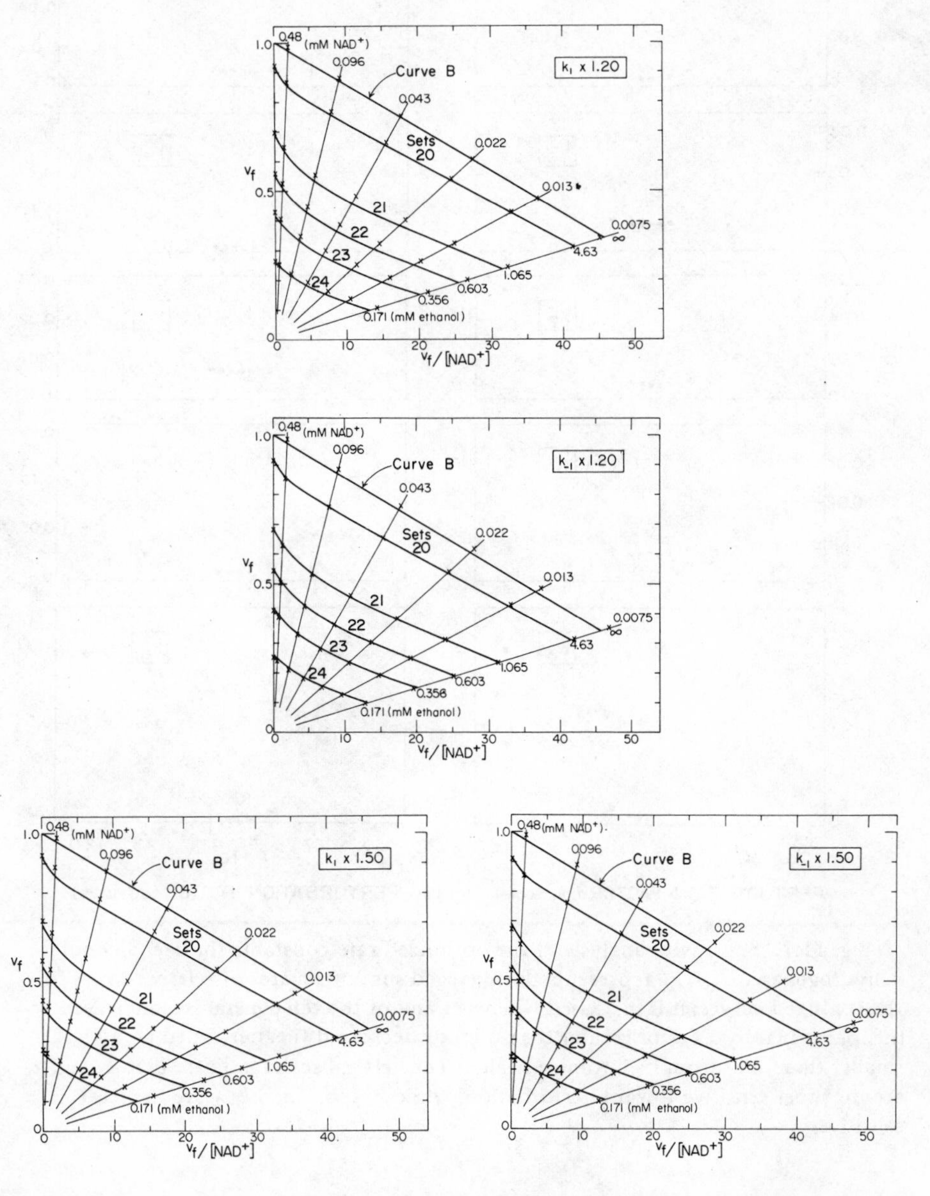


Fig. 11.6. Deterioration in the fitted curves for liver alcohol dehydrogenase when either $k_1$ and $k_{-1}$ was perturbed by a factor of 1.20 or 1.50 from its optimized value.

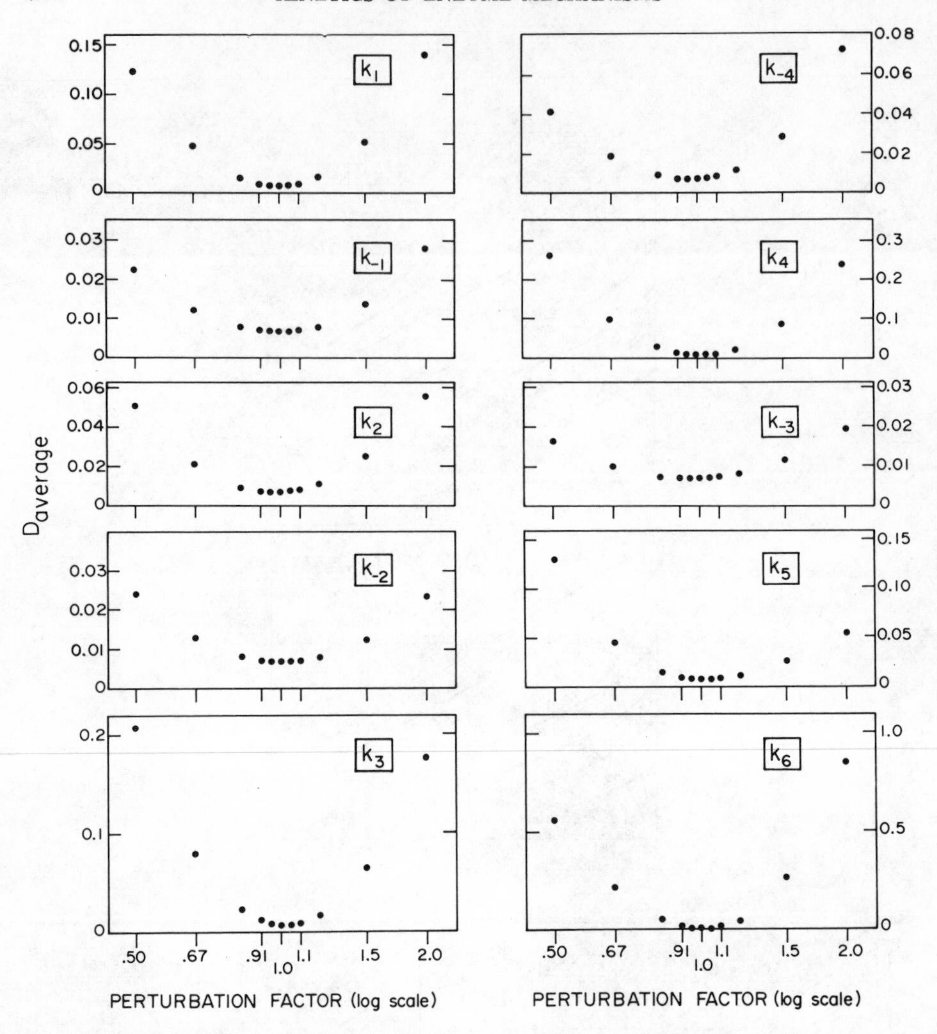


Fig. 11.7. Sensitivity analysis of the optimized rate constants for liver alcohol dehydrogenase. $D_{\text{average}}$ represents the averaged sum of squares of relative errors for the six fitted curves; it is increased whenever any of the ten constants is increased (i.e. perturbed by a factor greater than one) or decreased (i.e. perturbed by a factor smaller than one) from its optimized value. The vertical scales indicate that $k_6$ and $k_4$ are the most sensitive toward perturbation, while $k_{-3}$, $k_{-1}$ and $k_{-2}$ are the least sensitive.

## 11.4. Weight and Precision

In using the least-squares criterion as the basis for linear or nonlinear regression, the assumption is made that the random experimental error is unrelated to the magnitude of the independent or dependent variable. This is often unsupported, and for technical reasons the errors of velocity measurements at high ligand concentrations may be quite different in character from those at low ligand concentrations. An understanding of the error structure would enable more weight to be assigned to the reliable measurements, and less to the unreliable ones. In general, if we measure a reaction velocity at some fixed ligand concentrations a number of times, the replicate measurements are not identical. Due to random scatter, they are spread over a discrete distribution. The basic statistics of such a distribution are:

*Number of replicate measurements* $= N$

*Sample average* $\bar{v} = \dfrac{\Sigma v_i}{N}$

*Sample variance* $s^2 = \dfrac{\Sigma (v_i - \bar{v})^2}{N - 1}$

*Sample standard deviation* $s = \sqrt{\dfrac{\Sigma (v_i - \bar{v})^2}{N - 1}}$

When $N$ becomes infinitely large, the discrete distribution approaches a continuous *normal distribution*, the sample average $\bar{v}$ approaches the *mean* $\mu$, and the sample variance $s^2$ approaches the *variance* $\sigma^2$. The variance is the true indicator of experimental scatter. The smaller the scatter, the smaller $\sigma^2$ will be; and the larger the scatter, the larger $\sigma^2$ will be. However, as long as we are dealing with a finite number of replicates, we only have $s^2$, which may be accepted as a provisional estimate of $\sigma^2$. The greater the $N$, the better this provisional estimate will be. Therefore replicated measurements are immensely helpful toward an assessment of experimental scatter.

When velocity measurements obtained at different ligand concentrations do not have the same scatter, more weight should be given to the more reliable measurements, and less weight to the less reliable measurements. The least-squares criterion stated in Eqn (11.5) should be amended to include weights:

$$\text{Best fit} = \text{Min } \Sigma W_i (v_i - v_{\text{th}})^2. \tag{11.11}$$

The weight $W_i$ should be inversely proportional to the variance, namely,

$$\text{Best fit} = \text{Min } \Sigma \frac{1}{\sigma_i^2} (v_i - v_{\text{th}})^2. \tag{11.12}$$

If the different $v_i$ have a *constant error*, $\sigma_i^2$ is constant, and Eqn (11.12) is reduced back to (11.5). If the different $v_i$ have a *constant relative error*, $\sigma_i$ is proportional to $v_i$, and Eqn (11.12) is transformable to (11.13):

$$\text{Best fit} = \text{Min } \Sigma \frac{1}{v_i^2} (v_i - v_{\text{th}})^2. \tag{11.13}$$

In this equation, the weight employed is shown as $1/v_i^2$. However, experimental $v_i$ is in fact the "true" $v_i$ plus a finite amount of scatter. Perhaps the weight should be $1/v_{th}^2$ instead? Actually, some combination of $v_i$ and $v_{th}$ might be best. Both $1/(v_i + v_{th})^2$ (Ottaway, 1973) and $1/v_i v_{th}$ (Endrenyi and Kwong, 1973) were found to yield parameters of increased accuracy and precision. In general, there are two ways to decide if the experimental system is characterized by constant error, or constant relative error, or some intermediate position. First, replicate determinations of several or all of the $v_i$ could be performed to provide estimates of the sample variances $s_i^2$, which in turn will indicate the nature of the variances $\sigma_i^2$. Secondly, in the absence of replicates the use of residual plots (Anscombe and Tukey, 1960; Daniel and Wood, 1971) as well as other routines (Box and Cox, 1964; Goldfeld and Quandt, 1965) also have been proposed. The use of replicates are on the whole more straightforward, prompting Ottaway (1973) to suggest that *replicate measurements* should always be made if *weighted regression* is to be performed.

Experimental scatter always affects the precision of the parameters obtained from regression. In linear regression, the total variance of the experimental points about the fitted line is:

$$s^2 = \frac{\Sigma(v_i - v_{th})^2}{N - j} \tag{11.14}$$

where $N$ is the number of points, $j$ the number of $\theta$-parameters; and $N - j$ is referred to as the *degree of freedom*. Knowing $s^2$, the variance of the $\theta$-parameters of Eqn (11.2) are:

$$s_{\theta_2}^2 = \frac{s^2}{\Sigma(\mathrm{X}_i - \bar{\mathrm{X}})^2} \tag{11.15}$$

$$s_{\theta_1}^2 = s^2 \left[\frac{1}{N} + \frac{\bar{\mathrm{X}}^2}{\Sigma(\mathrm{X}_i - \bar{\mathrm{X}})^2}\right] \tag{11.16}$$

where $\bar{\mathrm{X}}$ is the average concentration. The variance of parameters obtained from nonlinear regression likewise could be analysed, but requires the processing of an *information matrix* (Draper and Smith, 1967; Reich *et al.*, 1972). A less sophisticated approach is to employ the method of *sensitivity analysis* developed in operations research. Each optimized parameter is perturbed away from its optimized position, and the sum of squares of deviation between theory and experiment is analysed as a function of the perturbation. This defines how sensitive the parameter estimate is toward perturbation, which in turn defines the precision of the parameter estimate itself.

For example, Fig. 10.7 shows the satisfactory fit between theory and experiment after the ten rate constants $k_1$, $k_{-1}$, $k_2$, $k_{-2}$, $k_3$, $k_{-3}$, $k_4$, $k_{-4}$,

$k_5$ and $k_6$ of liver alcohol dehydrogenase have been optimized by nonlinear regression. Figure 11.6 shows the distinctly poorer fit obtained when either $k_1$ or $k_{-1}$ were perturbed from their optimized values. Figure 11.7 shows the relative sensitivities of the ten constants toward such perturbation; either increasing or decreasing the value of each constant brings about an increase in $D_{\text{average}}$, namely the averaged sum of squares of relative errors for the six experimental curves. It might be noted from these two figures that, although $k_{-1}$ was one of the least sensitive of the constants, perturbation by a factor of 1.20 already brought visible deterioration in the fit. Thus the precision of fit between theory and experiment was such that a 20% deviation in any of the estimated rate constant would have brought about an unmistakable deterioration in the fit.

## 11.5. Linearizations

Enzymic rate laws, being mostly a ratio between two polynomials in substrate concentration, are nonlinear functions. Their parameters may be estimated by various nonlinear regression methods. In addition, a number of short-cut, or linearization, methods also have been developed for this purpose. This is especially the case for the simple rectangular hyperbola, i.e.

$$v_0 = \frac{V_S \cdot \mathrm{S}}{\mathrm{S} + K_m}. \tag{11.17}$$

This equation has the same form as Eqn (11.4). It may be *linearized* by three different transformations (the new variables are placed inside brackets):

*Lineweaver-Burk*

$$\left(\frac{1}{v_0}\right) = \frac{1}{V_S} + \frac{K_m}{V_S}\left(\frac{1}{\mathrm{S}}\right) \tag{11.17a}$$

*Eadie*

$$(v_0) = V_S - K_m\left(\frac{v_0}{\mathrm{S}}\right) \tag{11.17b}$$

*Hanes*

$$\left(\frac{\mathrm{S}}{v_0}\right) = \frac{K_m}{V_S} + \frac{1}{V_S}\,(\mathrm{S}) \tag{11.17c}$$

Graphically, these linear transforms provide a basis for placing the data into linear plots to yield estimates of $V_S$ and $K_m$. Statistically also, they provide a basis for estimating $V_S$ and $K_m$ by means of linear regression.

However, these transformed equations are not truly the same form as Eqn (11.2), where the dependent and independent variables are cleanly *separated* from one another. Consequently, when these transformed equations are employed for the purpose of linear regression, bias is introduced, and different $V_S$ and $K_m$ estimates are obtained with the different transforms. Wilkinson (1961) found the Hanes transform to be more reliable than the Lineweaver-Burk. Dowd and Riggs (1965) concluded from 500 computer experiments that the Lineweaver-Burk transform was by far the least reliable. The Hanes transform was slightly superior to the Eadie transform when the errors were small, but the reverse was true when the errors were large. Again on the basis of computer experiments, Endrenyi and Kwong (1972) found the Hanes transform, especially in its inverse form, to out-perform the other two when there were constant errors, or when there were constant relative errors and substrate concentrations were spaced harmonically or geometrically. The Eadie transform was preferred only when there were constant relative errors and substrate concentrations were spaced arithmetically. The Lineweaver-Burk transform consistently gave the poorest results. Thus the relative merits of the three linear transforms are not entirely the same when used statistically as when used graphically. For the graphical detection of nonlinearity, the Eadie plot is superior (Section 3.2).

Rate laws more complex than the first-degree rectangular hyperbola are not as readily treated by linear regression. An important exception emerges, however, from Bunting and Murphy's (1972) treatment of the *second-degree* rate law of

$$v_0 = \frac{m_2 S^2 + m_1 S}{d_2 S^2 + d_1 S + d_0} . \tag{11.18}$$

In this treatment, the rate expression is divided through by the coefficient $d_1$ to give

$$v_0 = \frac{a_2 S^2 + a_1 S}{b_2 S^2 + S + b_0} . \tag{11.18a}$$

This equation can be rearranged into two alternate forms:

$$S + b_2 S^2 - \frac{a_2 S^2}{v_0} = \frac{a_1 S}{v_0} - b_0 \tag{11.19}$$

$$b_2 v_0 = \frac{-v_0(b_0 + S)}{S^2} + \frac{a_1}{S} + a_2. \tag{11.20}$$

At this point, three substitute variables $X$, $Y$ and $Z$ are introduced:

$$X = \frac{a_1}{S} - \frac{v_0(b_0 + S)}{S^2}$$

$$Y = S + b_2S^2 - \frac{a_2S^2}{v_0}$$

$$Z = \frac{S}{v_0}.$$

Use of these substitute variables converts Eqns (11.19) and (11.20) into a linearized form:

$$Y = a_1Z - b_0 \tag{11.19a}$$

$$v_0 = \frac{X}{b_2} + \frac{a_2}{b_2}. \tag{11.20a}$$

If some initial estimates of $a_1$ and $b_0$ and therefore of $X$ can be made, least-squares linear regression of the velocity measurements on the basis of Eqn (11.20a) will yield estimates of $a_2$ and $b_2$. Utilizing these latter estimates to calculate $Y$, linear regression on the basis of Eqn (11.19a) will in turn yield improved estimates of $a_1$ and $b_0$, and so on. Thus Eqns (11.19a) and (11.20a) are mutually complementary, and continued iteration between them will lead to optimized estimates of all four parameters. The optimized estimates were found to be independent of the initial estimates in an application of this technique to the hydrolysis of 2-hippuroxy-DL-isovaleric acid by carboxypeptidase A. In Fig. 11.8, part (a) portrays the second-degree rate profile with a maximum, part (b) the linear relationship (11.19a), and part (c) the linear relationship (11.20a). In part (a), the fit between the optimized curve and the experimental points is visibly satisfactory and free of bias.

In principle, even more complex rate functions than the second-degree one might be linearized. Thus the general rate law (11.21) might be linearized into (11.21a):

$$v_0 = \frac{m_zS^z + \cdots + m_1S}{d_zS^z + \cdots + d_1S + d_0} \tag{11.21}$$

$$d_z(v_0S^z) + \cdots + d_1(v_0S) + d_0v_0 - m_zS^z - \cdots - m_1S = 0. \tag{11.21a}$$

As suggested by J. Z. Hearon (cited in Botts, 1958), an equation such as (11.21a) is open to treatment by multiple linear regression. However, nonlinear methods are much more attractive for such higher-degree rate-laws. The nonlinear programs of Arihood and Trowbridge (1970), Atkins (1971) and Reich *et al.* (1972) are all applicable to rate laws of any degree. Pettersson and Pettersson (1970) have devised a general strategy whereby velocity measurements can be tested, using nonlinear regression, against rate laws of increasingly higher degree, in order to establish the *lowest-degree* rate law that is compatible with the data.

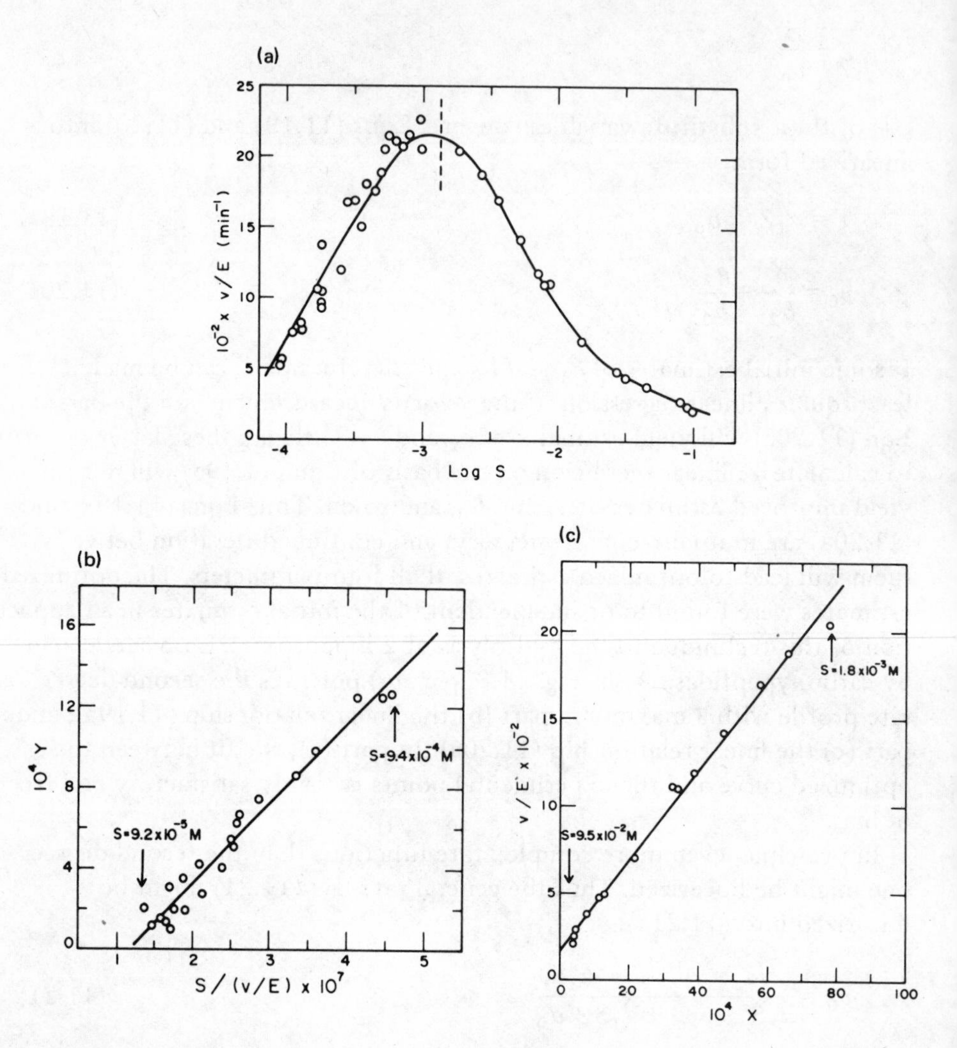


Fig. 11.8. Parameter evaluation for second-degree equations by linearization. (a) Fit between the optimized second-degree curve and experimental points of the hydrolysis of 2-hippuroxy-isovaleric acid by carboxypeptidase *A*. (b) Fit of linearized Eqn (11.19a) to experimental points. (c) Fit of linearized Eqn (11.20a) to experimental points. (From Bunting and Murphy, 1972.)

## 11.6. Goodness of Fit

For any model mechanism, an optimized least-squares fit between its rate law and experimental observations can be obtained by regression methods, but the least-squares criterion by itself may not be a sufficient basis for clearly establishing the relative merits of different model mechanisms. Additional criteria for goodness of fit are required. Haarhoff (1969), Reich (1970) and Bartfai and Mannervik (1972) have examined a number of useful criteria, which were divided by the latter workers into five major categories:

(a) *Success of linear regression*: The rate laws for some model mechanisms in principle could be linearized. Failure of linear regression in such cases might indicate that the model carries redundant parameters.

(b) *Success of nonlinear regression*: If nonlinear regression methods converge to an optimized fit for one model, but fail to do so for another, the latter model would be rejected in favour of the former.

(c) *Parameter values*: Models which yield unreasonable (e.g. negative) parameters or unreliable (e.g. large standard deviation) parameters would be rejected.

(d) *Distribution of residuals*: Models which yield residual deviations between theory and experiment that are not normally distributed or do not have zero mean would be rejected.

(e) *Sum of squares of residuals*: The sum of squares of residual deviations should not be excessive relative to the variance of the experimental observations. The model that yields a significantly smaller sum of squares would be the best model.

Systematic application of these criteria by Mannervik, Gorna-Hall and Bartfai (1973) to the kinetics of porcine glyoxalase I led to the proposal of Mechanism 11.I for the enzyme:

$G$ = glutathione
$M$ = methylglyoxal
$A$ = preformed hemimercaptal of glutathione and methylglyoxal

Some of the results relating to criterion (d) are shown in Fig. 11.9. The residuals calculated for Mechanism 11.I fell mostly within the 95% confidence band and did not vary with glutathione concentration in any pronounced trend. By comparison, the residuals calculated for three of the alternative model mechanisms were both larger and definitely more trendy.

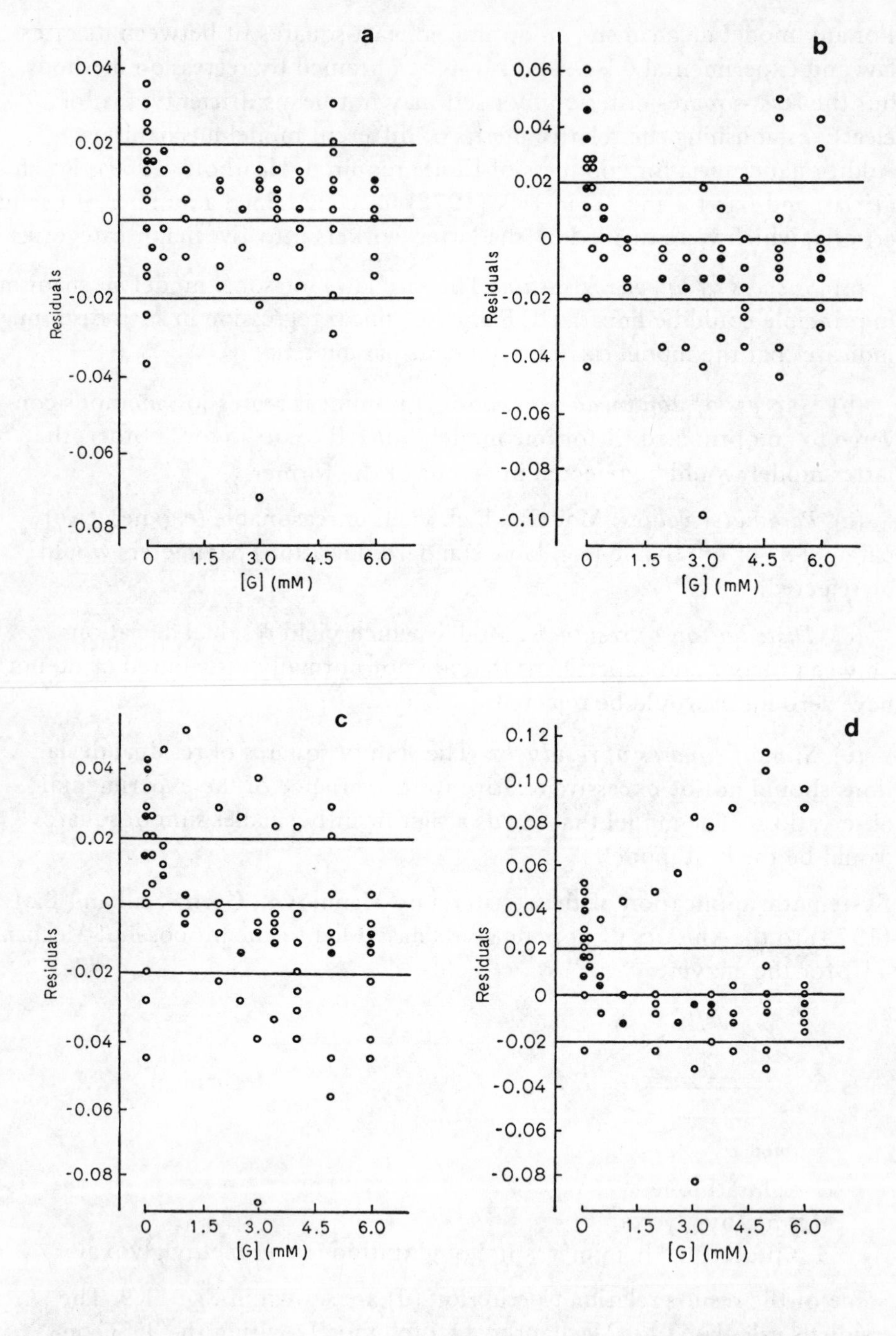


Fig. 11.9. Variation of the residual error with glutathione ($G$) concentration in the glyoxalase *I*-system. Part (a) shows the variation calculated for Mechanism 11.I. Parts (b–d) show the variation calculated for three other plausible mechanisms. (From Mannervik, Gorna-Hall and Bartfai, 1973.)

## Problem 11.1

The following rate measurements are obtained for a hyperbolic system with constant errors that are independent of the concentration and velocity values:

| S (mM) | 0.25 | 0.30 | 0.40 | 0.50 | 0.70 | 1.00 | 1.40 | 2.00 |
|---|---|---|---|---|---|---|---|---|
| $v$ ($\mu M/min$) | 2.4 | 2.6 | 4.2 | 3.8 | 6.2 | 7.4 | 10.2 | 11.4 |

Evaluate by linear regression the saturation velocity $V$ and its standard error $s_V$ on the basis of the Lineweaver-Burk, Eadie and Hanes linear transforms. Compare the three sets of results. It might be noted that all three linear transforms (Eqn 11.17a–c) can be represented in the general form of

$$y = \theta_1 + \theta_2 x \tag{11.22}$$

In the Lineweaver-Burk case, $y$ is $1/v$ and $x$ is $1/S$. In the Eadie case, $y$ is $v$ and $x$ is $v/S$. In the Hanes case, $y$ is $S/v$ and $x$ is S. Accordingly, from the S and $v$ measurements given, some useful sums for each of these transforms can be calculated as follows:

| Linear Transform | $\Sigma y$ | $\Sigma x$ | $\Sigma y^2$ | $\Sigma x^2$ | $\Sigma xy$ |
|---|---|---|---|---|---|
| Lineweaver-Burk | 1.785 | 15.47 | 0.50937 | 41.138 | 4.5497 |
| Eadie | 48.20 | 65.61 | 371.80 | 554.09 | 367.49 |
| Hanes | 1.006 | 6.550 | 0.1309 | 8.0125 | 0.92048 |

Also, in calculating the error of the reciprocal of the parameter $\theta_1$ or $\theta_2$, the following approximate relationship can be employed:

$$s_{1/\theta} = s_\theta / \theta^2. \tag{11.23}$$

On this basis, the standard error for $1/\theta$ can be determined from $\theta$ and its standard error. A few additional formulae that will help to simplify the statistical calculations are:

$$\Sigma(y_i - y_{\text{th}})^2 = \Sigma(y_i - \bar{y})^2 - \theta_2 \Sigma(x_i - \bar{x})(y_i - \bar{y}) \tag{11.24}$$

$$\Sigma(y_i - \bar{y})^2 = \Sigma y^2 - (\Sigma y)^2/N \tag{11.25}$$

$$\Sigma(x_i - \bar{x})^2 = \Sigma x^2 - (\Sigma x)^2/N \tag{11.26}$$

$$\Sigma(x_i - \bar{x})(y_i - \bar{y}) = \Sigma xy - \Sigma x \Sigma y/N \tag{11.27}$$

# 12 Across the Membrane into the Cell

## 12.1. Membrane Carriers

Generally, solutes may be transported across cell membranes by either *passive diffusion* or *carrier-mediated transport.* Passive diffusion obeys Fick's first law of diffusion, such that the rate of transport is proportional to the concentration gradient of the solute across the membrane. The rate increases indefinitely with increasing gradient, until the process is finally limited by the solubility of the solute, aggregation of solute molecules, or other physical factors. Although passive diffusion is widely encountered, no less widely encountered is the phenomenon of *saturation,* with the rate of transport levelling off to a finite asymptote at high gradients of solute. This is suggestive of a process other than passive diffusion, namely a combination between solute and some carrier structure on the membrane, analogous to the combination between substrate and enzyme in enzymic reactions. The analogy is further reinforced by the observation that the combination between solute and carrier is usually highly specific and inhibited by competing analogues of the solute. Consequently the concepts and equations of enzyme kinetics have been widely applied to carrier-transport reactions, even though the nature and mechanism of such reactions are only beginning to be understood.

### 12.1.1. Monovalent carrier

The use of mobile carriers to explain the kinetics of transport reactions was initiated by Widdas (1952) and Wilbrandt and Rosenberg (1961), but it remained for Jacquez (1961) and Regen and Morgan (1964) to remove some of the simplifying assumptions employed in these earlier studies. The basic carrier model mechanism that emerges is:

$$
\text{cis-side} \quad \left|
\begin{array}{ccc}
\multicolumn{3}{c}{\longleftarrow \text{Membrane} \longrightarrow} \\
E_cS_c & \underset{k_{-2}}{\overset{k_2}{\rightleftharpoons}} & E_tS_t \\
k_1S_c \Big\uparrow\Big\downarrow k_{-1} & & k_{-3}S_t \Big\uparrow\Big\downarrow k_3 \\
E_c & \underset{k_4}{\overset{k_{-4}}{\rightleftharpoons}} & E_t
\end{array}
\right| \quad \text{trans-side}
$$

Mechanism 12.I

In this mechanism, $E$ is a monovalent transport carrier with a single binding site for the transport substrate $S$. It combines reversibly with $S$ from the liquid phase on either side of the cell membrane. The two liquid phases may be designated as *cis* and *trans*, and the velocity of the transport reaction is measured on the basis of substrate transport from the *cis* to the *trans* side. Both free $E$ and the $ES$ complex can move between the two sides, the *mobility* being unrestrictedly translational, rotational or conformational in nature. All that is needed of the movement is to bring the carrier into contact alternatively with the two liquid phases. Simple as it is, this model provides a plausible basis for examining some of the fundamental questions concerning the transport reaction.

### 12.1.2. Facilitated diffusion vs. active transport

In Mechanism 12.I, when transport from *cis* to *trans* is measured, $S_c$ represents the substrate and $S_t$ the product. The steady-state rate law is readily obtained by application of the structural rule:

$$v_{\text{carrier}} = \frac{k_1 S_c k_2 k_3 k_4 - k_{-1} k_{-2} k_{-3} S_t k_{-4}}{\Delta}. \quad (12.1)\dagger$$

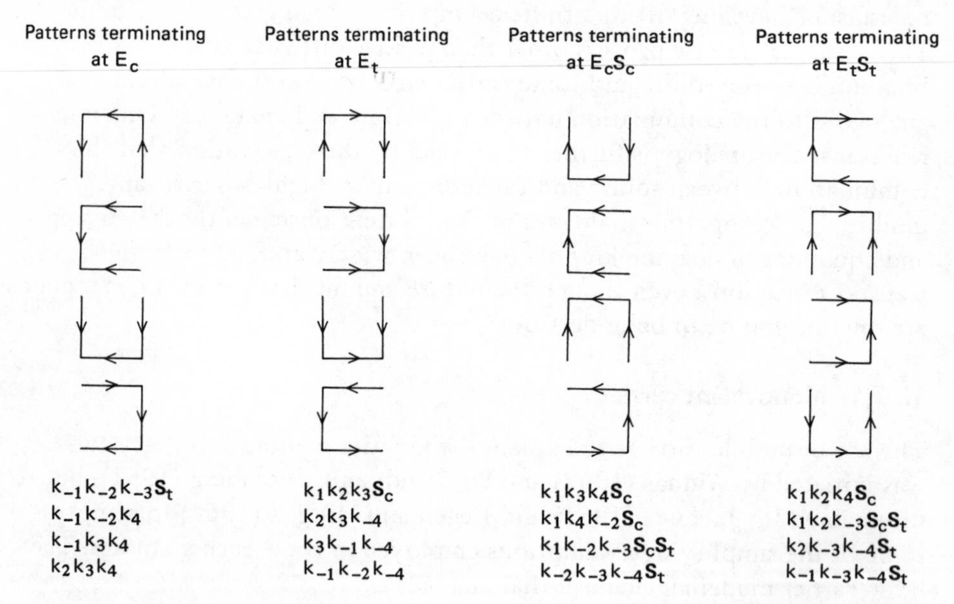


Fig. 12.1. Arrow patterns in the expression for $\Delta$ in rate law (12.1) (From Schachter, 1972.)

† In Eqn (12.1), $v_{\text{carrier}}$ stands for the rate of carrier-mediated transport per unit carrier. If passive diffusion occurs to a significant extent in the system, it must be first deducted from the total rate of transport:

$$v_{\text{carrier}} = v_{\text{total}} - v_{\text{diffusion}}$$

The arrow patterns that constitute $\Delta$ are shown in Fig. 12.1. According to this equation, net transport will cease whenever

$$k_1 S_c k_2 k_3 k_4 - k_{-1} k_{-2} k_{-3} S_t k_{-4} = 0. \tag{12.2}$$

In the case of *facilitated diffusion*, the transport system is incapable of transporting against a concentration gradient, i.e. net transport ceases when $S_t$ becomes equal to $S_c$. It follows from Eqn (12.2) that, for facilitated diffusion,

$$k_1 k_2 k_3 k_4 = k_{-1} k_{-2} k_{-3} k_{-4}. \tag{12.3}$$

In the case of *active transport,* the system is capable of transporting against a gradient, and net transport ceases only when $S_t$ becomes significantly greater than $S_c$. It follows from Eqn (12.2) that, for active transport,

$$k_1 k_2 k_3 k_4 > k_{-1} k_{-2} k_{-3} k_{-4} \tag{12.4}$$

Thus active transport occurs if and only if the condition described by Eqn (12.4) is fulfilled, and Jacquez (1961) distinguished four different ways by which this may be accomplished:

1. *Active transport of free carrier:*

   $k_1 = k_{-3}, \quad k_3 = k_{-1}, \quad k_2 = k_{-2}$

   $k_4 > k_{-4}$ (either $k_4$ is increased by energy coupling, or $k_{-4}$ is decreased by energy coupling)

2. *Active transport of carrier-substrate complex:*

   $k_1 = k_{-3}, \quad k_3 = k_{-1}, \quad k_4 = k_{-4}$

   $k_2 > k_{-2}$ (either $k_2$ is increased by energy coupling, or $k_{-2}$ is decreased by energy coupling)

3. *Active association between carrier and substrate:*

   $k_2 = k_{-2}, \quad k_4 = k_{-4}$

   $k_1/k_{-1} > k_{-3}/k_3$ (the former is increased by energy coupling)

4. *Active dissociation of carrier-substrate complex:*

   $k_2 = k_{-2}, \quad k_4 = k_{-4}$

   $k_1/k_{-1} > k_{-3}/k_3$ (the latter is decreased by energy coupling)

In each instance, the basic postulate is that some form of coupling of the carrier system to an energy-generating system brings about an asymmetry in the *apparent* rate constants of the carrier mechanism, and hence active transport. On this basis, if energy coupling were abolished, the possibility arises that an active-transport system might be transformed into a facilitated-diffusion system. Such an apparent transformation was observed for example by Winkler and Wilson (1966) when they abolished active transport of $^{14}$C-lactose into *Escherichia coli* cells, but found that the system remained capable of counter-transport of $^{14}$C-lactose into cells preloaded with $^{12}$C-lactose or thio-$\beta$-digalactoside.

### 12.1.3. *Cis* and *trans* stimulation and inhibition

The transport velocity can be affected by ligands present on either the *cis* or *trans* side of the membrane, and the rate effect due to any ligand $X$ falls into one of four categories:

*cis-stimulation:* When added on the *cis* side, $X$ stimulates transport of $S$ from *cis* to *trans.*

*cis-inhibition:* When added on the *cis* side, $X$ inhibits transport of $S$ from *cis* to *trans.*

*trans-stimulation:* When added on the *trans* side, $X$ stimulates transport of $S$ from *cis* to *trans.*

*trans-inhibition:* When added on the *trans* side, $X$ inhibits transport of $S$ from *cis* to *trans.*

Experimentally, *cis* effects by $X$ can be studied by adding $X$ and $S$ to the external medium, and measuring influx of $S$ into the cells; or by preloading the cells with $X$ and $S$, and measuring efflux of $S$ from the cells. Trans effects by $X$ can be studied by preloading the cells with $X$, and measuring influx of $S$ into the cells; or by adding $X$ to the external medium, and measuring efflux of $S$ from preloaded cells.

In Eqn (12.1) which describes the net flux of $S_c$ from *cis* to *trans*, addition of $S_t$ increases both the denominator and the negative part of the numerator. Accordingly an increase in $S_t$ invariably brings about a reduction of velocity, i.e. *trans*-inhibition. The situation is different if $S_c$ and $S_t$ were isotopically distinguishable, so that the *unidirectional flux* of $S_c$ from *cis* to *trans*, rather than just the net flux, could be measured. Thus, using labelled $S_c^*$ and unlabelled $S_t$, the initial rate of transport of label per unit carrier is described by Eqn (12.5), which was derived by Schachter (1972) on the basis of the structural rule for steady-state rate laws:

$$v^* = \frac{k_1 S_c^* k_2 k_3 k_{-1} k_{-2} k_{-3} S_t + p k_1 k_2 k_3 k_4 S_c^*}{p \cdot \Delta} \qquad (12.5)$$

$$p = k_{-1} k_3 + k_{-1} k_{-2} + k_2 k_3.$$

The expression for $\Delta$ is the same as that given in Fig. 12.1, except $S_c$ is replaced by $S_c^*$. If $\mathbf{S}_c^*$ is fixed, and only $\mathbf{S}_t$ varied, the dependence of $v^*$ on $\mathbf{S}_t$ according to Eqn (12.5) has a first-degree modifier form:

$$v^* = \frac{m_1\mathbf{S}_t + m_0}{d_1\mathbf{S}_t + d_0}. \tag{12.6}$$

Consequently the effect of $S_t$ on the unidirectional flux of $S_c^*$ can be either *trans*-stimulation (if $m_1 d_0 > m_0 d_1$) or *trans*-inhibition (if $m_1 d_0 < m_0 d_1$).

Structural analogues of $S$ also can bring about either *trans*-stimulation or *trans*-inhibition. Consider for example Mechanism 12.II.

$$\begin{array}{ccc} E_cS_c & \rightleftharpoons & E_tS_t \\ S_c \Big\Updownarrow & & S_t \Big\Updownarrow \\ E_c & \rightleftharpoons & E_t \\ X_c \Big\Updownarrow & & X_t \Big\Updownarrow \\ E_cX_c & \rightleftharpoons & E_tX_t \end{array}$$

Mechanism 12.II

This mechanism describes the competition between substrate $S$ and an analogue $X$ for the same binding site on a monovalent carrier. Its initial rate law for the transport of $S_c$ in the absence of $S_t$ has been derived by Jacquez (1961) and Schachter (1972), and has the following form:

$$v_0 = \frac{m_{110}\mathbf{S}_c\mathbf{X}_t + m_{100}\mathbf{S}_c}{d_{110}\mathbf{S}_c\mathbf{X}_t + d_{100}\mathbf{S}_c + d_{011}\mathbf{X}_t\mathbf{X}_c + d_{010}\mathbf{X}_t + d_{001}\mathbf{X}_c + d_{000}}. \tag{12.7}$$

$S_c$ and $X_t$, since they react with different carrier species in the mechanism, can appear in the same terms in the rate law; on the other hand, $S_c$ and $X_c$, since they react with the same carrier species, cannot appear together. Accordingly, $v_0(\mathbf{X}_t)$ is first-degree modifier type, and $X_t$ can bring about either *trans*-stimulation or *trans*-inhibition. In contrast, $v_0(\mathbf{X}_c)$ is first-degree inhibitory type, and $X_c$ can bring about *cis*-inhibition but *not cis*-stimulation.

### 12.1.4. Divalent carrier

The transports of chemically unrelated solutes across the cell membrane often exhibit interactions which are largely inexplicable on the basis of the monovalent carrier. For example, the intestinal transports of $Na^+$ and of sugars profoundly influence one another (Fig. 12.2). Since $Na^+$ and sugars are not structural analogues expected to compete for the same carrier site, Crane, Miller and Bihler (1961) had suggested that divalent carrier with separate sites for $Na^+$ and sugars might furnish a more plausible explanation

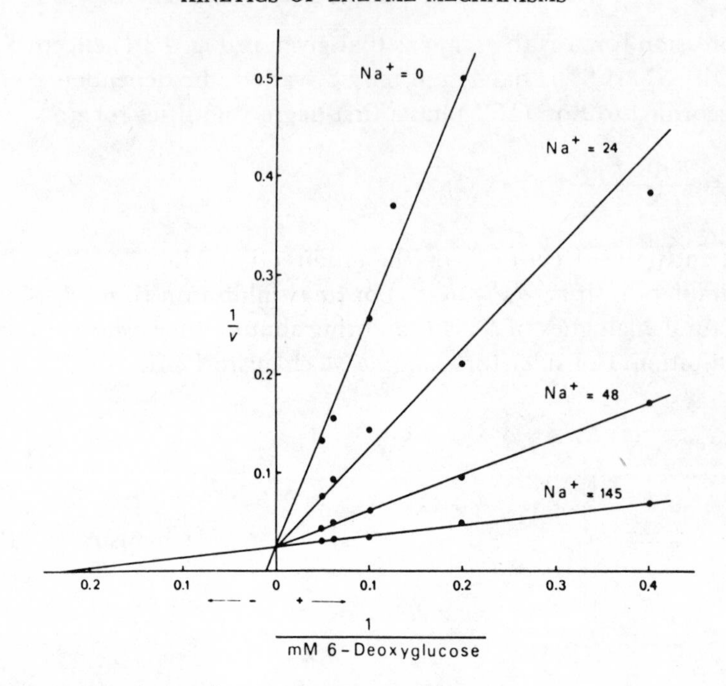


Fig. 12.2. Effects of $Na^+$ on deoxyglucose transport by hamster intestinal tissue. (From Crane, Forstner and Eicholz, 1965.)

for the observed interaction. Such heterotropic interactions between dissimilar solutes binding to separate sites on the carrier would be analogous to the interactions between dissimilar ligands binding to separate sites on an allosteric enzyme. The former can be just as important to the regulation of carrier systems as the latter are for enzymic ones.

Fig. 12.3 shows four divalent carrier mechanisms. In Mechanism 12.III, one of the two sites specifically binds $S$, and the other $X$. In Mechanism 12.IV, both sites bind $S$, but only one of them binds $X$. In Mechanism 12.V, both sites bind $X$, but only one of them binds $S$. Finally, in Mechanism 12.VI, both sites bind $S$ as well as $X$. The steady-state rate predictions by these mechanisms, which have been analysed with the use of the structural rule, are all topologically evident (Wong, 1965b). They indicate that the divalent carrier is far more diversified in the properties predictable for it than the monovalent carrier:

(a) The $v_0(S_c)$ function must be first-degree, hyperbolic for the monovalent carrier, but can be higher-degree for the divalent carrier. Thus the latter but not the former can explain a sigmoidal rate curve, or a rate curve that passes through a maximum.

(b) The $v_0(X_c)$ function is first-degree inhibitor type for the monovalent carrier, so that $X_c$ can exert *cis*-inhibition but not *cis*-stimulation on the transport of $S_c$ from *cis* to *trans*. However, the variation of $v_0$ with

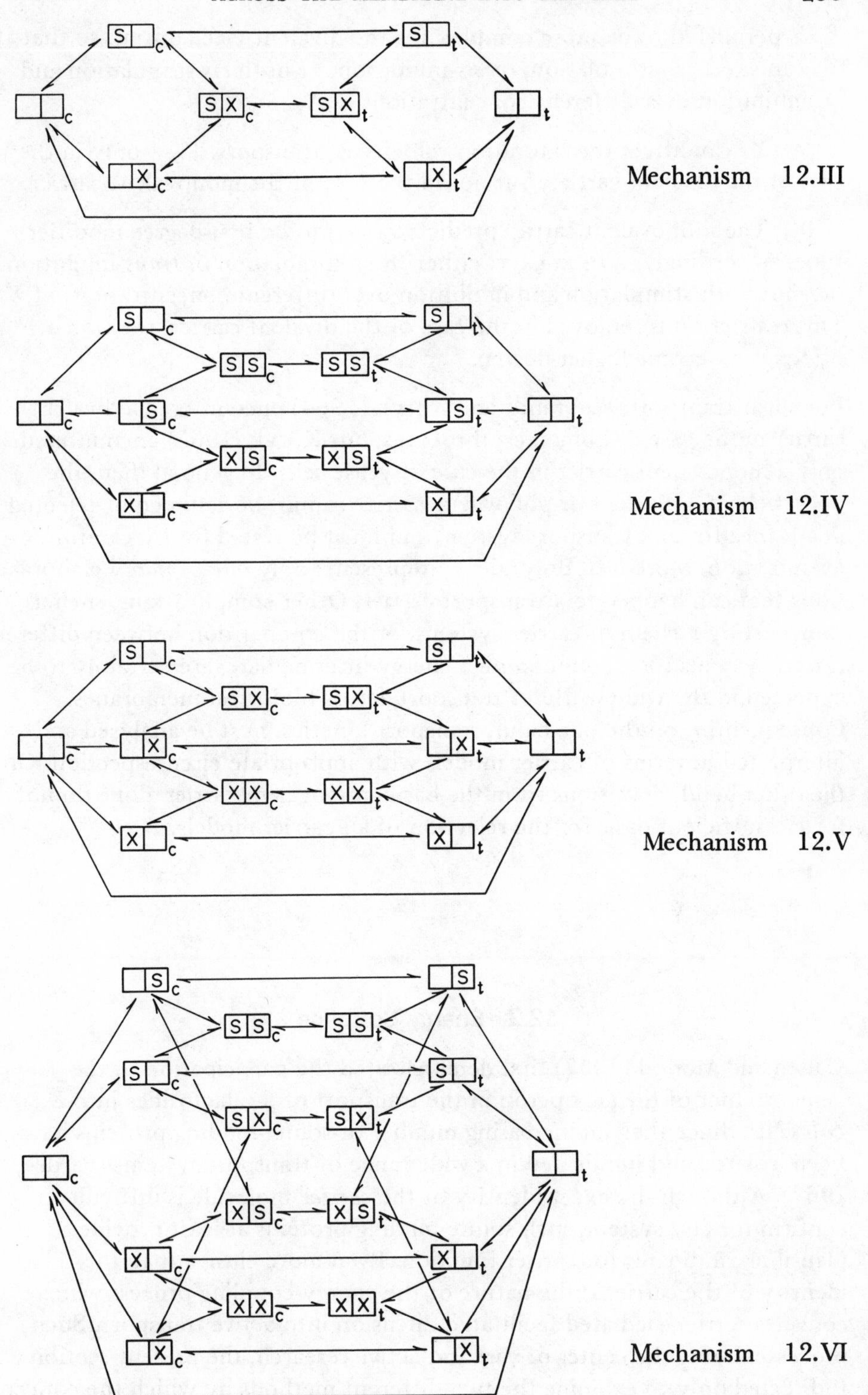


Fig. 12.3. Divalent carrier mechanisms.

$X_c$ is permitted to be more complex for the divalent mechanisms, so that $X_c$ can exert *cis*-stimulation, or *cis*-inhibition, or both *cis*-stimulation and *cis*-inhibition over different concentrations of $X_c$.

(c) $X_c$ can affect the saturation velocity of transport, $V_{(S)}$, only in the case of the divalent carrier, but not in the case of the monovalent carrier.

(d) The monovalent carrier predicts $v_0(\mathbf{X}_t)$ to be first-degree modifier type. Accordingly, $X_t$ can exert either *trans*-stimulation or *trans*-inhibition, but not both stimulation and inhibition over different concentrations of $X_t$. This restriction is removed in the case of the divalent carrier, for which $v_0(\mathbf{X}_t)$ can become higher-degree.

For sugar transport, Wilbrandt and Kotyk (1964) encountered a divalent carrier in the case of human erythrocytes, but Kotyk (1965) encountered only a monovalent carrier in the case of yeast cells. In general then, the operation of a divalent or polyvalent carrier cannot be assumed or rejected in advance for any transport system, and must be tested for by careful examination. Moreover, polyvalency represents only *one of many* elaborations that can happen to a transport carrier. Other complications, such as transport by a chain of carrier systems, or the competition between different carrier systems for a common high-energy intermediate, are certainly to be expected in the wide world of transport across biological membranes. Consequently, on the one hand, transport kinetics must be analysed and interpreted in terms of carrier models with appropriate circumspection. On the other hand, deviations from the basic monovalent carrier alone do not form a sufficient basis for the rejection of all carrier models.

## 12.2. Energy Coupling

Cohen and Monod (1957) first demonstrated the participation of the *y*-gene product of the *lac*-operon in the transport of β-galactosides into *E. coli* cells. Since then an increasing number of solute-binding proteins have been isolated and implicated in a wide range of transport systems (Pardee, 1967). Although the exact identity of the *carrier* molecule is difficult to confirm for any system, such solute-binding proteins at least provide plausible candidates for carrier function. Even more elusive than the identity of the carrier is the nature of the energy-coupling process which converts carrier-mediated facilitated diffusion into active transport. Such processes being the center of vast and active research, the present section is directed only to examine the two different methods by which the concepts and approaches of kinetic analysis may contribute toward their elucidation.

### 12.2.1. Analysis of transport kinetics

In the basic carrier model, it is postulated that active transport results from some step in the carrier mechanism being coupled to an energy generating system. This energy coupling alters the *apparent* rate constant for the reaction step, and creates the necessary asymmetry in the mechanism. On this basis, inhibition of energy metabolism is expected to affect only the coupled step, and not the other steps in the mechanism. Therefore analysis of transport kinetics both in the presence and in the absence of an inhibitor of energy metabolism will help to define the identity of the coupled step. This approach may be illustrated by the application of Mechanism 12.I to β-galactoside transport.

The transport, or permease, system for β-galactosides in *E. coli* cells could transport galactosidase-hydrolysable substrates such as *o*-nitrophenylgalactoside (NPG), as well as nonhydrolysable substrates such as thiomethylgalactoside (TMG). Cyanide did not inhibit NPG transport (Koch, 1964), but actively inhibited the *initial rates* of TMG transport (Fig. 12.4a). This provided an initial basis for the following analysis of the cyanide-sensitive step in TMG transport (Wong, Pincock and Bronskill, 1971):

(a) Since the initial rate was measured under the zero-time condition of $(\mathrm{TMG})_t = 0$, $k_{-3}(\mathrm{TMG})_t$ would be absent from the rate law, and $k_{-3}$ could not be the cyanide-sensitive step.

(b) TMG inhibited NPG transport competitively (Fig. 12.4b) and this inhibition was not alleviated by cyanide (Fig. 12.4c). Clearly, cyanide did not inhibit the binding of TMG to the carrier, and $k_1$ and $k_{-1}$ were not the cyanide-sensitive rate constant.

(c) Furthermore, cyanide inhibited TMG transport even in the presence of NPG (Fig. 12.4d). Accordingly, the resistance of NPG transport to cyanide was not due to any counter-transport (*trans*-stimulation) by hydrolytic products of NPG, which would have stimulated TMG transport as well. Thus the resistance of NPG transport to cyanide suggested that the act of hydrolysis of the transport substrate could substitute for energy coupling in driving the transport process.

These observations ruled out $k_{-3}$, $k_1$ and $k_{-1}$, and suggested $k_3$ as an important site of energy coupling. It was proposed that the dissociation of β-galactoside, or *G—X*, from the *E/G—X* complex at this step might be accelerated by any one of a number of possible reactions, e.g. hydrolysis of the *G—X* molecule, energized conformational change in the *E/G—X* complex, phosphorylation of the *G—X* molecule, or displacement of *G—X* by another metabolite (Fig. 12.5). If none of these accelerating reactions for $k_3$ was operative, then only simple dissociation of the *E/G—X* complex remained to permit the occurrence of un-energized facilitated diffusion.

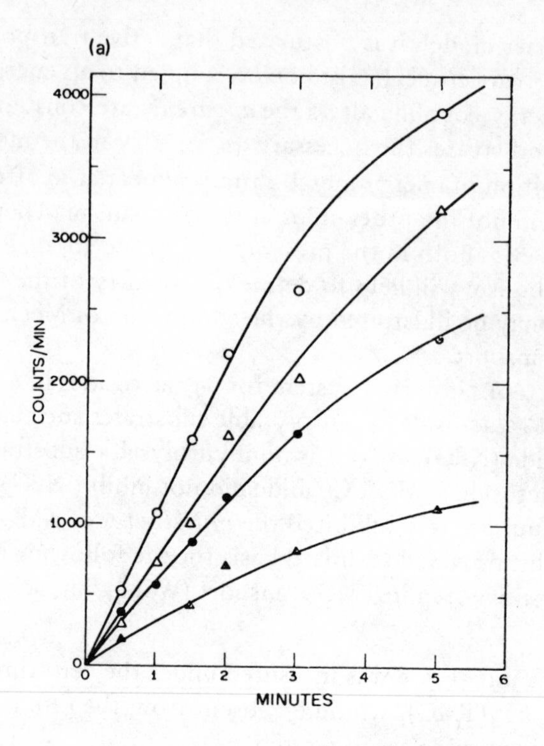


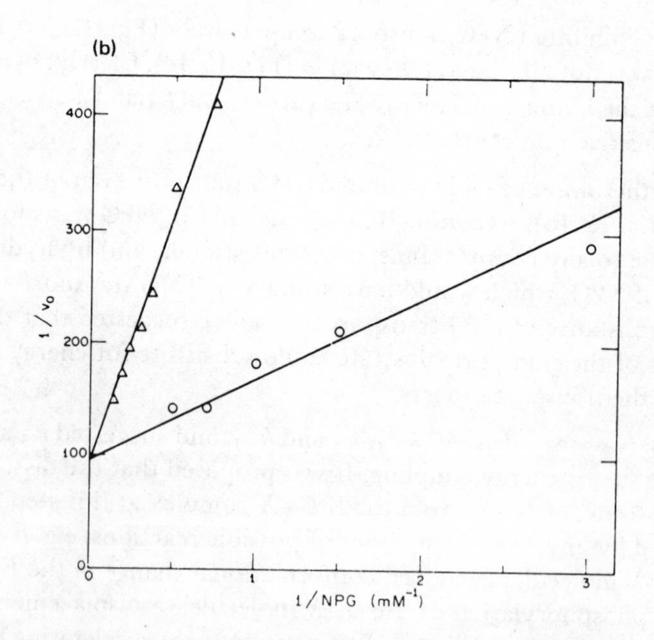


Fig. 12.4. Cyanide inhibition of TMG transport into *Escherichia coli.* (a) Increasin concentrations of cyanide progressively lowered the initial rate of TMG transport. (b) Competitive inhibition of NPG transport by TMG (○—no TMG added; △—TMG

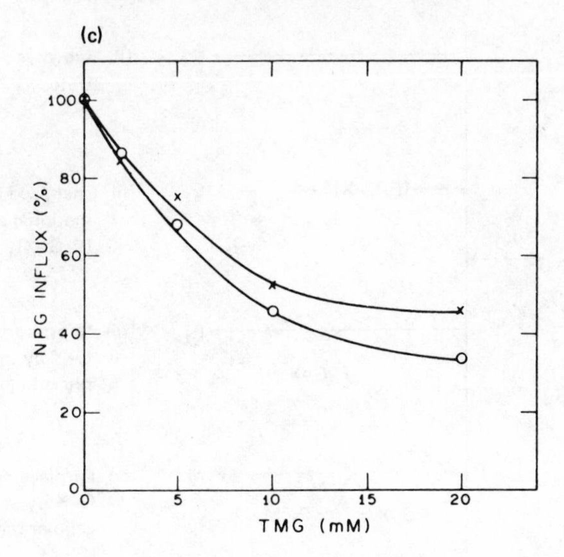


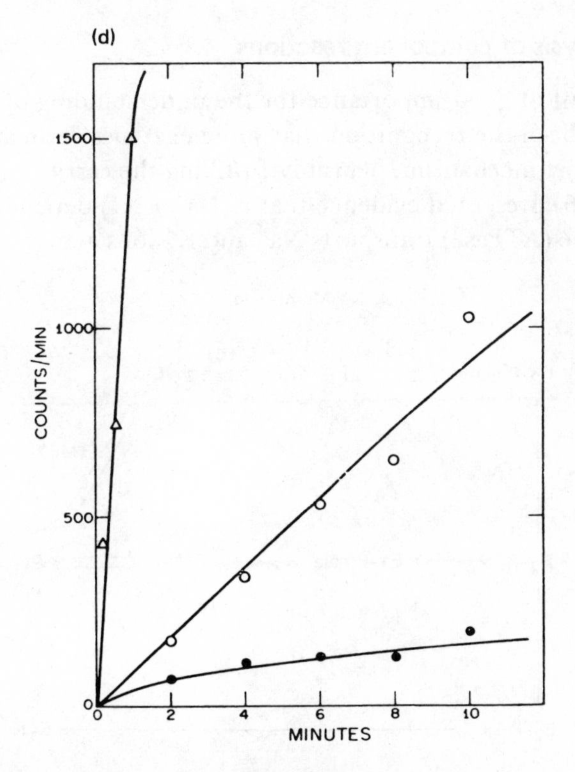


added). (c) Inhibition of NPG transport by TMG in the absence (x) or presence (○) of cyanide. (d) TMG transport in the absence of cyanide and NPG (△), in the presence of NPG (○), and in the presence of both NPG and cyanide (●). (From Wong, Pincock and Bronskill, 1971.)

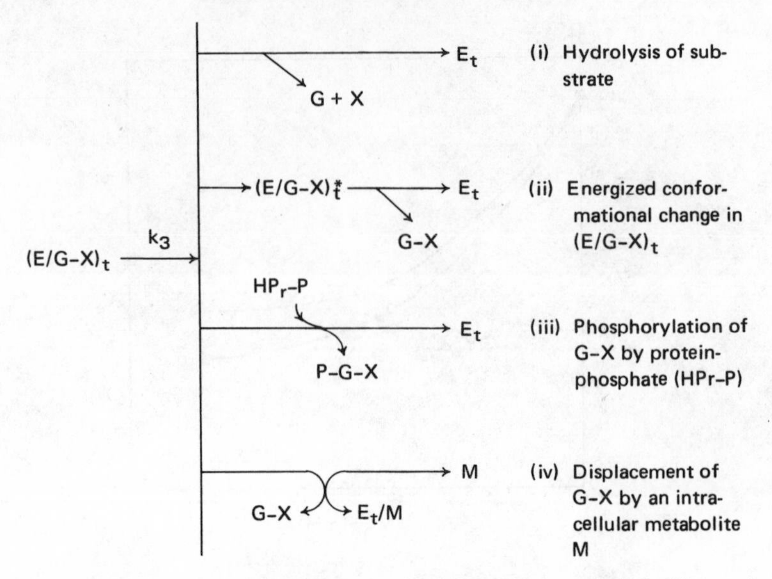


Fig. 12.5. Possible pathways for the energized dissociation of *E/G–X* complex in the TMG transport system. (After Wong, Pincock and Bronskill, 1971.)

## 12.2.2. Analysis of component reactions

A development of great importance for the understanding of energy coupling has been the recognition that some enzymes form an integral part of the transport mechanism, plausibly fulfilling the carrier function itself.

Skou (1965) presented evidence that a ($Na^+ + K^+$)-dependent adenosine triphosphatase (ATPase) transports $Na^+$ and $K^+$ ions across cell membranes

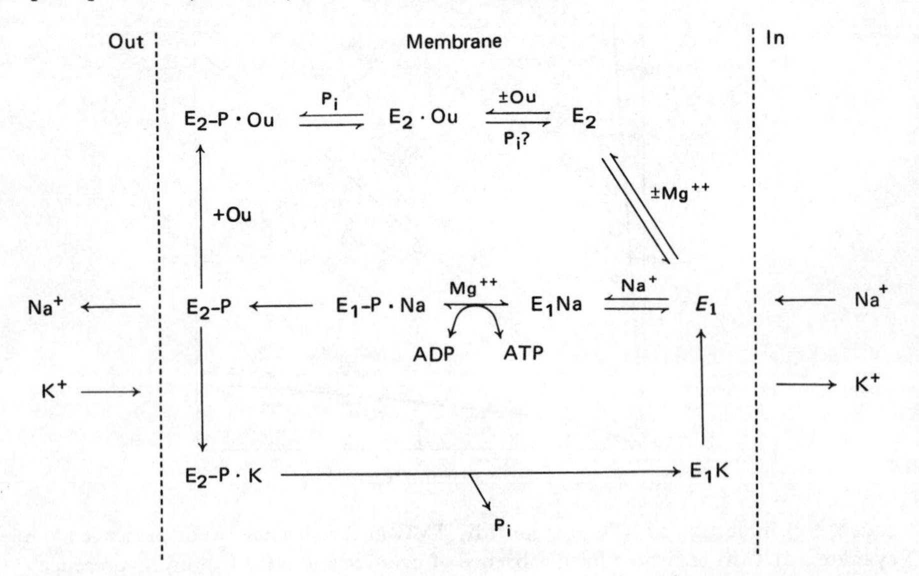


Fig. 12.6. Possible mechanism of ion transport by ($Na^+ + K^+$)-ATPase. (After Sen, Tobin and Post, 1969.)

by a mechanism which is sensitive to ouabain inhibition. Sen, Tobin and Post (1969) analysed the mechanism of interaction between the enzyme and its various ligands, and some of the observations that led to the proposed mechanism shown in Fig. 12.6 were as follows:

(a) Ouabain slowly reacted with the enzyme, $E_1$, to generate a new enzyme conformation, $E_2 \cdot$Ou, which was less readily phosphorylated by $^{32}$P-ATP to form the $E$-$^{32}$P intermediate; this reaction was stimulated by $Mg^{++}$ (Fig. 12.7a).

(b) The reaction between ouabain and enzyme could be arrested by $Na^+$, i.e. forming $E_1 \cdot$Na (Fig. 12.7b).

(c) Once the enzyme was phosphorylated by ATP, the $E_2$-P intermediate was rapidly dephosphorylated in the presence of $K^+$, i.e. forming $E_1$K; this reaction could be inhibited by ouabain, i.e. forming $E_2$-P · Ou (Fig. 12.7c).

(d) Whereas the reaction between $E_1$ and ouabain was stimulated by $Mg^{++}$, the reaction between $E_2$-P and ouabain did not require $Mg^{++}$ (Fig. 12.7d), thus underlining the different reactivities of the two enzyme conformations.

Since the *conformational shift* of the carrier (enzyme) between $E_1$ and $E_2$ could produce a change in carrier orientation with respect to the two sides of the membrane as well as a change in its ligand affinities, the proposed mechanism is capable of accomplishing the active transports of $Na^+$ and $K^+$ in opposite directions. In this case, all four aspects of the carrier mechanism, viz. transport of free carrier, transport of carrier-ligand complex, and formation and dissociation of the carrier-ligand complex, all stand to be energized by the phosphorylation of the carrier by ATP.

Another important example of enzyme participation in transport function is the proposal by Simoni and Roseman (1973) that the phosphotransferase system from *Staphylococcus aureus* catalyses the transport of galactose and lactose. This enzyme system consists of four proteins: HPr, Enzyme I, Enzyme II and Factor III$^{lac}$:

$$\text{P-enolpyruvate} + \text{HPr} \underset{Mg^{++}}{\overset{\text{Enzyme I}}{\rightleftharpoons}} \text{P-HPr} + \text{pyruvate}$$

$$\text{P-HPr} + 1/3\text{III}^{lac} \rightleftharpoons 1/3[\text{P}_3\text{-III}^{lac}] + \text{HPr}$$

$$1/3[\text{P}_3\text{-III}^{lac}] + \text{lactose} \xrightarrow{\text{Enzyme II}} \text{lactose-P} + 1/3\text{III}^{lac}$$

Enzyme I or HPr mutants could not transport either galactosides or glucosides. In contrast, Factor III$^{lac}$ and Enzyme II$^{lac}$ mutants could transport glucosides but not galactosides. Thus the two former proteins were required for all sugar transports, whereas the two latter proteins were

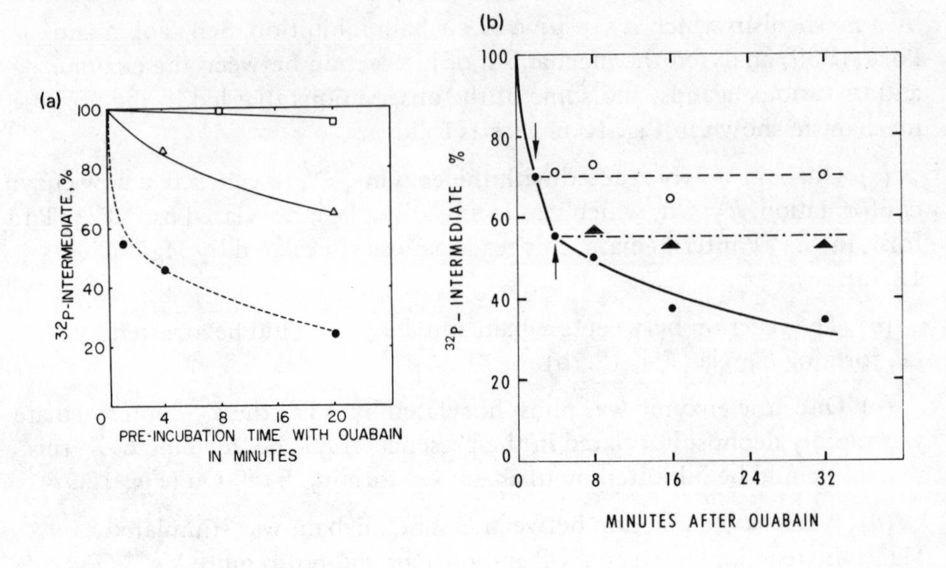


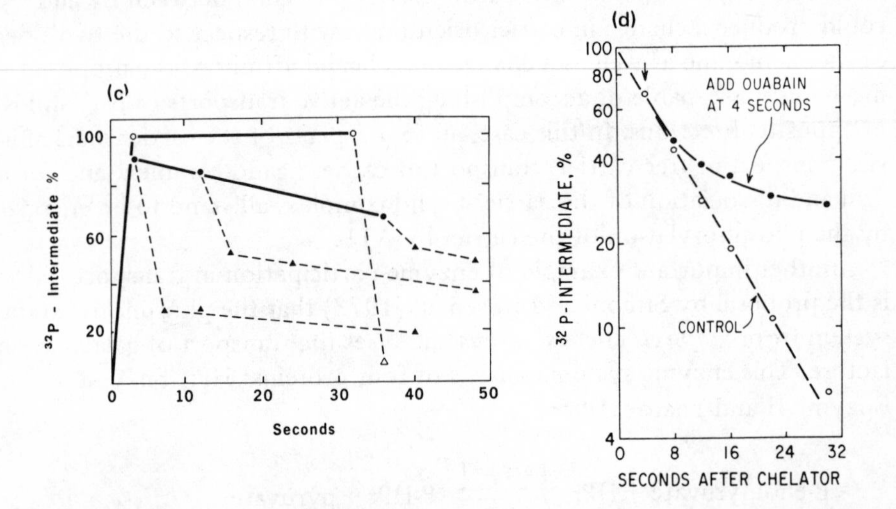


Fig. 12.7. Interaction of $(Na^+ + K^+)$-ATPase with its ligands. (a) Preincubation of enzyme with ouabain decreased the subsequent phosphorylation of enzyme by $^{32}$P-ATP (□—chelator added; △—$Mg^{++}$ added; ●—$Mg^{++}$ and phosphate added). (b) Reaction between enzyme and ouabain was arrested when $Na^+$ was added at the indicated times (●—no $Na^+$ added; ○, ▲—$Na^+$ added). (c) Dephosphorylation of the phosphoenzyme by $K^+$ (open symbols—in the absence of ouabain; solid symbols—in the presence of ouabain; $K^+$ was added in each instance at the start of the dotted line). (d) Dephosphorylation of phosphoenzyme after the addition of chelator to remove $Mg^{++}$ and inhibit re-formation of phosphoenzyme; thus dephosphorylation was inhibited by ouabain even in the absence of $Mg^{++}$. (From Sen, Tobin and Post, 1969.)

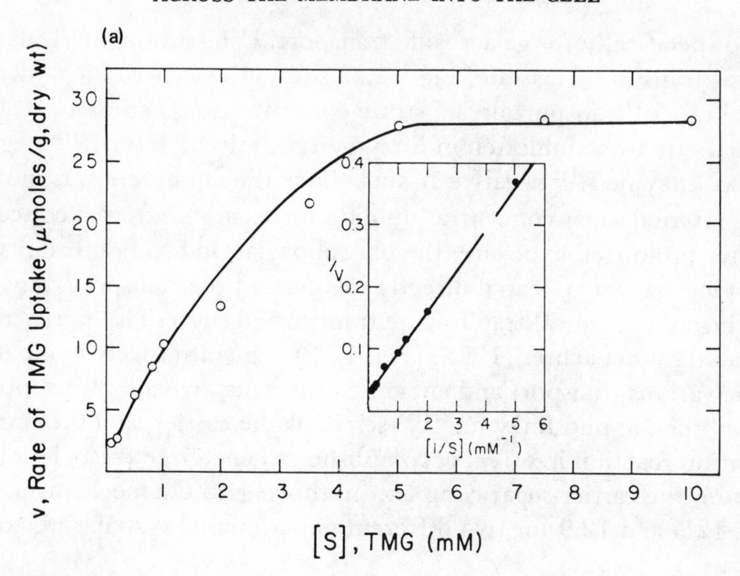


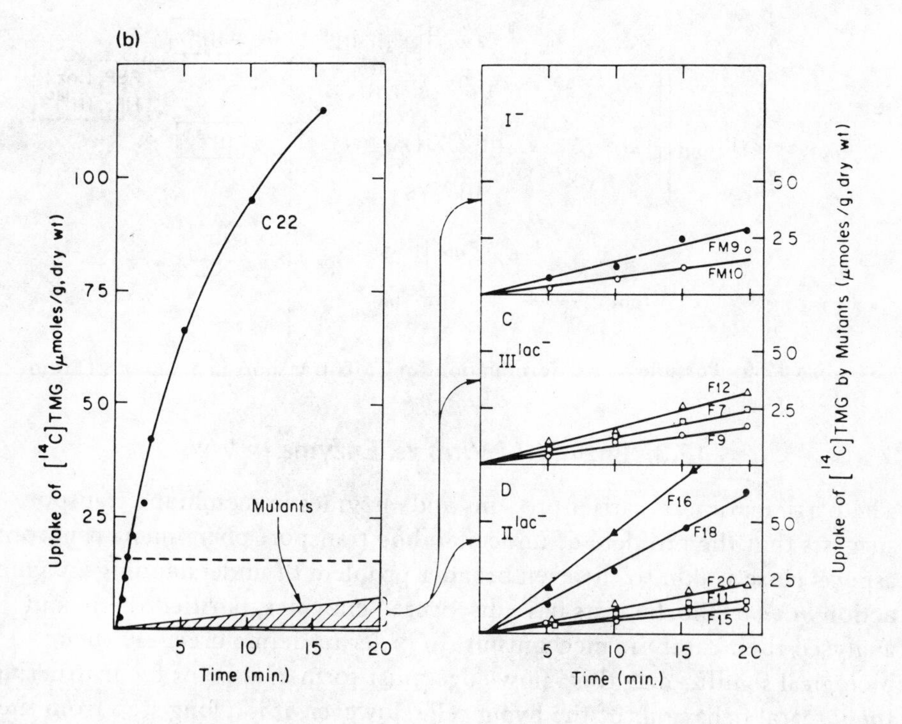


Fig. 12.8. TMG transport into *Staphylococcus aureus.* (a) Dependence of initial rate of transport on TMG concentration. (b) Inhibited initial rate in Enzyme I-negative, Factor $III^{lac}$-negative, and Enzyme $II^{lac}$-negative mutants. (From Simoni and Roseman, 1973.)

required specifically for galactoside transport. Using thiomethylgalactoside (TMG) as transport substrate, Fig. 12.8a shows the hyperbolic dependence of initial rate of transport on substrate concentration, and Fig. 12.8b shows the initial rate to be inhibited in Enzyme I-negative, Factor $III^{lac}$-negative, as well as Enzyme $II^{lac}$-negative strains. Since the *initial rate* was inhibited, and the internal sugar concentration did not reach its external concentration even after prolonged exposure, the phosphorylation reaction catalysed by the enzyme system appeared directly involved in the transport rather than just acting as a chemical trap for the transported sugar. The carrier mechanism formulated by Schachter (1973) (Fig. 12.9) is quantitatively compatible with the various transport and enzymic rate measurements. It proposes that the membrane-bound Enzymic $II^{lac}$ serves as the carrier, and the transphosphorylation reaction in effect accomplishes an energized dissociation of the sugar from the carrier-sugar complex. In this regard the mechanisms shown in Figs. 12.5 and 12.9 for two different experimental systems are completely analogous.

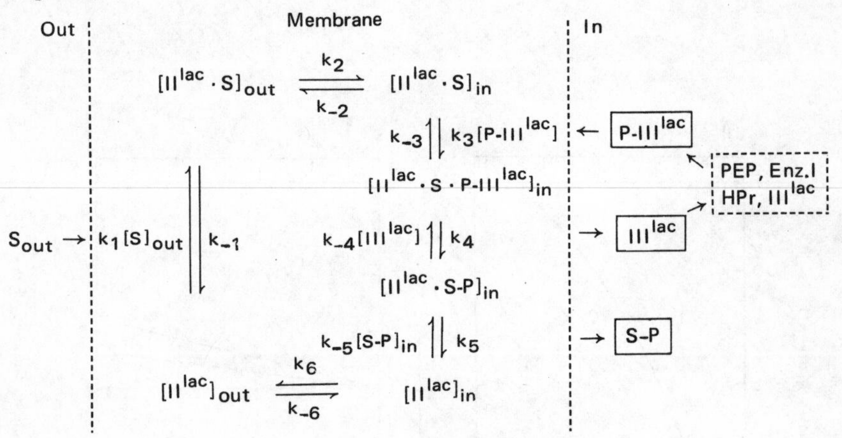


Fig. 12.9. Possible carrier formulation for TMG transport in *S. aureus.* (From Schachter, 1973.)

## 12.3. Enzyme *In Vitro* vs. Enzyme *In Vivo*

The participation of carrier-proteins and enzymes in membrane transport suggests that the problem of understanding transport phenomena is in some aspects closely akin to the even broader problem of understanding enzyme action *in vivo*. Biochemists have discovered enzymes, purified them, and analysed their catalytic mechanisms. In order to define precisely their biological significance, this knowledge must form a basis for reconstructing the integral behaviour of the living cell. However, it is a long step from the test-tube to the living cell, and the microenvironments of any enzyme molecule *in vitro* and *in vivo* may be entirely dissimilar. There is no easy solution to this problem, but there are workable approaches.

First, the difference between *in vitro* and *in vivo* conditions need not always result in irreconcilable discrepancy. For example, for a number of glycolytic enzymes, the intracellular flux rate calculated on the basis of their performance *in vitro* was found to be in satisfactory agreement with the flux rate measured *in vivo:*

| *Enzyme* | *Calculated flux* | *Measured flux* | *Reference* |
|---|---|---|---|
| Muscle enolase | 0.33–0.38 | 0.3 | Bücher and Sies (1969) |
| Yeast phosphoglycomutase | 7–10 | 7.5–15 | Wurster and Schneider (1970) |
| Yeast pyruvate kinase | 57–100 | 52–65 | Barwell and Hess (1972) |

Likewise, Chance, Hess and Betz (1964) observed that the *glycolytic oscillations* of yeast cells, a highly-structured multienzyme phenomenon, could be reproduced by carefully prepared cell extracts. In fact the oscillation parameters were so sensitive to ambient conditions that similarity between the *in vitro* and *in vivo* parameters provided a valuable criterion for the comparability of *in vitro* and *in vivo* conditions (Betz and Selkov, 1969).

Secondly, the study of enzyme mechanisms *in vitro* depends on the analysis of velocity-concentration, or $v(S)$, profiles. Such profiles are difficult but not impossible to establish for enzymes functioning within the cell. Thus, Kohen, Kohen and Thorell (1968) controlled the entry of glucose-1-phosphate into ascites cells by means of microelectrophoresis, and obtained the $v(S)$ profile for the reduction of $NAD^+$ by glyceraldehyde phosphate dehydrogenase (Fig. 12.10a). Nazar, Tyfield and Wong (1972) varied the intracellular concentrations of nucleoside triphosphates by means of phosphate depletion, and obtained the $v(S)$ profile for RNA polymerization (Fig. 12.10b). In the case of glycolysis in human erythrocytes, Rose and Warms (1966) could even determine that normal cells and pyruvate-kinase deficient cells utilized different regions of the $v(S)$ profile of the kinase. Furthermore, it is feasible to analyse experimental systems *midway* between the extracted enzyme and the intact cell. Reeves and Sols (1973) treated bacterial cells with toluene and freezing and thawing, and obtained "permeabilized" cells which were permeable to small substrates but retained their enzymes within the intracellular space. Such cells may be employed to define the kinetic behaviour of enzymes working at least *in situ* if not *in vivo* (Fig. 12.10c).

Finally, the problem of integrating kinetic information on a wide range of enzymes into a single multienzyme unit is vastly aided by the development of computer simulation. With ever faster computers having ever greater memory capacity, it has been possible to construct sophisticated kinetic models for a number of multienzyme systems:

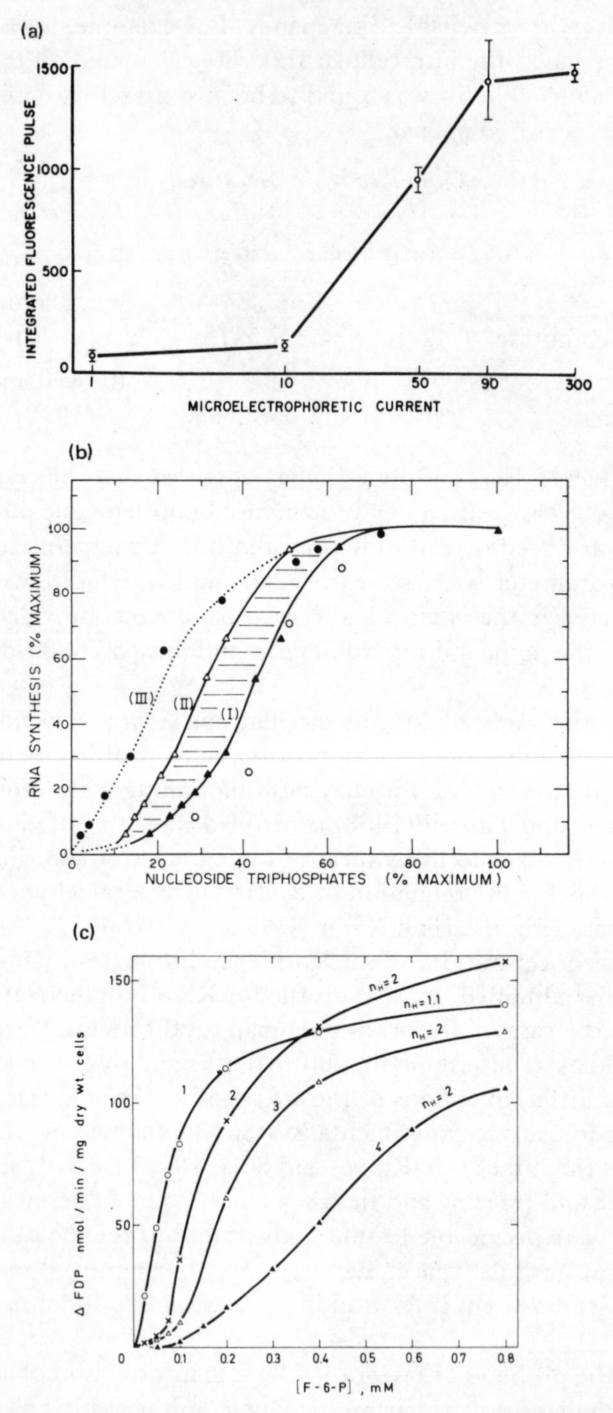


Fig. 12.10. Rate profile of enzymes functioning inside cells. (a) Rate of $NAD^+$ reduction by glyceraldehyde phosphate dehydrogenase (fluorescence pulse) as a function of glucose-1-phosphate entry into ascites cells (microelectrophoretic current). (From Kohen, Kohen and Thorell, 1968.) (b) Rate of stable RNA synthesis as a function of

| *Multienzyme system* | *Reference* |
|---|---|
| Glycolysis in ascites cells | Garfinkel and Hess (1964) |
| Cytochrome chain in mitochondria | Wagner, Erecinska and Pring (1971) |
| Glycolysis in human erythrocytes | Rapoport, Heinrich, Jacobasch and Rapoport (1974) |
| Metabolism of perfused rat heart | Garfinkel, Achs and Dzubow (1974) |

Such models can define on the one hand the regulatory significance of individual enzymes, and on the other hand the predicted behaviour of multienzyme systems. For example, one of the key regulatory enzymes of intermediary metabolism is phosphofructokinase. Lowry and Passoneau (1966) discovered that this enzyme responds to a remarkably wide range of allosteric effectors (Fig. 12.11). The interplay between so many effectors

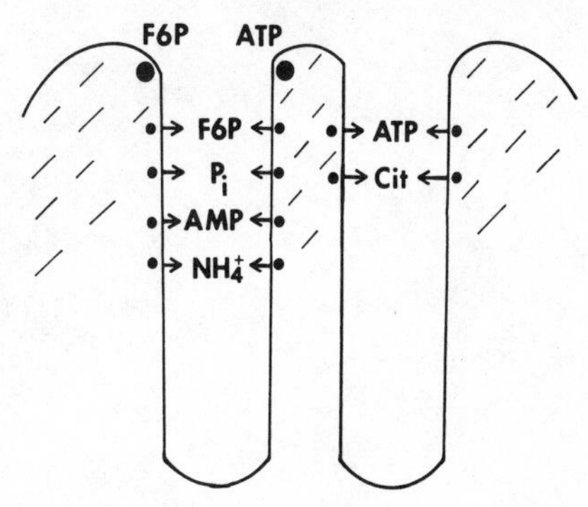


Fig. 12.11. Substrate and effector sites postulated for phosphofructokinase: large circles show substrate sites for fructose-6-phosphate and ATP; small circles on the right show inhibitor sites for ATP and citrate; small circles on the left show deinhibitor sites for fructose-6-phosphate, inorganic phosphate, AMP and $NH_4^+$. (From Lowry and Passoneau, 1966.)

can be delineated, and its regulatory significance defined, only on the basis of an adequate kinetic model (Garfinkel and Hess, 1964). The perfused-heart model contains 68 such submodels of enzyme and membrane-carrier mechanisms in a system of 334 chemical reactions among 275 chemicals, representing the biochemical pathways of energy metabolism, glycolysis, the Krebs cycle, fatty acid metabolism and glycogen metabolism. Within such a model, every detail of the phosphofructokinase rate profile will become meaningful. In this manner, kinetic models of multienzyme

---

nucleoside triphosphate concentrations inside *E. coli* cells. (curve I—rate vs. averaged concentration of ATP, GTP, CTP and UTP; curve II—rate vs. concentrations of UTP; curve III—*in vitro* rate for RNA polymerase included for comparison). (From Nazar, Tyfield and Wong, 1972). (c) Kinetics of *E. coli* phosphofructokinase in cells treated with toluene and freezing and thawing (curve 1—1 mM GDP added; curve 2 no effector added; curve 3—0.2 mM phosphoenolpyruvate added; curve 4—0.8 mM phosphoenolpyruvate added). (From Reeves and Sols, 1973.)

systems help to bridge the gulf between our understanding of enzyme behaviour *in vitro* and *in vivo*. By comparing the predictions of a model to observations on the experimental system, the adequacy or inadequacy of the model will be revealed. If adequate, the model may be extended and more revealing experiments may be designed. If inadequate, the structure of the model as well as the properties of its component enzymes would have to be reexamined. Either way the foundation is laid for the achievement of newer and more profound insight.

## Problem 12.1

What are the rate consequences if the interconversion of $E_c$ and $E_t$ in Mechanism 12.II is prohibited?

# Appendix

## Solutions to Problems

### Problem 1.1

Since 1 mM of the ester substrate is far below its $K_m$ value of 100 mM, the reaction velocity will decrease continuously during the reaction as the substrate is being consumed. Accordingly the progress curve will show a continuous decrease in slope (velocity) right from the start of the reaction. In contrast, 1 mM of the amide substrate is far above its $K_m$ value of 0.01 mM, and the enzyme will remain saturated by substrate over the greater part of the reaction. Accordingly the progress curve will show a near constant slope until toward the end of the reaction, when the amide concentration will be reduced to well below 0.1 mM.

### Problem 2.1

The rate law consists of three groups of terms:

$$\frac{v}{(E)_0} = \frac{(\textit{Forward numerator terms}) - (\textit{Reverse numerator terms})}{(\textit{Denominator terms})}$$

There are four admissible arrow patterns for the *forward numerator terms.* Each of the five enzyme species in the mechanism contributes one arrow to every pattern, which must contain a complete cyclic reaction pathway for the forward reaction (those arrows which form part of the cyclic pathway are shown as heavy arrows):

$k_1 k_{-2} k_3 k_6 k_8 \mathbf{AB}$

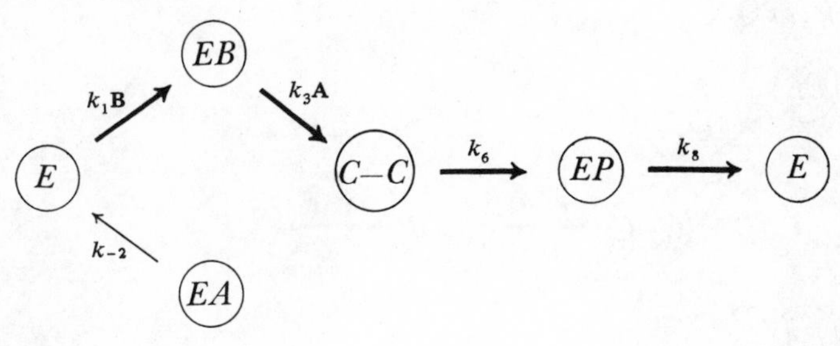

$k_1k_3k_4k_6k_8AB^2$

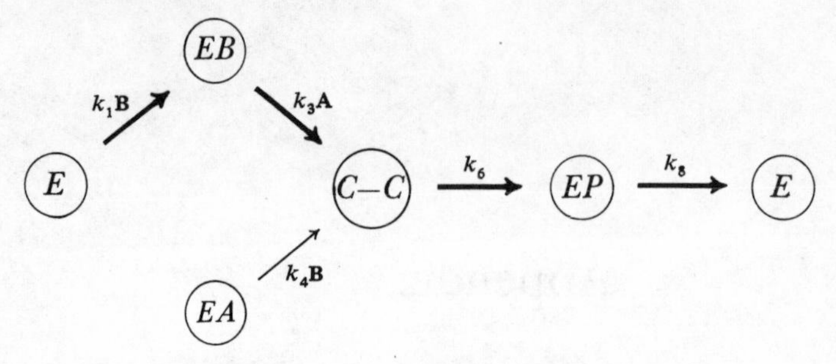


$k_{-1}k_2k_4k_6k_8AB$

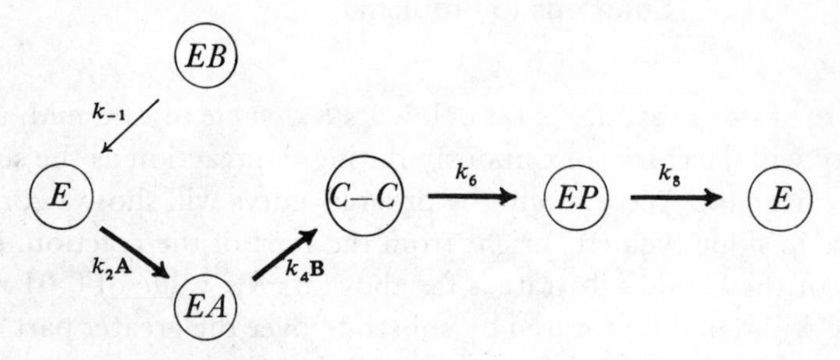


$k_2k_3k_4k_6k_8A^2B$

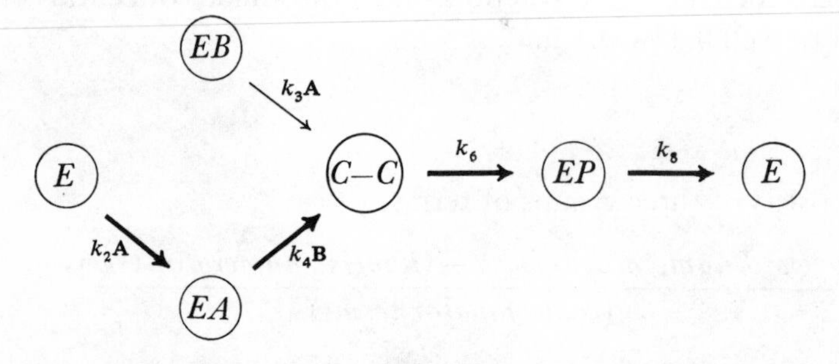


Similarly, there are four admissible arrow patterns for the *reverse numerator terms*. Each of the five enzyme species in the mechanism contributes one arrow to every pattern, which must contain a complete cyclic reaction pathway for the reverse reaction (those arrows which form part of the cyclic pathway are again shown as heavy arrows):

$k_{-1}k_{-2}k_{-3}k_{-6}k_{-8}PQ$

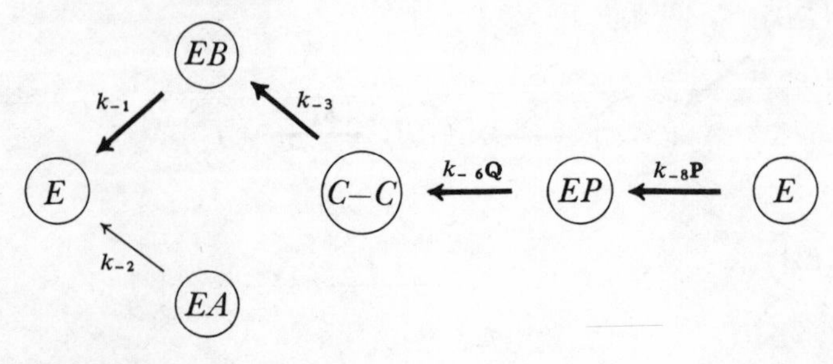

$k_{-1}k_{-3}k_4k_{-6}k_{-8}\mathbf{PQB}$

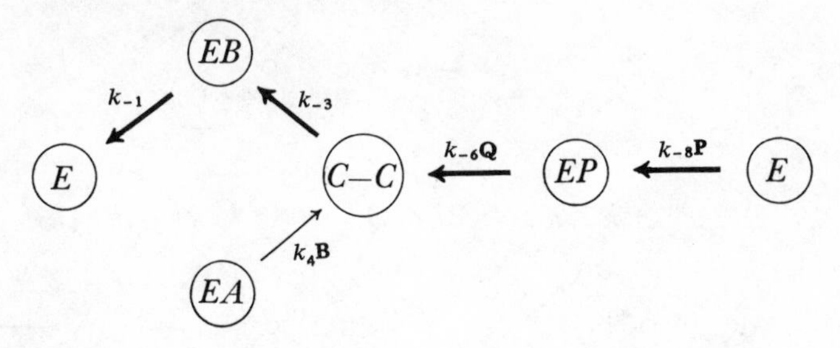


$k_{-1}k_{-2}k_{-4}k_{-6}k_{-8}\mathbf{PQ}$

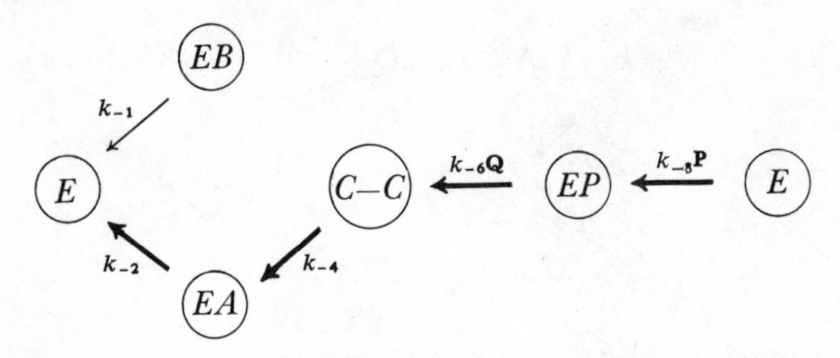


$k_{-2}k_3k_{-4}k_{-6}k_{-8}\mathbf{PQA}$

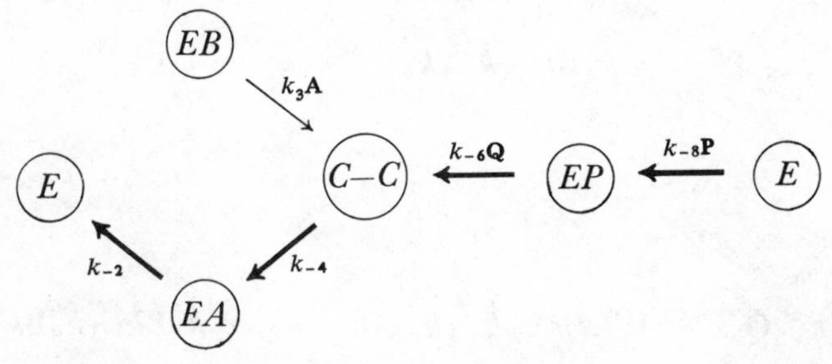


Finally, all the arrow patterns in the *denominator terms* contain four arrows, one from a different enzyme species. The four arrows form a convergent pattern that terminates at the sole enzyme species not contributing an arrow to the pattern. There are altogether sixty admissible patterns, twelve of which terminate at each of the five enzyme species in the mechanism. They can be derived by exhausting all the noncyclic four-arrow combinations, aided by tabulating all the arrows coming from each enzyme species (Hanes *et al.*, 1972), and are as follows:

$$\Delta_E = \text{Patterns terminating at } E$$

$$= \not{C}(k_{-1} + k_3\mathbf{A})(k_{-2} + k_4\mathbf{B})(k_{-3} + k_{-4} + k_6)(k_8 + k_{-6}\mathbf{Q})$$

$$= k_{-1}k_{-2}k_{-3}k_8 \qquad + k_{-1}k_{-2}k_{-4}k_8 \qquad + k_{-1}k_{-2}k_6k_8$$

$+k_{-1}k_{-2}k_{-3}k_{-6}Q \quad + k_{-1}k_{-2}k_{-4}k_{-6}Q \quad + k_{-1}k_{-3}k_4Bk_8$

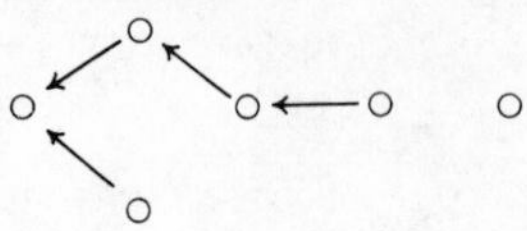

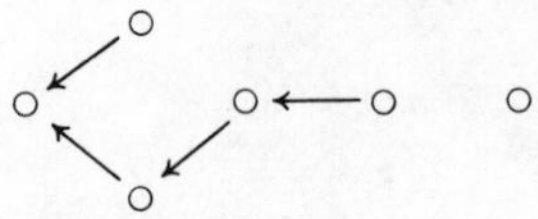

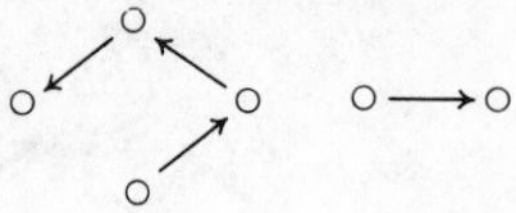

$+k_{-1}k_4Bk_6k_8 \quad + k_{-1}k_{-3}k_4Bk_{-6}Q \quad k_{-2}k_3Ak_{-4}k_8$

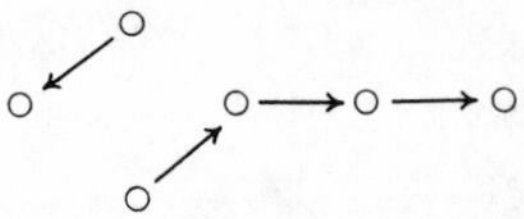

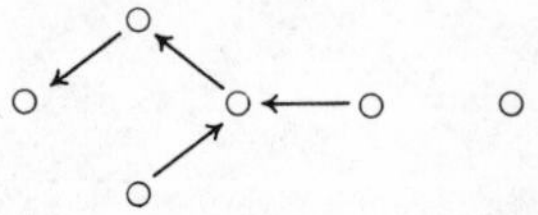

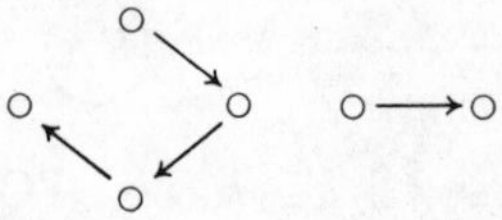

$+k_{-2}k_3Ak_6k_8 \quad + k_{-2}k_3Ak_{-4}k_{-6}Q \quad + k_3Ak_4Bk_6k_8$

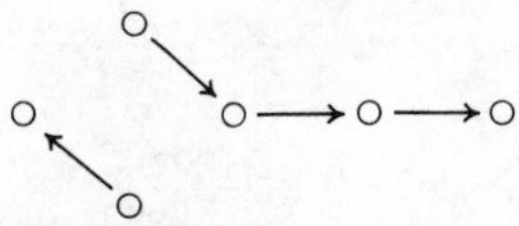

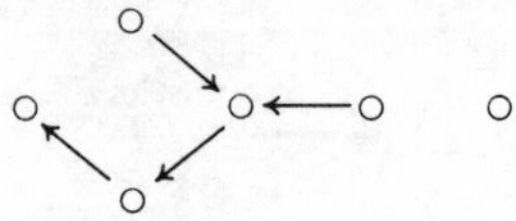

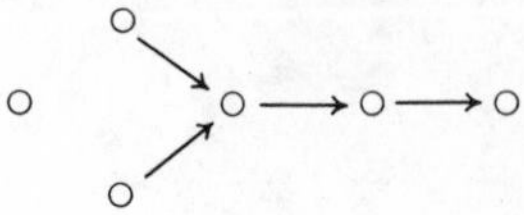

$\Delta_{EB}$ = Patterns terminating at $EB$

$$= \not{C}(k_1B + k_2A + k_{-8}P)(k_{-2} + k_4B)(k_{-3} + k_{-4} + k_6)(k_8 + k_{-6}Q)$$

$= k_1Bk_{-2}k_{-3}k_8 \quad + k_1Bk_{-2}k_{-4}k_8 \quad + k_1Bk_2k_6k_8$

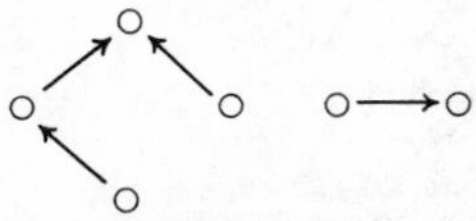

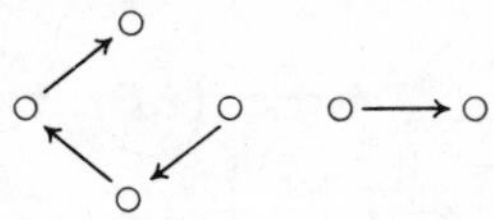

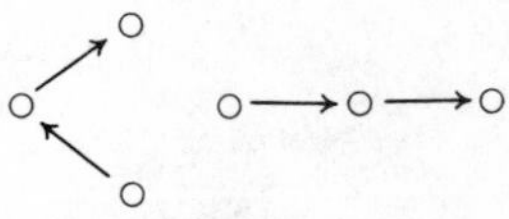

$+k_1Bk_{-2}k_{-3}k_{-6}Q \quad + k_1Bk_{-2}k_{-4}k_{-6}Q \quad + k_1Bk_{-3}k_4Bk_8$

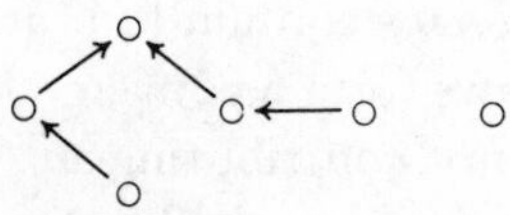

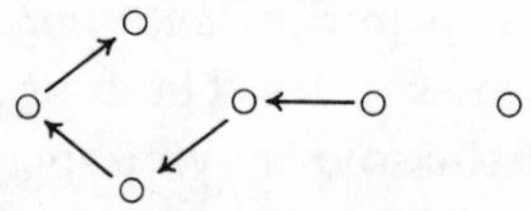

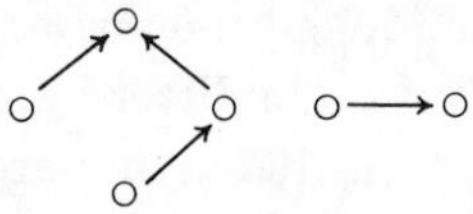

$+k_1Bk_4Bk_6k_8 \quad + k_1Bk_{-3}k_4Bk_{-6}Q \quad + k_2Ak_{-3}k_4Bk_8$

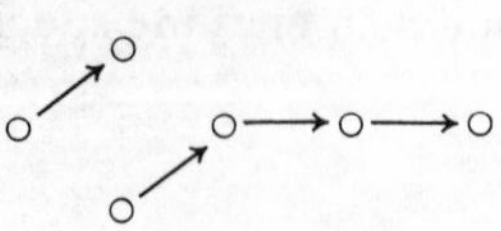

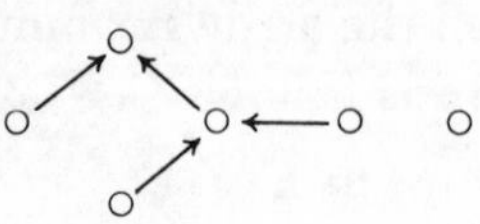

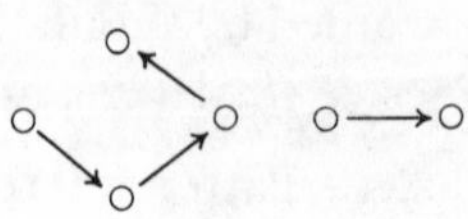

$+k_{-2}k_{-3}k_{-6}Qk_{-8}P \quad + k_2Ak_{-3}k_4Bk_{-6}Q \quad + k_{-3}k_4Bk_{-6}Qk_{-8}P$

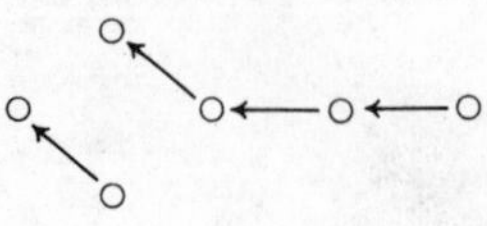

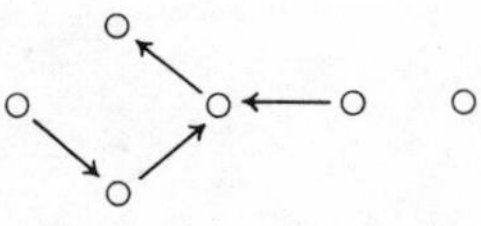

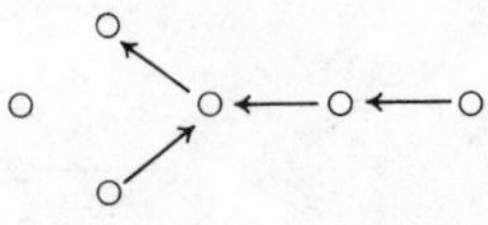

$$\Delta_{EA} = \text{Patterns terminating at } EA$$

$$= \text{₵}(k_1\mathbf{B} + k_2\mathbf{A} + k_{-8}\mathbf{P})(k_{-1} + k_3\mathbf{A})(k_{-3} + k_{-4} + k_6)(k_8 + k_{-6}\mathbf{Q})$$

$$= k_{-1}k_2\mathbf{A}k_{-3}k_8 \quad + k_{-1}k_2\mathbf{A}k_{-4}k_8 \quad + k_{-1}k_2\mathbf{A}k_6k_8$$

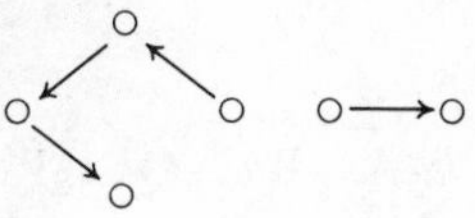

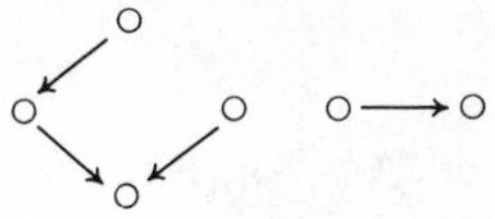

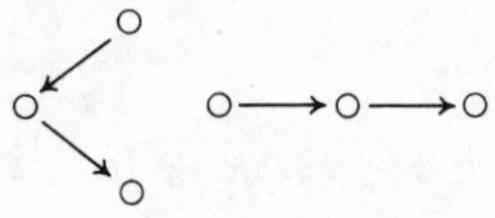

$$+ k_{-1}k_2\mathbf{A}k_{-3}k_{-6}\mathbf{Q} \quad + k_{-1}k_2\mathbf{A}k_{-4}k_{-6}\mathbf{Q} \quad + k_1\mathbf{B}k_3\mathbf{A}k_{-4}k_8$$

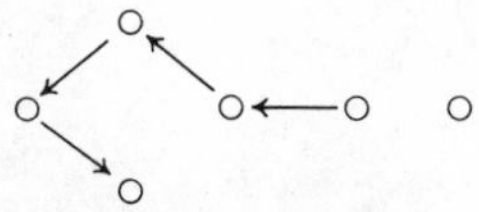

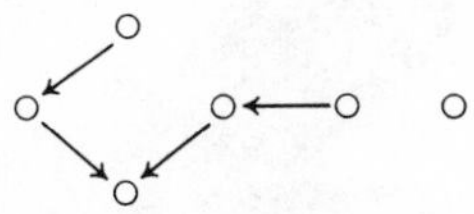

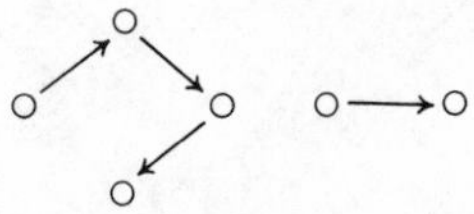

$$+ k_{-1}k_{-4}k_{-6}\mathbf{Q}k_{-8}\mathbf{P} \quad + k_1\mathbf{B}k_3\mathbf{A}k_{-4}k_{-6}\mathbf{Q} \quad + k_2\mathbf{A}k_3\mathbf{A}k_{-4}k_8$$

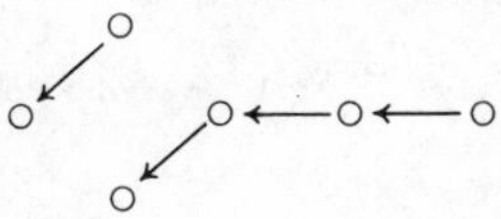

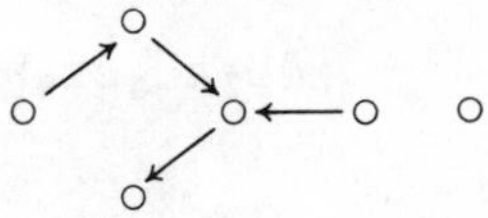

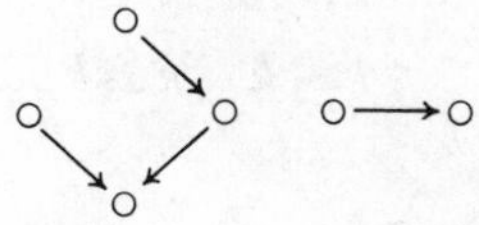

$$+ k_2\mathbf{A}k_3\mathbf{A}k_6k_8 \quad + k_2\mathbf{A}k_3\mathbf{A}k_{-4}k_{-6}\mathbf{Q} \quad + k_3\mathbf{A}k_{-4}k_{-6}\mathbf{Q}k_{-8}\mathbf{P}$$

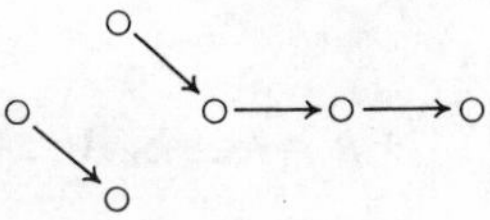

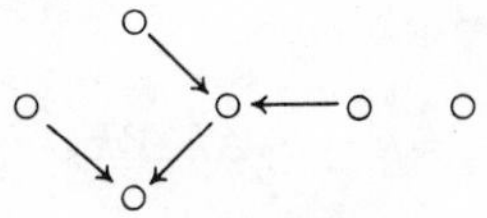

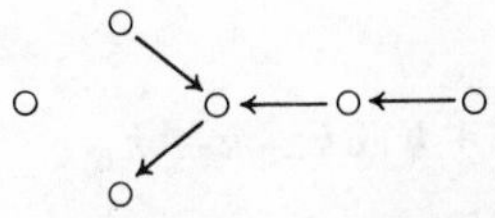

$$\Delta_{C-C} = \text{Patterns terminating at } C\text{—}C$$

$$= \text{₵}(k_1\mathbf{B} + k_2\mathbf{A} + k_{-8}\mathbf{P})(k_{-1} + k_3\mathbf{A})(k_{-2} + k_4\mathbf{B})(k_8 + k_{-6}\mathbf{Q})$$

$$= k_1\mathbf{B}k_{-2}k_3\mathbf{A}k_8 \quad + k_{-1}k_2\mathbf{A}k_4\mathbf{B}k_8 \quad + k_{-1}k_{-2}k_{-6}\mathbf{Q}k_{-8}\mathbf{P}$$

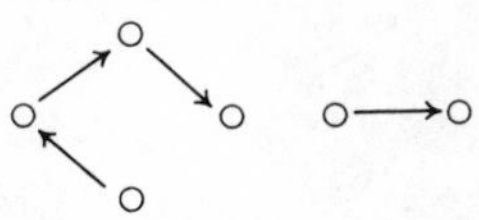

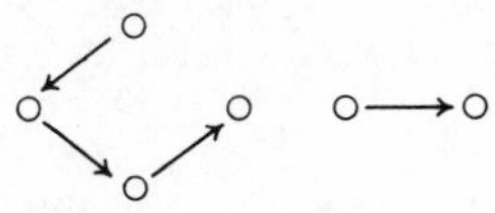

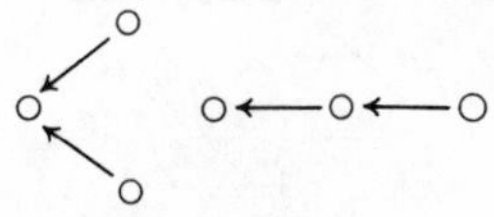

$$+ k_1\mathbf{B}k_{-2}k_3\mathbf{A}k_{-6}\mathbf{Q} \quad + k_{-1}k_2\mathbf{A}k_4\mathbf{B}k_{-6}\mathbf{Q} \quad + k_1\mathbf{B}k_3\mathbf{A}k_4\mathbf{B}k_8$$

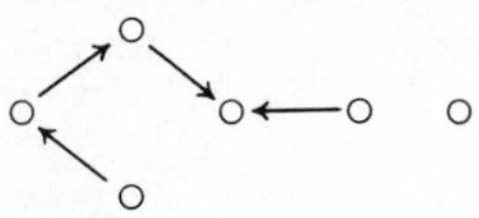

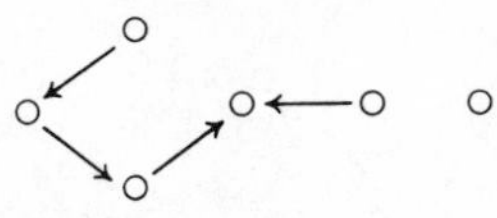

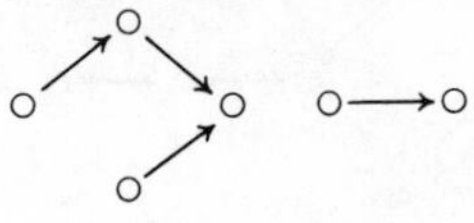

$+ k_{-1}k_4\mathbf{B}k_{-6}\mathbf{Q}k_{-8}\mathbf{P} \quad + k_1\mathbf{B}k_3\mathbf{A}k_4\mathbf{B}k_{-6}\mathbf{Q} \quad + k_2\mathbf{A}k_3\mathbf{A}k_4\mathbf{B}k_8$

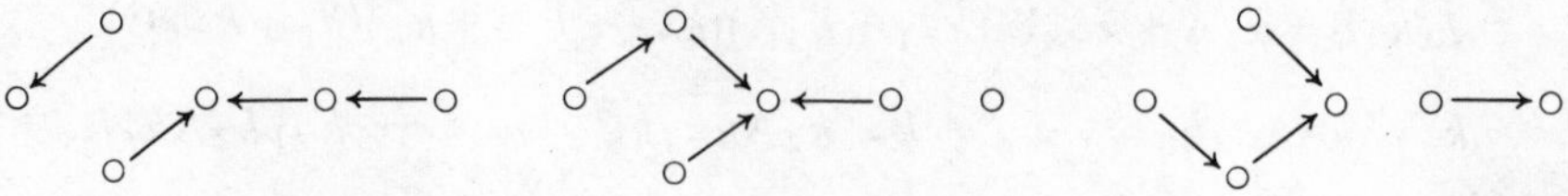

$+ k_{-2}k_3\mathbf{A}k_{-6}\mathbf{Q}k_{-8}\mathbf{P} \quad + k_2\mathbf{A}k_3\mathbf{A}k_4\mathbf{B}k_{-6}\mathbf{Q} \quad + k_3\mathbf{A}k_4\mathbf{B}k_{-6}\mathbf{Q}k_{-8}\mathbf{P}$

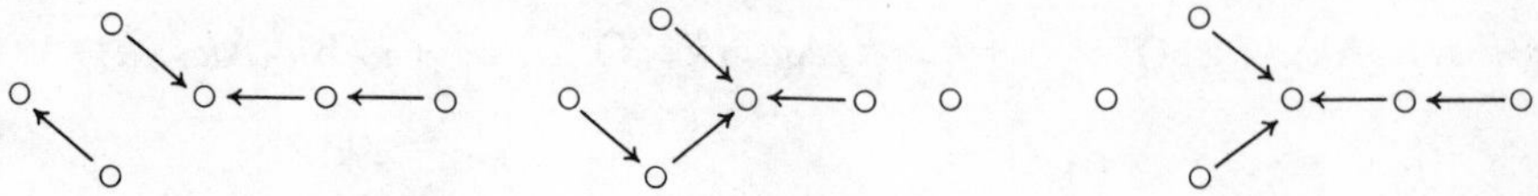

$\Delta_{EP}$ = Patterns terminating at $EP$

$= \mathbb{C}(k_1\mathbf{B} + k_2\mathbf{A} + k_{-8}\mathbf{P})(k_{-1} + k_3\mathbf{A})(k_{-2} + k_4\mathbf{B})(k_{-3} + k_{-4} + k_6)$

$= k_{-1}k_{-2}k_{-3}k_{-8}\mathbf{P} \quad + k_{-1}k_{-2}k_{-4}k_{-8}\mathbf{P} \quad + k_{-1}k_{-2}k_6k_{-8}\mathbf{P}$

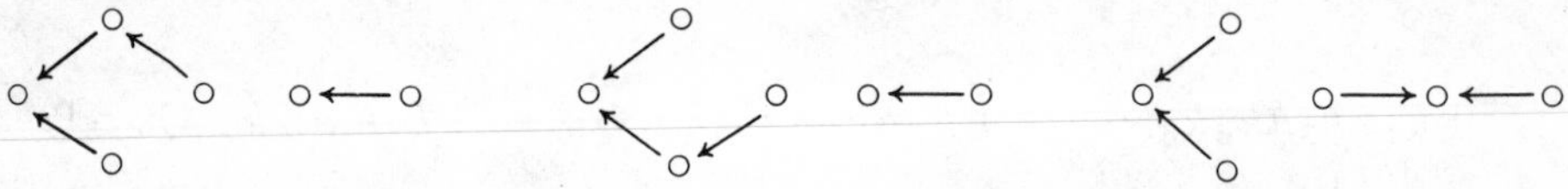

$+ k_1\mathbf{B}k_{-2}k_3\mathbf{A}k_6 \quad + k_{-1}k_2\mathbf{A}k_4\mathbf{B}k_6 \quad + k_{-1}k_{-3}k_4\mathbf{B}k_{-8}\mathbf{P}$

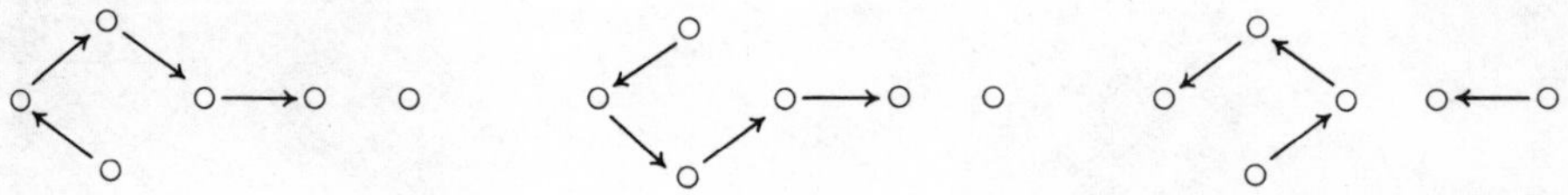

$+ k_{-1}k_4\mathbf{B}k_6k_{-8}\mathbf{P} \quad + k_1\mathbf{B}k_3\mathbf{A}k_4\mathbf{B}k_6 \quad + k_{-2}k_3\mathbf{A}k_{-4}k_{-8}\mathbf{P}$

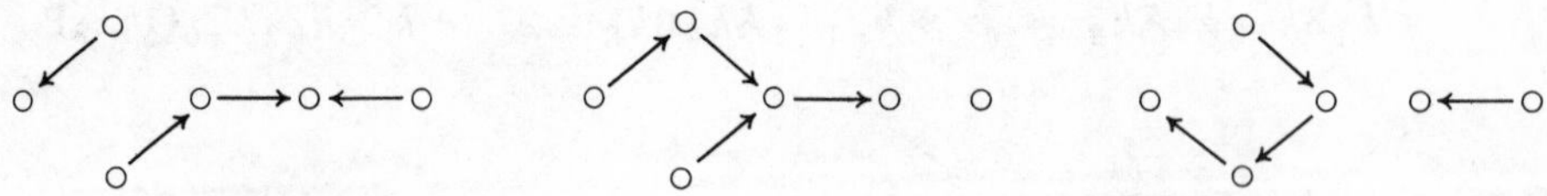

$+ k_{-2}k_3\mathbf{A}k_6k_{-8}\mathbf{P} \quad + k_2\mathbf{A}k_3\mathbf{A}k_4\mathbf{B}k_6 \quad + k_3\mathbf{A}k_4\mathbf{B}k_6k_{-8}\mathbf{P}$

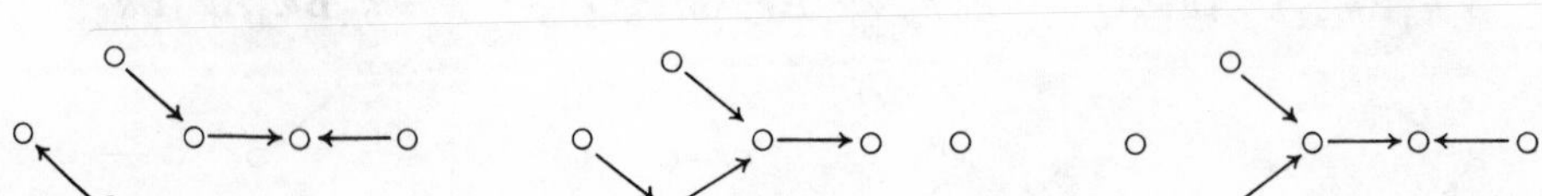

**Problem 3.1**

The formation of multiple dead-ends in a mechanism involving $X$ does not necessarily give rise to Eqn (3.11a) or (3.11b). For this purpose, it is essential that $X$ reacts with more than one enzyme species within the same dead-end (cf. Section 4.7.2). For example, Mechanism 3.II is a model mechanism that predicts Eqn (3.11a), and Mechanism 3.III is a model mechanism that predicts Eqn (3.11b):

$$\begin{array}{lcl} E \overset{S}{\rightleftharpoons} ES \longrightarrow E + P \\ \Updownarrow X \\ EX \overset{S}{\rightleftharpoons} EXS \longrightarrow EX + P \\ \Updownarrow X \\ EX_2 \overset{X}{\rightleftharpoons} EX_3 \end{array} \qquad \text{Mechanism 3.II}$$

$$\begin{array}{l} E \overset{S}{\rightleftharpoons} ES \longrightarrow E + P \\ \Updownarrow X \\ EX \overset{X}{\rightleftharpoons} EX_2 \overset{X}{\rightleftharpoons} EX_3 \end{array} \qquad \text{Mechanism 3.III}$$

In Mechanism 3.II, $EX$ can take part in a productive reaction cycle. Accordingly, the $X$-arrow that leads to the formation of $EX$ from $E$ can appear in the numerator of the rate law, whereas the other two $X$-arrows can only appear in the denominator. In Mechanism 3.III, all three $X$-arrows can only appear in the denominator, because $EX$, $EX_2$ and $EX_3$ are all non-productive enzyme species in a dead-end branch.

**Problem 4.1**

There are three enzyme species in Mechanisms 4.II and 4.IV, but four enzyme species in Mechanism 4.III. Accordingly, there are three arrows per pattern in the numerator of the rate law for 4.II and 4.IV, but four arrows per pattern for 4.III. In fact, only one numerator pattern can be constructed in each case to conform to the structural rule. As usual, the denominator patterns all contain one less arrow than the numerator ones, and terminate at one of the enzyme species in the mechanism.

$$v_0 = \frac{k_{-i}k_1 S k_2}{k_{-i}k_1 S + k_{-i}(k_{-1} + k_2) + k_i I(k_{-1} + k_2)}$$

$$= \frac{S \cdot V_S}{S + K_m^S\left(1 + \dfrac{I}{K_I}\right)}$$

$$V_S = k_2 \qquad K_m^S = (k_{-1} + k_2)/k_1 \qquad K_I = k_{-i}/k_i \qquad \text{Mechanism} \quad 4.II$$

$$v_0 = \frac{k_{-i}k_T k_1 S k_2}{k_{-i}k_T k_1 S + k_{-i}(k_R + k_T)(k_{-1} + k_2) + k_i I k_R (k_{-1} + k_2)}$$

$$= \frac{S \cdot V_S}{S + K_m^S\left(1 + \dfrac{I}{K_I + K_I/L}\right)}$$

$$V_S = k_2 \quad K_m^S = \frac{k_{-1} + k_2}{k_1} \cdot (1 + L) \quad L = k_R/k_T \quad K_I = k_{-i}/k_i \qquad \text{Mechanism} \quad 4.III$$

$$v_0 = \frac{k_{-i}k_1 S k_2}{k_1 S(k_{-i} + k_i I) + k_{-i}(k_{-1} + k_2)}$$

$$= \frac{S \cdot V_S/(1 + I/K_I')}{S + K_m^S/(1 + I/K_I')}$$

$$V_S = k_2 \quad K_m^S = (k_{-1} + k_2)/k_1 \quad K_I' = k_{-i}/k_i \qquad \text{Mechanism} \quad 4.IV$$

## Problem 5.1

The rate observations recorded in Fig. 5.7 point to the following mechanism:

$$E \underset{k_{-1}}{\overset{k_1 A}{\rightleftharpoons}} EA; \quad EA \underset{k_{-2}}{\overset{k_2 B}{\rightleftharpoons}} EAB \underset{k_{-4} Q}{\overset{k_4}{\rightleftharpoons}} EP; \quad EA \underset{k_{-3} Q}{\overset{k_3}{\rightleftharpoons}} E\text{—}G \underset{k_{-5}}{\overset{k_5 B}{\rightleftharpoons}} EP; \quad EP \underset{k_{-6} P}{\overset{k_6}{\rightleftharpoons}} E; \quad E \underset{k_{-7}}{\overset{k_7 Q}{\rightleftharpoons}} EQ$$

The reasons for this proposal are:

(a) The Eadie plots for **A** and **P** are linear, suggesting that $A$ and $P$ each react with a single enzyme species. That they react with the same species, namely the free enzyme, is indicated by the lack of effect of **P** on $V_A$ and the lack of effect of **A** on $V_P$.

(b) The dependences of $v_f$ on **B**, of $V_A$ on **B**, and of $V_P$ on **B** are all higher degree. Therefore *B* must react with two (or more) enzyme species other than the free enzyme.

(c) The $v_r$(Q) function is substrate-inhibitor type, but Q fails to inhibit $V_P$. Thus the inhibitory action of *Q* likely involves the formation of a dead-end complex with the free enzyme. Such a dead-end would not be significant if the system is saturated by either *P* or *A*, both of which react with the free enzyme.

(d) Even though Q fails to inhibit $V_P$, the $V_P$(Q) function is still higher-degree. Therefore *Q* also reacts with two (or more) enzyme species other than the free enzyme. This is confirmed by the higher-degree effects of Q on $V_A$.

(e) The proposed mechanism provides two species, *EA* and *E–G*, to react with *B*. It likewise provides two species, *EP* and *E–G*, to react with *Q* even when the dead-end formation of *EQ* is insignificant.

(f) Since the $V_B$(P) relationship is first-degree inhibitory, *B* must not react with the free enzyme. Otherwise the relationship would be zero-degree. The first-degree $V_B$(A) dependence confirms this.

(g) The Dixon plot of $1/V_B$ versus Q appears linear, and suggests that $V_B$(Q) is first-degree inhibitory. From the proposed mechanism, this function is expected to be second-degree inhibitory, because *Q* reacts with two enzyme species (namely *E* and *EP*) that do not react with *B*. However, this apparent inconsistency does not undermine the proposed mechanism. Instead, it is a useful reminder that, whereas a seemingly higher-degree observation rules out a first-degree function, a seemingly first-degree observation by no means rules out a higher-degree function.

### Problem 6.1

Adding *P* to the forward reaction system brings into the denominator of the rate law two new arrow patterns, namely $k_2\mathrm{B}k_{-3}\mathrm{P}$ and $k_{-1}k_{-3}\mathrm{P}$. This has no effect on $\phi_B$, but causes both $\phi_A$ and $\phi_{AB}$ to be increased by the same factor of $(k_3 + k_{-3}\mathrm{P})/k_3$. Accordingly the Dalziel relation of $\phi_{AB} = V_r\phi_A\phi_B$ is unaltered.

Adding *Q* to the forward reaction system brings into the denominator of the rate law two new arrow patterns, namely $k_1\mathrm{A}k_{-2}\mathrm{Q}$ and $k_{-1}k_{-2}\mathrm{Q}$. This has no effect on $\phi_A$, but causes both $\phi_B$ and $\phi_{AB}$ to be increased by the same factor of $(k_3 + k_{-2}\mathrm{Q})/k_3$. Accordingly the Dalziel relation of $\phi_{AB} = V_r\phi_A\phi_B$ is once again unaltered.

**Problem 7.1**

The protein under consideration has $n$ independent and equivalent binding sites for $S$. They all react with $S$ with the same intrinsic association rate constant $k_a$, and the same intrinsic dissociation rate constant $k_d$. The ratio $k_a/k_d$ is $\bar{K}$. The apparent stability constant $U_j$ pertains to the addition of $S$ to $E_nS_{j-1}$ to form $E_nS_j$. We note that:

Since there are $n - (j - 1)$ empty sites on $E_nS_{j-1}$, the rate of conversion of $E_nS_{j-1}$ to $E_nS_j$ must be $[n - (j - 1)]k_a \cdot \mathrm{S} \cdot (E_nS_{j-1})$.

Since there are $j$ molecules of $S$ on $E_nS_j$, the rate of conversion of $E_nS_j$ back to $E_nS_{j-1}$ must be $jk_d \cdot (E_nS_j)$.

It follows that:

$$U_j = \frac{(E_nS_j)}{(E_nS_{j-1}) \cdot \mathrm{S}} = \frac{[n - (j - 1)]k_a}{jk_d} = \frac{n - j + 1}{j} \times \bar{K}.$$

Entirely similar considerations give us

$$U_{j+1} = \frac{n - j}{j + 1} \times \bar{K}.$$

Dividing $U_{j+1}$ by $U_j$ yields Eqn (7.9).

**Problem 8.1**

The mechanism furnishes two enzyme species, namely $E_R$ and $E_T$, to react with $S$. Since arrow patterns containing both of the S-arrows can be constructed for the numerator and the denominator of the rate law, the $v(\mathrm{S})$ function is second-degree. Therefore, as pointed out by Rabin, it represents a plausible model mechanism for nonhyperbolic rate behaviour even though there is neither interactions between multiple enzyme protomers nor branching pathways for the additions of multiple substrates.

With only a single substrate site, the $y(\mathrm{S})$ function has to be first-degree, i.e. substrate binding must be hyperbolic. In addition, if the interconversion between $E_RS$ and $E_TS$ is disallowed, the kind of topological constraint discussed in Section 4.7.2 becomes operative. As a result, the $v(\mathrm{S})$ function is first-degree even though $S$ still reacts with two enzyme species in the mechanism. The mechanism thus loses the capability of predicting non-hyperbolic rate behaviour.

**Problem 9.1**

It is possible that the bulk of the group-transfer reaction flows through the ternary complexes $E/GX/Y$ and $E/X/GY$, and only a trickle through the substituted enzyme $E{-}G$. It is also possible that not even a trickle flows through $E$–$G$, in which case $E$–$G$ is not a normal reaction intermediate at all, but is induced to form by experimental conditions chosen to entrap the $G$ moiety on the enzyme surface. For example, Schellenberg (1967) discovered that a hydrogenated enzyme could be isolated when some dehydrogenases were inactivated by appropriate means in the presence of a reduced substrate. The hydrogenation involved the conversion of a tryptophan residue from the indolenine to the indole form. However, in the case of alcohol dehydrogenases, there was no significant exchange either between the reduced and oxidized coenzymes in the absence of substrates (Kaplan, Colowick and Neufeld, 1952), or between the reduced and oxidized substrates in the absence of coenzymes (Wong and Williams, 1968). Consequently, although the hydrogenation of enzyme-bound indolenine represents fascinating chemistry, the hydrogenated enzyme cannot be considered a normal reaction intermediate. Another example is the succinyl-coenzyme A synthetase from *E. coli*. Even though evidence pointed to the catalytic importance of a phosphorylated enzyme in this system, Moffet and Bridger (1970) found that a quaternary complex between enzyme, ATP, succinate and coenzyme A had to be formed before a significant release of any reaction product occurred. Consequently $E$–$G$ (i.e. a freely existing phosphorylated enzyme) was not a significant enzyme species in the kinetic mechanism.

## Problem 10.1

The Alberty-Massey derivation runs as follows. In the case of the bell-shaped curve, $H_1^+E_i$ is the only active enzyme species in the $k_i$-step, and it is apparent from Eqns (10.44) and (10.42) that

$$k_i = k_{i1}\psi_1 = k_{i1}\Sigma(E_i)\frac{(\mathrm{H}^+)K_1}{(\mathrm{H}^+)^2 + (\mathrm{H}^+)K_1 + K_1K_2}. \tag{10.64}$$

At the peak of the bell-shaped curve, if $\Sigma(E_i)$ does not vary with $(\mathrm{H}^+)$,

$$\frac{dk_i}{d(\mathrm{H}^+)} = k_{i1}\Sigma(E_i)\frac{K_1^2K_2 - (\mathrm{H}^+)^2K_1}{[(\mathrm{H}^+)^2 + (\mathrm{H}^+)K_1 + K_1K_2]^2} = 0 \tag{10.65}$$

which requires that

$$(\mathrm{H}^+)_{\text{peak}} = \sqrt{K_1K_2}. \tag{10.66}$$

Combination of Eqns (10.64) and (10.66) indicates that

$$k_i \text{ at } (\mathrm{H}^+)_{\text{peak}} = k_{i1}\Sigma(E_i)\frac{K_1\sqrt{K_1K_2}}{2K_1K_2 + K_1\sqrt{K_1K_2}}. \tag{10.67}$$

Since $k_i$ at $(\mathrm{H}^+)_a$ or $(\mathrm{H}^+)_b$ is equal to one half $k_i$ at $(\mathrm{H}^+)_{\text{peak}}$, we can write, on the basis of Eqns (10.64) and (10.67),

$$\frac{(\mathrm{H}^+)_a K_1}{(\mathrm{H}^+)_a^2 + (\mathrm{H}^+)_a K_1 + K_1 K_2} = \frac{(\mathrm{H}^+)_b K_1}{(\mathrm{H}^+)_b^2 + (\mathrm{H}^+)_b K_1 + K_1 K_2} \tag{10.68}$$

and

$$\frac{(\mathrm{H}^+)_a K_1}{(\mathrm{H}^+)_a^2 + (\mathrm{H}^+)_a K_1 + K_1 K_2} = \frac{K_1\sqrt{K_1 K_2}}{4K_1 K_2 + 2K_1\sqrt{K_1 K_2}}\,. \tag{10.69}$$

Rearrangement of Eqn (10.68) gives

$$K_1[(\mathrm{H}^+)_a(\mathrm{H}^+)_b - K_1 K_2][(\mathrm{H}^+)_a - (\mathrm{H}^+)_b] = 0. \tag{10.70}$$

Since $(\mathrm{H}^+)_a$ and $(\mathrm{H}^+)_b$ are not the same, Eqn (10.70) leads to the first part of Eqn (10.46):

$$K_1 K_2 = (\mathrm{H}^+)_a(\mathrm{H}^+)_b.$$

Rearrangement of Eqn (10.69) gives

$$(\mathrm{H}^+)_a K_1 = (\mathrm{H}^+)_a^2 + K_1 K_2 - (\mathrm{H}^+)_a \cdot 4\sqrt{K_1 K_2}. \tag{10.71}$$

Substitution of $K_1 K_2$ is this equation by $(\mathrm{H}^+)_a(\mathrm{H}^+)_b$ finally leads to the second part of Eqn (10.46):

$$K_1 = (\mathrm{H}^+)_a + (\mathrm{H}^+)_b - 4\sqrt{(\mathrm{H}^+)_a(\mathrm{H}^+)_b}.$$

In deriving Eqn (10.65), it is assumed that $\Sigma(E_i)$ does not vary with pH. Such would be the case if the free enzyme and the various enzyme-ligand complexes in the mechanism have essentially the same p*K*s, so that pH has little effects on the distribution of total enzyme amongst them. Such also would be the case in considering the variation of $V_S$ with pH for Mechanism 10.VIII, because upon saturation by substrate all the enzyme would be in the form of the *ES* complex. However, such would not be the case in considering the variation of $V_S$ with pH for Mechanism 10.III, because upon saturation by substrate all the enzyme would be distributed between *ES* and *E–Q*, a distribution that could be affected by changing pH.

**Problem 11.1**

For the Lineweaver-Burk plot, we can begin with Eqn (11.10). Replacing $v_i$ by $y$, and X by $x$, and noting that there are a total of eight observations ($N = 8$), this yields

$$\theta_2 = \frac{K_m}{V} = \frac{4.5497 - 1.785 \times 15.47/8}{41.138 - 15.47 \times 15.47/8} = 0.09786\ (\mathrm{min}/\mu\mathrm{M})/(1/\mathrm{mM})$$

$$= 97.86\ \mathrm{min}.$$

Having obtained $\theta_2$, Eqn (11.9) can be used to calculate $\theta_1$:

$$\theta_1 = \frac{1}{V} = \frac{1.785 - 0.09786 \times 15.47}{8}$$

$$= 0.03389 \text{ min}/\mu\text{M}.$$

Therefore the saturation velocity is

$$V = 1/\theta_1 = 29.5\ \mu\text{M/min}.$$

Having the least-squares estimates of both $\theta_1$ and $\theta_2$, it is possible to define by means of Eqn (11.22) the theoretical $y_{th}$ that corresponds to any value of $x$, and in turn to calculate the total variance on the basis of Eqn (11.14), which becomes

$$s^2 = \frac{\Sigma(y_i - y_{th})^2}{N - j}.$$

Instead of defining $y_{th}$, a more attractive alternative is to make use of Eqn (11.24) at this point. Remembering that $N$ is 8 and $j$, the number of $\theta$-parameters, is 2, we have

$$s^2 = \frac{\Sigma(y_i - \bar{y})^2 - \theta_2 \Sigma(x_i - \bar{x})(y_i - \bar{y})}{8 - 2}$$

$$= \frac{0.11109 - 0.09786 \times 1.098}{6}$$

$$= 0.000607 \text{ min}^2/\mu\text{M}^2.$$

Application of Eqn (11.16) then yields the standard error for $\theta_1$:

$$s_{\theta_1} = \left\{ s^2 \left[ \frac{1}{N} + \frac{\bar{x}^2}{\Sigma(x_i - \bar{x})^2} \right] \right\}^{1/2}$$

$$= \left\{ 0.000607 \left[ \frac{1}{8} + \frac{(15.47/8)^2}{11.223} \right] \right\}^{1/2}$$

$$= 0.01668 \text{ min}/\mu\text{M}.$$

Since $V$ is equal to the reciprocal of $\theta_1$, substitution of the standard error for $\theta_1$ into Eqn (11.23) gives the standard error for $V$:

$$s_V = s_{\theta_1}/\theta_1^2 = 0.01668/0.03389^2$$

$$= 14.5\ \mu\text{M/min}.$$

Similarly, $V$ and its error can be evaluated from the Eadie and the Hanes transforms, as well as from nonlinear regression. The results are summarized as follows:

| | |
|---|---|
| Lineweaver-Burk: | $V = 29.5 \pm 14.5\ \mu M/min$ |
| Eadie: | $V = 20.3 \pm 4.9\ \mu M/min$ |
| Hanes: | $V = 27.4 \pm 5.4\ \mu M/min$ |
| Nonlinear regression: | $V = 25.4 \pm 4.0\ \mu M/min$ |

Thus the error estimated from the Lineweaver-Burk transform far exceeds the errors from the other two transforms. Nonlinear regression gives a lower error than any of the linear transforms. The Eadie transform gives a lower error than the Hanes transform, but the Hanes transform gives a $V$-value closer to that from nonlinear regression than the Eadie transform.

For any two variables $y$ and $x$, the extent of their linear association can be described by a *correlation coefficient* $r$:

$$r = \frac{\Sigma(x_i - \bar{x})(y_i - \bar{y})}{\{\Sigma(x_i - \bar{x})^2\,\Sigma(y_i - \bar{y})^2\}^{1/2}}\,.$$

Application of this formula to the three linear transforms yields:

| | |
|---|---|
| Lineweaver-Burk: | $r = 0.983$ |
| Eadie: | $r = -0.771$ |
| Hanes: | $r = 0.899$ |

The absolute value of $r$ is close to unity in all three instances. In fact, the extent of linear correlation is highest in the Lineweaver-Burk plot, giving no indication at all of the very poor capability of this linearizing method for parameter estimation. Consequently the nice appearance of a linear plot is no assurance for accurate parameter estimation.

**Problem 12.1**

If the interconversion of $E_c$ and $E_t$ is prohibited, the carrier can traverse the membrane only when it is liganded to a substrate. Accordingly, a complete forward reaction cycle that brings about the uptake of $S_c$ and release of $S_t$ must include the $X_t$-arrow. Since the structural rule requires all arrow patterns in the numerator of the forward initial rate law to contain a complete forward reaction cycle, the rate dependence of $S_c$-transport on $X_t$ is substrate (essential) type. This kind of absolute dependence of $S_c$-transport on $X_t$ has been observed, for example, in the case of transport of anions across the mitochondrial membrane (Ferguson and Williams, 1966; Chappell and Robinson, 1968). Mitchell (1967) has discussed various model mechanisms capable of explaining the phenomenon, including such a prohibited interconversion of $E_c$ and $E_t$.

# References

Adair, G. S. (1925). *J. biol. Chem.* **63**, 529.
Alberty, R. A. (1953). *J. Am. chem. Soc.* **75**, 1928.
Alberty, R. A. (1958). *J. Am. chem. Soc.* **80**, 1777.
Alberty, R. A. and Massey, V. (1954). *Biochim. biophys. Acta.* **13**, 347.
Alberty, R. A., Bloomfield, V., Peller, L. and King, E. L. (1962). *J. Am. chem. Soc.* **84**, 4381.
Albery, W. J. and Knowles, J. R. (1974). Personal communication.
Anscombe, F. J. and Tukey, J. W. (1960). *Technometrics*, **5**, 141.
Arihood, S. A. and Trowbridge, C. G. (1970). *Archs Biochem. Biophys.* **141**, 131.
Atkins, G. L. (1971). *Biochim. biophys. Acta* **252**, 405.
Atkins, G. L. (1973). *Eur. J. Biochem.* **33**, 175.
Atkinson, D. E. (1970). *In* "The Enzymes" (P. D. Boyer, ed.), Vol. I, 3rd Edition, p. 461. Academic Press, New York.
Atkinson, D. E. and Walton, G. M. (1965). *J. biol. Chem.* **240**, 757.
Bailey, J. M. and French, D. (1957). *J. biol. Chem.* **226**, 1.
Barber, E. D. and Bright, H. J. (1968). *Proc. natn. Acad. Sci. U.S.A.* **60**, 1363.
Bártfai, T. and Mannervik, B. (1972). *FEBS Letters* **26**, 252.
Barwell, C. J. and Hess, B. (1972). *J. physiol. Chem.* **353**, 1178.
Bender, M. L., Clement, G. E., Kezdy, F. J. and Heck, H. D'A. (1964). *J. Am. chem. Soc.* **86**, 3680.
Benesch, R. and Benesch, R. E. (1969). *Nature* **221**, 618.
Berger, R. L., Antonini, E., Brunori, M., Wyman, J. and Rossi-Fanelli, A. (1967). *J. biol Chem.* **242**, 4841.
Bernhard, S. A., Dunn, M. F., Luisi, P. L. and Schack, J. (1970). *Biochemistry* **9**, 185.
Betz, A. and Selkov, E. (1969). *FEBS Letters* **3**, 5.
Bjerrum, J. (1941). "Metal Amine Formation in Aqueous Solution." P. Haase and Son, Copenhagen.
Bjorksten, F. (1968). *Eur. J. Biochem.* **5**, 133.
Blangy, D., Buc, H. and Monod, J. (1968). *J. molec. Biol.* **31**, 13.
Bloomfield, V. and Alberty, R. A. (1963). *J. biol. Chem.* **238**, 2811.
Blow, D. M., Birktoft, J. J. and Hartley, B. S. (1969). *Nature* **221**, 337.
Booman, K. A. and Niemann, C. (1956). *J. Am. chem. Soc.* **78**, 3642.
Botts, J. (1958). *Trans. Faraday Soc.* **54**, 593
Botts, J. and Morales, M. (1953). *Trans. Faraday Soc.* **49**, 696.
Box, G. E. P. (1957). *Appl. Statist.* **2**, 81.
Box, G. E. P. and Cox, D. R. (1964). *Jl. R. statist. Soc.* **B26**, 211.
Box, M. J., Davies, D. and Swann, W. H. (1969). "Non-Linear Optimization Techniques." Mathematical and Statistical Techniques for Industry Monograph No. 5. Published for Imperial Chemical Industries Ltd. by Oliver and Boyd, Edinburgh.

Boyer, P. D. (1959). *Archs Biochem. Biophys.* **82**, 387.
Boyer, P. D. and Silverstein, E. (1963). *Acta chem. scand.* **17**, S195.
Boyer, P. D., Cross, R. L. and Momsen, W. (1973). *Proc. natn. Acad. Sci. U.S.A.* **70**, 2837.
Brandt, K. G., Himoe, A. and Hess, G. P. (1967). *J. biol. Chem.* **242**, 3973.
Briggs, G. E. and Haldane, J. B. S. (1925). *Biochem. J.* **19**, 338.
Britton, H. G. (1964). *J. Physiol.* **170**, 1.
Britton, H. G. (1966). *Archs Biochem. Biophys.* **117**, 167.
Britton, H. G. (1973). *Biochem. J.* **133**, 255.
Brown, A. J. (1902). *J. chem. Soc.* **81**, 373.
Bruice, T. C. (1970). *In* "The Enzymes" (P. Boyer, ed.), Vol. II, 3rd Edition, p. 217. Academic Press, New York and London.
Buc, H. (1967). *Biochem. biophys. Res. Commun.* **28**, 59.
Buc, M. H. and Buc, H. (1967). *In* "Regulation of Enzyme Activity and Allosteric Interactions" (E. Kvamme and A. Pihl, eds.), p. 109. (4th meeting FEBS Symp. Oslo). Academic Press, London.
Bücher, T. and Sies, H. (1969). *Eur. J. Biochem.* **8**, 273.
Bunting, J. W. and Murphy, J. (1972). *Can. J. Biochem.* **50**, 1369.
Cennamo, C. (1969). *J. theor. Biol.* **23**, 53.
Cha, S. (1968). *J. biol. Chem.* **243**, 820.
Chan, W. W. C. and Mort, J. S. (1973). *J. biol. Chem.* **248**, 7614.
Chance, B. (1943). *J. biol. Chem.* **151**, 553.
Chance, B. (1954). *Discuss. Faraday Soc.* **17**, 120.
Chance, B. and Williams, G. R. (1955). *Adv. Enzymol.* **17**, 65.
Chance, B., Hess, B. and Betz, A. (1964). *Biochem. biophys. Res. Commun.* **16**, 182.
Changeux, J.-P., Gerhart, J. C. and Schachman, H. K. (1968). *Biochemistry* **7**, 531.
Chappell, J. B. and Robinson, B. H. (1968). *Biochem. Soc. Symp.* **27**, 123.
Clark, A. G. (1970). *Biochem. J.* **117**, 997.
Cleland, W. W. (1963a). *Biochim. biophys. Acta* **67**, 104.
Cleland, W. W. (1963b). *Biochim. biophys. Acta* **67**, 188.
Cleland, W. W. (1967a). *A. Rev. Biochem.* **36**, 77.
Cleland, W. W. (1967b). *Adv. Enzymol.* **29**, 1.
Cohen, G. N. and Monod, J. (1957). *Bact. Rev.* **21**, 169.
Cornish-Bowden, A. and Koshland, D. E., Jr. (1970a). *Biochemistry* **9**, 3325.
Cornish-Bowden, A. J. and Koshland, D. E., Jr. (1970b). *J. biol. Chem.* **245**, 6241.
Coryell, C. D. (1939). *J. phys. Chem.* **43**, 841.
Craine, J. E., Hall, E. S. and Kaufman, S. (1972). *J. biol. Chem.* **247**, 6082.
Crane, R. K., Miller, D. and Bihler, I. (1961). *In* "Membrane Transport and Metabolism " (A. Kleinzeller and A. Kotyk, eds.), pp. 439-449. Academic Press, New York.
Crane, R. K., Forstner, G. and Eicholz, A. (1965). *Biochim. biophys. Acta* **109**, 467
Cumme, G. A., Hoppe, H., Horn, A. and Bornig, H. (1972). *Acta biol. med. germ.* **28**, 727.
Cunningham, L. W. (1957). *Science* **125**, 1145.
Czerlinski, G. H. (1968). *J. theor. Biol.* **21**, 408.
Dalziel, K. (1957). *Acta chem. scand.* **11**, 1706.
Dalziel, K. (1958). *Trans. Faraday Soc.* **54**, 1247.
Dalziel, K. (1968). *FEBS Letters* **1**, 346.
Dalziel, K. (1969). *Biochem. J.* **114**, 547.
Dalziel, K. and Dickinson, E. M. (1966). *Biochem. J.* **100**, 34.
Daniel, C. and Wood, F. S. (1971). "Fitting Equations to Data", p. 27. Wiley-Interscience, New York.
Darvey, I. G. (1968). *J. theor. Biol.* **19**, 215.

Darvey, I. G. (1972). *Biochem. J.* **128**, 383.
Darvey, I. G. and Williams, J. F. (1964). *Biochim. biophys. Acta* **85**, 1.
Davidon, W. C. (1959). A.E.C. Research and Development Report, ANL-5990 (Rev).
DeMoss, J. A. and Novelli, G. D. (1956). *Biochim. biophys. Acta* **22**, 49.
Dixon, M. (1953a). *Biochem. J.* **55**, 170.
Dixon, M. (1953b). *Biochem. J.* **55**, 161.
Dixon, M. and Webb, E. C. (1958). "Enzymes". Longmans, Green and Co., London, New York and Toronto.
Dixon, M. and Webb, E. C. (1959). *Nature* **184**, 1296.
Dodgson, K. S., Spencer, B. and Williams, K. (1955). *Biochem. J.* **61**, 374.
Doudoroff, M., Barker, H. A. and Hassid, W. Z. (1947). *J. biol. Chem.* **168**, 725.
Dowd, J. E. and Riggs, D. S. (1965). *J. biol. Chem.* **240**, 863.
Draper, N. R. and Smith, H. (1967). "Applied Regression Analysis", p. 17. Wiley, New York.
Eadie, G. S. (1942). *J. biol. Chem.* **146**, 85.
Ebersole, E. R., Guttentag, C. and Wilson, P. W. (1943). *Archs Biochem.* **3**, 399.
Edelstein, S. J. (1971). *Nature* **230**, 224.
Edsall, J. T. (1943). *In* "Proteins, Amino Acids and Peptides" (E. J. Cohn and J. T. Edsall, eds.), p. 445. Reinhold Corp., New York.
Edsall, J. T. and Wyman, J. (1958). "Biophysical Chemistry", Vol. I. Acadmic Press, New York.
Eigen, M. (1954). *Discuss. Faraday Soc.* **17**, 194.
Eigen, M. and De Maeyer, L. (1963). "Technique of Organic Chemistry" (A. Weissberger, ed.), Vol. VIII, Part II, p. 895, Interscience Publishers Inc., New York.
Endrenyi, L. and Kwong, F. H. F. (1972). *In* "Analysis and Stimulation of Biochemical Systems" (H. C. Hemker and B. Hess, eds.), FEBS 8th Meeting Symposium, Vol. 25, p. 219. North Holland, Amsterdam.
Endrenyi, L. and Kwong, F. H. F. (1973). *Acta biol. med. germ.* **31**, 495.
Endrenyi, L., Chan, M.-S. and Wong, J. T.-F. (1971). *Can. J. Biochem.* **49**, 581.
Endrenyi, L., Fajszi, C. and Kwong, F. H. F. (1975). *Eur. J. Biochem.* (In press).
Engers, H. D., Bridger, W. A. and Madsen, N. B. (1970). *Biochemistry* **9**, 3281.
Epand, R. M. and Wilson, I. B. (1963). *J. biol. Chem.* **238**, 1718.
Everett, D. H. (1967). Adsorption hysteresis. *In* "The Solid-Gas Interface", Vol. 2, (E. A. Flood, ed.), pp. 1055-1113. M. Dekker, New York.
Eyring, H. (1935). *J. Chem. Phys.* **3**, 107.
Ferdinand, W. (1966). *Biochem. J.* **98**, 278.
Ferguson, S. M. F. and Williams, G. R. (1966). *J. biol Chem.* **241**, 3969.
Fisher, D. D. and Schulz, A. R. (1969). *Math. Bioscience* **4**, 189.
Fletcher, R. and Powell, M. J. D. (1963). *Comput. J.* **6**, 163.
Florini, J. R. and Vestling, C. S. (1957). *Biochim. biophys. Acta* **25**, 575.
Flossdorf, J. and Kula, M. R. (1972). *Eur. J. Biochem.* **30**, 325.
Folk, J. E., Cole, P. W. and Mullooly, J. P. (1967). *J. biol. Chem.* **242**, 2615.
Folk, J. E., Mullooly, J. P. and Cole, P. W. (1967). *J. biol. Chem.* **242**, 1838.
Frieden, C. (1959). *J. biol. Chem.* **234**, 2891.
Frieden, C. (1967). *J. biol. Chem.* **242**, 4045.
Frieden, C. (1970). *J. biol. Chem.* **245**, 5788.
Frieden, C. and Colman, R. D. (1967). *J. biol. Chem.* **242**, 1705.
Frieden, C., Wolfe, R. G. and Alberty, R. A. (1957). *J. Am. chem. Soc.* **79**, 1523.
Gaffney, T. J. and O'Sullivan, W. J. (1964). *Biochem. J.* **90**, 177.
Garfinkel, D. and Hess, B. (1964). *J. biol. Chem.* **239**, 971.
Garfinkel, D. H., Achs, M. J. and Dzubow, L. (1974). *Fedn Proc. Fedn Am. Socs. exp. Biol.* **33**, 176.

Gerhardt, J. C. and Pardee, A. B. (1962). *J. biol. Chem.* **237**, 891.

Gibson, Q. H. (1969). *Meth. Enzym.* **16**, 187.

Gibson, Q. H. (1970). *J. biol. Chem.* **245**, 3285.

Gibson, Q. H. (1973). *Proc. natn. Acad. Sci. U.S.A.* **70**, 1.

Gilbert, W. and Müller-Hill, B. (1966). *Proc. natn. Acad. Sci. U.S.A.* **56**, 1891.

Glende, M., Reich, J. G. and Wangermann, G. (1972). *Acta. biol. med. germ.* **29**, 793.

Goldfeld, S. M. and Quandt, R. E. (1965). *J. Am. Statist. Ass.* **60**, 539.

Griffith, J. S. (1967). "A View of the Brain." Oxford University Press, London.

Grisolia, S. and Cleland, W. W. (1968). *Biochemistry* **7**, 1115.

Guidotti, G. (1967). *J. biol. Chem.* **242**, 3704.

Gurr, P. A., Wong, J. T.-F. and Hanes, C. S. (1973). *Analyt. Biochem.* **51**, 584.

Gutfreund, H. (1955). *Discuss. Faraday Soc.* **20**, 167.

Gutfreund, H. and Sturtevant, J. M. (1956). *Biochem. J.* **63**, 656.

Haarhoff, K. N. (1969). *J. theor. Biol.* **22**, 117.

Haldane, J. B. S. (1930). "Enzymes." Longmans, Green and Co. Ltd, London.

Hammes, G. G. (1964). *Nature* **204**, 342.

Hammes, G. G. and Haslam, J. L. (1969). *Biochemistry* **8**, 1591.

Hammes, G. G. and Hurst, J. K. (1969). *Biochemistry* **8**, 1083.

Hammes, G. G. and Schimmel, P. R. (1967). *J. phys. Chem.* **71**, 917.

Hanes, C. S. (1932). *Biochem. J.* **26**, 1406.

Hanes, C. S., Bronskill, P. M., Gurr, P. A. and Wong, J. T.-F. (1972). *Can. J. Biochem.* **50**, 1385.

Hanson, K. R., Ling, R. and Havir, E. (1967). *Biochem. biophys. Res. Commun.* **29**,194.

Harada, K. and Wolfe, R. G. (1968). *J. biol. Chem.* **243**, 4131.

Hartley, B. S. and Kilby, B. A. (1954). *Biochem. J.* **56**, 288.

Hartridge, H. and Roughton, F. J. R. (1923). *Proc. R. Soc.* Ser. A, **104**, 376.

Hearon, J. Z. (1952). *Physiol. Rev.* **32**, 499.

Hearon, J. Z. (1953). *Bull. math. Biophys.* **15**, 121.

Hearon, J. Z., Bernhard, S. A., Friess, S. L., Botts, D. J. and Morales, M. F. (1958). *In* "The Enzymes", (P. D. Boyer, H. Lardy and K. Myrback, eds.), 2nd Edition, p. 49. Academic Press, New York.

Hebb, D. O. (1949). "The Organization of Behaviour", Wiley, New York.

Henderson, J. F. (1968). *Can. J. Biochem.* **46**, 1381.

Henri, V. (1901). *C. r. hebd. Seanc. Acad. Sci., Paris* **133**, 891.

Hess, G. P. and Rupley, J. A. (1971). *A. Rev. Biochem.* **40**, 1013.

Hess, B., Chance, E. M., Busse, H. and Wurster, B. (1972). *FEBS* **25**, 119.

Hijazi, N. H. and Laidler, K. J. (1973a). *Biochim, biophys. Acta* **315**, 209.

Hijazi, N. H. and Laidler, K. J. (1973b). *Can. J. Biochem.* **51**, 806.

Hill, A. V. (1910). *J. Physiol.* **40**, 4.

Hill, T. L. (1969). *Proc. natn. Acad. Sci. U.S.A.* **64**, 267.

Hoagland, M. B., Keller, E. B. and Zamecnik, P. C. (1956). *J. biol. Chem.* **218**, 345.

Hofstee, B. H. J. (1952). *Science* **116**, 329.

Hofstee, B. H. J. (1954). *J. biol. Chem.* **207**, 219.

Hofstee, B. H. J. (1959). *Nature* **184**, 1296.

Hogness, T. R. (1942). *In* "A Symposium on Respiratory Enzymes", p. 134. University of Wisconsin Press, Madison.

Hopfield, J. J., Schulman, R. G. and Ogawa, S. (1971). *J. molec. Biol.* **61**, 425.

Horn, A. and Börnig, H. (1969). *FEBS Letters* **3**, 325.

Hurst, R. O. (1969). *Can. J. Biochem.* **47**, 941.

Illingworth, J. A. and Tipton, K. F. (1969). *Biochem. J.* **115**, 511.

Ince, E. L. (1944). "Ordinary Differential Equations", p. 115. Dover Publications, New York.

Jacquez, J. A. (1961). *Proc. natn. Acad. Sci. U.S.A.* **47**, 153.
Janin, J. and Cohen, G. N. (1969). *Eur. J. Biochem.* **11**, 520.
Jencks, W. P. (1966). *In* "Current Aspects of Biochemical Energetics" p. 273. (N. O. Kaplan and E. P. Kennedy, eds.), Academic Press, New York.
Johannes, K. J. and Hess, B. (1973). *J. molec. Biol.* **76**, 181.
Jones, J. B. and Wigfield, D. C. (1967). *J. Am. chem. Soc.* **89**, 5294.
Josse, J. (1966). *J. biol. Chem.* **241**, 1948.
Kaplan, N. O., Colowick, S. P. and Neufeld, E. F. (1952). *J. Biol. Chem.* **195**, 107.
Katchalsky, A. and Neumann, E. (1972). *Int. J. Neuroscience* **3**, 175.
Keech, B. and Barritt, G. J. (1967). *J. biol. Chem.* **242**, 1983.
Keilin, D. and Mann, T. (1937). *Proc. R. Soc. B.* **122**, 119.
Keleti, T. (1972). *FEBS Letters* **28**, 287.
Keleti, T. and Fajszi, C. (1971). *Math. Biosci.* **12**, 197.
King, E. L. and Altman, C. (1956). *J. phys. Chem.* **60**, 1375.
Kirschner, K. (1971a). *J. molec. Biol.* **58**, 51.
Kirschner, K. (1971b). *In* "Current Topics in Cellular Regulations" (B. L. Horecker and E. R. Stadtman, eds.), Vol. 4, p. 167. Acadmic Press, New York.
Kirschner, K., Eigen, M., Bittman, R. and Voigt, B. (1966). *Proc. natn. Acad. Sci. U.S.A.* **56**, 1661.
Knowles, J. R. (1974). Personal communication.
Koch, A. L. (1964). *Biochim. biophys. Acta* **79**, 177.
Kohen, E., Kohen, C. and Thorell, B. (1968). *Biochim. biophys. Acta* **167**, 635.
Koshland, D. E., Jr. (1958). *Proc. natn. Acad. Sci. U.S.A.* **44**, 98.
Koshland, D. E., Jr. (1959). *J. Cellular Comparative Physiol.* **54**, suppl. 1, 245.
Koshland, D. E., Jr. (1962). *J. theor. Biol.* **2**, 75.
Koshland, D. E., Jr. (1970). *In* "The Enzymes" (P. O. Boyer, ed.), Vol. I, 3rd Edition, p. 342. Academic Press, New York.
Koshland, D. E., Jr., Nemethy, G. and Filmer, D. (1966). *Biochemistry* **5**, 365.
Kosow, D. P. and Rose, I. A. (1972). *Biochem. biophys. Res. Commun.* **48**, 376.
Kotyk, A. (1965). *Folia Microbiologica* **10**, 30.
Kowalik, J. and Morrison, J. F. (1968). *Math. Biosci.* **2**, 57.
Kuby, S. A., Noda, L. and Lardy, H. A. (1954). *J. biol. Chem.* **210**, 65.
Kuby, S. A. and Noltmann, E. A. (1962). *In* "The Enzymes" (P. D. Boyer, H. Lardy and K. Myrback, eds.), Vol. 6, 2nd Edition, p. 516. Academic Press, New York.
Kurganov, B. I. (1973). *Acta biol. med. germ.* **31**, 181.
Leininger, K. R. and Westley, J. (1968). *J. biol. Chem.* **243**, 1892.
Levitzki, A. and Koshland, D. E., Jr. (1969). *Proc. natn. Acad. Sci. U.S.A.* **62**, 1121.
Levy, H. M., Sharon, N. and Koshland, D. E. (1959). *Proc. natn. Acad. Sci. U.S.A.* **45**, 785.
Lineweaver, H. and Burk, D. (1934). *J. Am. chem. Soc.* **56**, 658.
London, W. P. (1968). *Bull. math. Biophys.* **30**, 253.
London, W. P. and Steck, T. L. (1969). *Biochemistry* **8**, 1767.
Longmuir, I. S. (1957). *Biochem. J.* **65**, 378.
Lorente de No, R. (1938). *J. Neurophysiol.* **1**, 207.
Loudon, G. M. and Koshland, D. E., Jr. (1972). *Biochemistry* **11**, 229.
Lowry, O. H. and Passoneau, J. V. (1966). *J. biol. Chem.* **241**, 2268.
Madsen, N. B. (1964). *Biochem. biophys. Res. Commum.* **15**, 390.
Madsen, N. B. and Shechosky, S. (1967). *J. biol. Chem.* **242**, 3301.
Magar, M. E. and Steiner, R. F. (1971). *J. theor. Biol.* **32**, 495.
Mahler, H. R., Baker, R. H., Jr. and Shiner, V. J., Jr. (1962). *Biochemistry* **1**, 47.
Mannervik, B., Gorna-Hall, B. and Bartfai, T. (1973). *Eur. J. Biochem.* **37**, 270.
Mantle, J. and Garfinkel, D. (1969). *J. biol. Chem.* **244**, 3884.

Markus, G., McClintock, D. K. and Bussel, J. B. (1971). *J. biol. Chem.* **246**, 762.
Marquardt, D. W. (1963). *J. Soc. ind. appl. Math.* **11**, 431.
Marshall, M. and Cohen, P. P. (1966). *J. biol. Chem.* **241**, 4197.
Melander, L. C. S. (1960). "Isotope Effects on Reaction Rates", p. 24. Ronald Press, New York.
Michaelis, L. and Davidsohn, H. (1911). *Biochem. Z.* **35**, 386.
Michaelis, L. and Menten, M. L. (1913). *Biochem. Z.* **49**, 333.
Mildvan, A. S. (1970). *In* "The Enzymes" (P. D. Boyer, ed.), Vol. II, 3rd Edition, p. 445. Acadmic Press, New York.
Mildvan, A. S. and Cohn, M. (1966). *J. biol. Chem.* **241**, 1178.
Minton, A. P. and Imai, K. (1974). *Proc. natn. Acad. Sci. U.S.A.* **71**, 1418.
Mitchell, P. (1967). *Adv. Enzymol.* **29**, 33.
Moffett, F. J. and Bridger, W. A. (1970). *J. biol. Chem.* **245**, 2758.
Monod, J., Changeux, J.-P. and Jacob, F. (1963). *J. molec. Biol.* **6**, 306.
Monod, J., Wyman, J. and Changeux, J.-P. (1965). *J. molec. Biol.* **12**, 88.
Morales, M. F., Horovitz, M. and Botts, J. (1962). *Archs Biochem. Biophys.* **99**, 258.
Morrison, J. F., O'Sullivan, W. J. and Ogston, A. G. (1961). *Biochim. biophys. Acta* **52**, 82.
Murray, D. R. P. (1930). *Biochem. J.* **24**, 1890.
Nachmansohn, D. and Wilson, I. B. (1951). *In* "Advances in Enzymology" (F. F. Nord, ed.), Vol. XII, p. 259. Interscience Publishers, Inc., New York.
Nazar, R. N., Tyfield, L. A. and Wong, J. T.-F. (1972). *J. biol. Chem.* **247**, 798.
Nichol, L. W., Jackson, W. J. H. and Winzor, D. J. (1967). *Biochemistry* **6**, 2449.
Noat, G., Ricard, J., Borel, M. and Got, C. (1970). *Eur. J. Biochem.* **13**, 347.
Noda, L., Kuby, S. A. and Lardy, H. A. (1954). *J. biol. Chem.* **210**, 83.
Ogawa, S. and McConnell, H. M. (1967). *Proc. natn. Acad. Sci. U.S.A.* **58**, 19.
Olive, C., Geroch, M. E. and Levy, H. R. (1971). *J. biol. Chem.* **246**, 2047.
O'Sullivan, W. J. and Cohn, M. (1966). *J. biol. Chem.* **241**, 3116.
Ottaway, J. H. (1973). *Biochem. J.* **134**, 729.
Ouellet, L. and Stewart, J. A. (1959). *Can. J. Chem.* **37**, 737.
Ouellet, L., Laidler, K. J. and Morales, M. (1952). *Archs Biochem. Biophys.* **39**, 37.
Pardee, A. B. (1967). *Science* **156**, 1627.
Patte, J. C., Truffa-Bachi, P. and Cohen, G. N. (1966). *Biochim. biophys. Acta* **128**, 426.
Pauling, L. (1935). *Proc. natn. Acad. Sci. U.S.A.* **21**, 186.
Peller, L. and Alberty, R. A. (1959). *J. Am. chem. Soc.* **81**, 5907.
Perutz, M. F. (1969). *Harvey Lect.* **63**, 213.
Pettersson, G. (1969). *Acta chem. scand.* **23**, 2717.
Pettersson, G. (1970). *Acta chem. scand.* **24**, 1271.
Pettersson, G. and Pettersson, I. (1970). *Acta chem. scand.* **24**, 1275 (1970).
Powell, M. J. D. (1964). *Comput. J.* **7**, 155.
Purich, D. L. and Fromm, H. J. (1972). *Archs Biochem. Biophys.* **149**, 307.
Rabin, B. R. (1967). *Biochem. J.* **102**, 22c.
Rapoport, T. A., Hohne, W. E., Reich, J. G., Heitmann, P. and Rapoport, S. M. (1972). *Eur. J. Biochem.* **26**, 237.
Rapoport, T. A., Hohne, W. E., Heitmann, P. and Rapoport, S. (1973). *Eur. J. Biochem.* **33**, 341.
Rapoport, T., Heinrich, R., Jacobasch, G. and Rapoport, S. (1974). *Eur. J. Biochem.* **42**, 107.
Raval, D. N. and Wolfe, R. G. (1962). *Biochemistry* **1**, 1118.
Raval, D. N. and Wolfe, R. G. (1963). *Biochemistry* **2**, 220.
Ray, W. J. and Roscelli, G. A. (1964). *J. biol. Chem.* **239**, 3935.

Reeves, R. E. and Sols, A. (1973). *Biochem. biophys. Res. Commun.* 50, 459.
Regen, D. M. and Morgan, H. E. (1964). *Biochim, biophys. Acta* 79, 151.
Reich, J. G. (1970). *FEBS Letters* 9, 245.
Reich, J. G., Wangermann, G., Falck, M. and Rohde, K. (1972). *Eur. J. Biochem.* 26, 368.
Reiner, J. M. (1959). "Behaviour of Enzyme Systems." Burgess Co., Minneapolis.
Reuben, J. (1971). *Proc. natn. Acad. Sci. U.S.A.* 68, 563.
Ricard, J., Noat, G., Got, C. and Borel, M. (1972). *Eur. J. Biochem.* 31, 14.
Ricard, J., Mouttet, C. and Nari, J. (1974). *Eur. J. Biochem.* 41, 479.
Rose, I. A. (1962). *Brookhaven Symp. Biol.* 15, 293.
Rose, I. A. (1970). *In* "The Enzymes" (P. D. Boyer, ed.), Vol. II, 3rd Edition, p. 281. Academic Press, New York.
Rose, I. A. and Warms, J. V. B. (1966). *J. biol. Chem.* 241, 4848.
Rossi Fanelli, A. and Antonini, E. (1958). *Archs Biochem. Biophys.* 77, 478.
Roughton, F. J. W. (1954). *Discuss. Faraday Soc.* 17, 116.
Roughton, F. J. W. (1963). *Clin. Chem.* 9, 682.
Roughton, F. J. W. (1964). "Oxygen in the Animal Organism," p. 5. Pergamon Press, London.
Roughton, F. J. W., DeLand, D. C., Kernohan, J. C. and Severinghaus, J. W. (1972). *In* "Oxygen Affinity and Red Cell Acid Base Status," (M. Rorth and P. Astrup, eds.). Academic Press, New York.
Rubin, M. M. and Changeux, J.-P. (1966). *J. molec. Biol.* 21, 265.
Rudolph, F. B. and Fromm, H. J. (1970). *Biochemistry* 9, 4660.
Rupley, J. A., Butler, L., Gerring, M., Hartdegen, F. A. and Pecoraro, R. (1967). *Proc. natn. Acad. Sci. U.S.A.* 57, 1088.
Sachs, E. (1961). *Fedn. Proc. Fedn. Am. Socs exp. Biol.* 20, 339.
Sanwal, B. D. (1970). *Bact. Rev.* 34, 20.
Sanwal, B. D. and Cook, R. A. (1966). *Biochemistry* 5, 886.
Sargent, D. F. and Taylor, C. P. S. (1971). *Anal. Biochem.* 42, 446.
Saroff, H. A. and Minton, A. P. (1972). *Science* 175, 1253.
Scatchard, G. (1949). *Ann. N. Y. Acad. Sci.* 51, 660.
Schachter, H. (1972). *In* "Metabolic Pathways" (L. E. Hokin, ed.), Vol. 6, 3rd Edition, p. 1. Academic Press, New York.
Schachter, H. (1973). *J. biol. Chem.* 248, 974.
Schellenberg, K. A. (1967). *J. biol. Chem.* 242, 1815.
Schwert, G. W. (1969). *J. biol. Chem.* 244, 1278.
Segal, H. L. (1958). *In* "The Enzymes," (P. D. Boyer, H. Lardy, and K. Myrback, eds.), Vol. 1, Second Edition, p. 1. Academic Press, New York.
Segal, H. L., Kachmar, J. F. and Boyer, P. D. (1952). *Enzymologia* 15, 187.
Sen, A. K., Tobin, T. and Post, R. L. (1969). *J. biol. Chem.* 244, 6596.
Seshagiri, N. (1972). *J. theor. Biol.* 34, 469.
Silanova, G. V., Livanova, N. B. and Kurganov, B. I. (1969). *Molek. Biol.* 3,768.
Sillén, L. G. (1956). *Acta chem. Scand.* 10, 186.
Silverstein, R., Vogt, J., Reed, D. and Abeles, R. H. (1967). *J. biol. Chem.* 242, 1338.
Simoni, R. D. and Roseman, S. (1973). *J. biol. Chem.* 248, 966.
Skou, J. C. (1965). *Physiol. Rev.* 45, 596.
Slater, E. C. and Bonner, W. D. (1952). *Biochem. J.* 52, 185.
Smith, R. L. and Shaw, E. (1969). *J. biol. Chem.* 244, 4704.
Spendley, W., Hext, G. R. and Himsworth, F. R. (1962). *Technometrics* 4, 441.
Stadtman, E. R. (1970). *In* "The Enzymes" (P. D. Boyer, ed.), Vol. 1, 3rd Edition, p. 397. Academic Press, New York.
Stallcup, W. B. and Koshland, D. E., Jr. (1973). *J. molec. Biol.* 80, 41.

Sweeny, J. R. and Fisher, J. R. (1968). *Biochemistry* 7, 561.
Swoboda, P. A. T. (1957). *Biochim. biophys. Acta* 23, 70.
Szabo, A. and Karplus, M. (1972). *J. molec. Biol.* 72, 163.
Taketa, K. and Pogell, B. M. (1965). *J. biol. Chem.* **240**, 651.
Tanford, C. (1962). *Adv. Protein. Chem.* **17**, 69.
Taniguchi, S. Theorell, H. and Akeson, A. (1967). *Acta chem. scand.* **9**, 1148.
Taraszka, M. and Alberty, R. A. (1964). *J. phys. Chem.* **68**, 3368.
Theorell, H. and Chance, B. (1951). *Acta chem. scand.* **5**, 1127.
Theorell, H., Nygaard, A. P. and Bonnichsen, R. (1955). *Acta chem. scand.* **9**, 1148.
Thompson, C. J. (1968). *Biopolymers* **6**, 1101.
Umbarger, H. E. (1956). *Science* 123, 848.
Van Slyke, D. D, and Cullen, G. E. (1914). *J. biol. Chem.* 19, 141.
Velick, S. F. and Vavra, J. (1962). *J. biol. Chem.* 237, 2109.
Volkenstein, M. V. and Goldstein, B. N. (1966). *Biochim. biophys. Acta* **115**, 478.
Wagner, M., Erecinska, M. and Pring, M. (1971). *Archs Biochem. Biophys.* **147**, 675.
Waley, S. G. (1953). *Biochim. biophys. Acta* **10**, 27.
Walker, A. C. and Schmidt, C. L. A. (1944). *Archs Biochem.* **5**, 445.
Wall, M. C. and Laidler, K. J. (1953). *Archs Biochem. Biophys.* 43, 299.
Walsh, K. A. (1959). Ph.D. Thesis, University of Toronto.
Walter, C. (1962). *Biochemistry* 1, 652.
Walter, C. F. and Morales, M. F. (1964). *J. biol. Chem.* 239, 1277.
Webb, J. L. (1963). "Enzyme and Metabolic Inhibitors", Vol. 1. Academic Press, New York.
Weber, G. (1972). *Proc. natn. Acad. Sci. U.S.A.* **69**, 3000.
Weber, G. and Anderson, S. R. (1965). *Biochemistry* **4**, 1942.
Wedler, F. C. and Boyer, P. D. (1973). *J. theor. Biol.* **38**, 539.
Westheimer, F. H. (1962). *In* "Advances in Enzymology" (F. F. Nord, ed.), Vol. 24, p. 441. Interscience Publishers, New York.
Whitehead, E. P. (1970). *Prog. Biophys. molec. Biol.* **21**, 321.
Whitehead, E. P. (1973). *Acta biol. med. germ.* **31**, 227.
Widdas, W. F. (1952). *J. Physiol.* **118**, 23.
Wieker, H.-J. and Hess, B. (1971). *Biochemistry* **10**, 1243.
Wieker, H.-J., Johannes, K.-J. and Hess, B. (1970). *FEBS Letters* **8**, 178.
Wilbrandt, W. and Kotyk, A. (1964). *Naunyn-Schmiedebergs Arch. exp. Path. u. Pharmak.* 249, 279.
Wilbrandt, W. and Rosenberg, T. (1961). *Pharmac. Rev.* **13**, 109.
Wilkinson, G. N. (1961). *Biochem. J.* **80**, 324.
Will, H. and Damaschun, G. (1973). *J. theor. Biol.* **38**, 579.
Wilson, I. B. and Cabib, E. (1956). *J. Am. chem. Soc.* 78, 202.
Winer, A. D. and Schwert, G. W. (1958). *J. biol. Chem.* 231, 1065.
Winkler, H. H. and Wilson, T. H. (1966). *J. biol. Chem.* 241, 2200.
Wolochow, H., Putman, E. W., Doudoroff, M., Hassid, W. Z. and Barker, H. A. (1949). *J. biol. Chem.* **180**, 1237.
Wong, J. T.-F. (1965a). *J. Am. chem. Soc.* **87**, 1788.
Wong, J. T.-F. (1965b). *Biochim. biophys. Acta* **94**, 102.
Wong, J. T.-F. and Endrenyi, L. (1970). Symp. Biochem. Regulation and Control Systems, 8th Int. Cong. Biochem. Abst., p. 230.
Wong, J. T.-F. and Endrenyi, L. (1971). *Can. J. Biochem.* **49**, 568.
Wong, J. T.-F. and Hanes, C. S. (1962). *Can. J. Biochem. Physiol.* **40**, 763.
Wong, J. T.-F. and Hanes, C. S. (1964). *Nature* **203**, 492.
Wong, J. T.-F. and Hanes, C. S. (1969). *Archs Biochem. Biophys.* 135, 50.
Wong, J. T.-F. and Hanes, C. S. (1973). *Acta biol. med. germ.* 31, 507.

Wong, J. T.-F. and Williams, G. R. (1968). *Archs Biochem. Biophys.* **124**, 344.
Wong, J. T.-F., Pincock, A. and Bronskill, P. M. (1971). *Biochim. biophys. Acta* 233, 176.
Wong, J. T.-F., Gurr, P. A., Bronskill, P. M. and Hanes, C. S. (1972). *In* "Analysis and Simulation of Biochemical Systems" (H. C. Hemker and B. Hess, eds.), FEBS Meeting Symp. Vol. 25, p. 327. North Holland/American Elsevier.
Wurster, B. and Hess, B. (1970). *Hoppe-Seyler's Z. Physiol. Chem.* **351**, 869.
Wurster, B. and Schneider, F. (1970). *Hoppe-Seyler's Z. Physiol. Chem.* **351**, 961.
Wyman, J. (1939). *J. biol. Chem.* **127**, 581.
Wyman, J. (1948). *Adv. Protein Chem.* **4**, 410.
Wyman, J. (1963). *Cold Spring Harb. Symp.* **28**, 483.
Wyman, J. (1967). *J. Am. chem. Soc.* **89**, 2202.
Yagi, K. and Ozawa, T. (1960). *Biochim. biophys. Acta* **42**, 381.
Yagil, G. and Hoberman, H. D. (1969). *Biochemistry* **8**, 352.
Yates, R. A. and Pardee, A. B. (1956). *J. biol. Chem.* **221**, 757.
Yon, R. J. (1972). *Biochem. J.* **128**, 311.
Yonetani, T. and Ray, G. S. (1966). *J. biol. Chem.* **241**, 700.
Yonetani, T. and Theorell, H. (1964). *Archs Biochem. Biophys.* **106**, 243.
Zerner, B., Bond, R. P. M. and Bender, M. L. (1964). *J. Am. chem. Soc.* **86**, 3674.

# Subject Index